Geometrical Drawing for Carpentry & Joinery

Geometrical Drawing for Carpentry & Joinery

John J. O'Connor

Gill & Macmillan

Gill & Macmillan Ltd
Hume Avenue
Park West
Dublin 12
with associated companies throughout the world www.gillmacmillan.ie

© John J. O'Connor 2009

978 07171 4448 8

Index compiled by Cover to Cover
Design and print origination by Macmillan Publishing Solutions

The paper used in this book is made from the wood pulp of managed forests.
For every tree felled, at least one tree is planted, thereby renewing natural resources.

All rights reserved. No part of this publication may be copied, reproduced or transmitted in any form or by any means without written permission of the publishers or else under the terms of any licence permitting limited copyright issued by the Irish Copyright Licensing Agency.

A CIP catalogue record for this book is available from the British Library.

For Michael and Joan
In Memoriam

Dynamic and easy-to-use online support material for this book is available at *www.gillmacmillan.ie*

*Provides **lecturers** with:*

- PowerPoint presentation slides, in colour, for selected drawings from all chapters
- Additional questions with solutions.

To access lecturer support material on our site:

1) Go to *www.gillmacmillan.ie*
2) Log on using your username and password. If you don't have a password, register online and we will email your password to you.

*Provides **students** with:*

- Additional questions with solutions
- Typical examination-type questions
- Additional drawing construction techniques for students without a foundation in geometrical drawing.

To access student support material:

1) Go to *www.gillmacmillan.ie*
2) Click on the link for Student Support material.

CONTENTS

Acknowledgments ix

Introduction x

1 ISOMETRIC PROJECTION 1
   Section I: Introduction To Isometric Projection 1
   Section II: Curves In Isometric Projection 5

2 PROJECTIONS OF LINES AND TRUE LENGTHS 10

3 AUXILIARY ELEVATIONS AND PLANS 17
   Section 1: Auxiliary Elevations 17
   Section II: Auxiliary Plans 21

4 GEOMETRIC SOLIDS 27
   Section I: Development Of Geometric Solids 27
   Section II: Rebatment Of Surfaces 40

5 SPLAYED WORK 43

6 GEOMETRIC CONSTRUCTION OF ARCHES 50
   Section I: Arch Outlines 50
   Section II: Soffit Development 57
   Section III: Skew Arches 61

7 CONIC SECTIONS AND CONIC DEVELOPMENT 66

8 ROOF GEOMETRY 78
   Section I: Conventional Pitched Roofs 78
   Section II: Pyramid Roofs 88
   Section III: Roofs To Bay Windows 95
   Section IV: Turrets And Domes 98
   Section V: Dormer Roofs 109

9 HANDRAILING 115
   Section I: Ramps And Knees 115
   Section II: Wreaths 118
   Section III: Scrolls 132

1 ISOMETRIC PROJECTION

SECTION I: INTRODUCTION TO ISOMETRIC PROJECTION

Orthographic projection shows drawings of an object in a two-dimensional format, with views given in plan, elevation and end elevation (as explained in Chapter 17 on constructions, Section II: geometric planes). Isometric projection gives a pictorial view of objects in a three-dimensional form. There are other pictorial forms such as oblique, axonometric and perspective projection, but isometric projection will be dealt with in this chapter.

In **Fig. 1.1b** the axis lines of isometric projection are shown. These consist of three lines: one is vertical and two are at an angle of 30° to the left and right. These may also be presented as the three 120° angles shown.

What does this mean? Quite simply that lines that are horizontal in orthographic projection become 30° lines in isometric, and lines that are vertical in orthographic will remain vertical in isometric; but more importantly, true measurements can only be applied (in isometric) to the axis lines, or lines parallel with them.

Fig. 1.1a shows the elevation and the isometric projection of an envelope. In both views the perimeter dimensions are the same, but dimensions **A-C** and **B-D** in isometric will not have the same dimensions as their counterparts in elevation.

In **Fig. 1.2** a 45° set square is shown in elevation and isometric. The longest edge of the set square will not have a true measurement in isometric, whereas the two shorter edges will, because they are on the axis lines, or parallel to them.

The elevation and end elevation of a *truncated* prism is shown in **Fig. 1.3**, and a completed isometric projection is also given. This solid could represent a roof. The dimensions **H**, **j**, **k**, **m**, **W** and **L** are all taken directly from the elevation and end elevation and applied to the isometric lines as shown. In the isometric view the lines that are not 'true' are: **A-B**, **A-C**, **D-E**, and **D-F**; in other words, they will be different from their counterparts in elevation and end elevation.

A useful method of transferring objects from orthographic to isometric is to 'box the object', as shown in dotted construction in isometric projection in **Fig. 1.3**.

In **Fig. 1.4** a square-based pyramid is shown in orthographic and isometric projection. Again the object is 'boxed-in' to facilitate easy transfer to isometric projection. Note that perimeter dimensions **a**, **b**, **c**, **d** and height **H** remain the same in both projections. It is also important to note here that when a square is drawn in isometric, its diagonals will end up horizontal and vertical, respectively (see diagonals **A-C** and **B-D** in the isometric projection in **Fig. 1.4**).

Fig. 1.5 shows orthographic and isometric views of an octagonal pyramid, the octagon in plan has a grid drawn on it, which takes in the eight corners.

GEOMETRICAL DRAWING FOR CARPENTRY & JOINERY

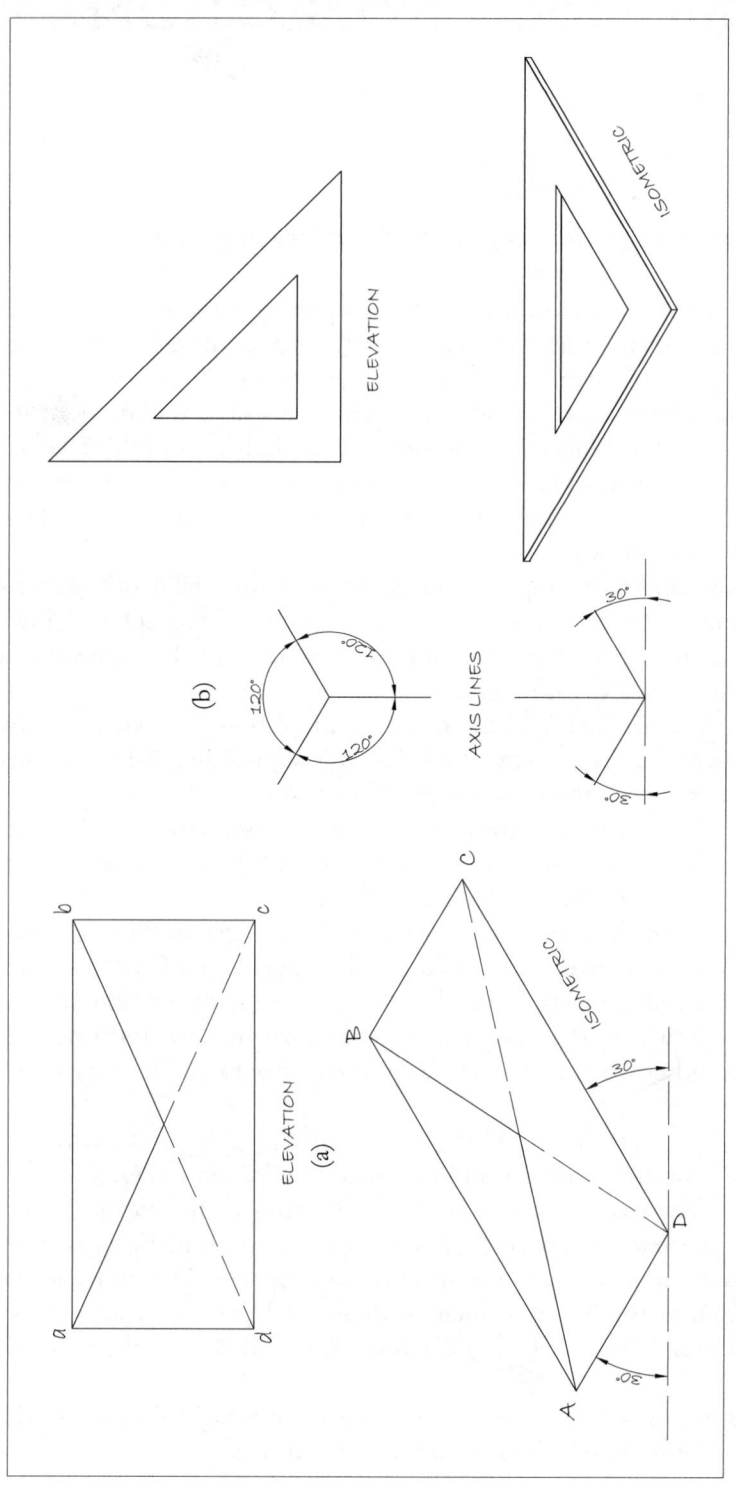

Fig. 1.1a
Fig. 1.1b
Fig. 1.2

1 ISOMETRIC PROJECTION

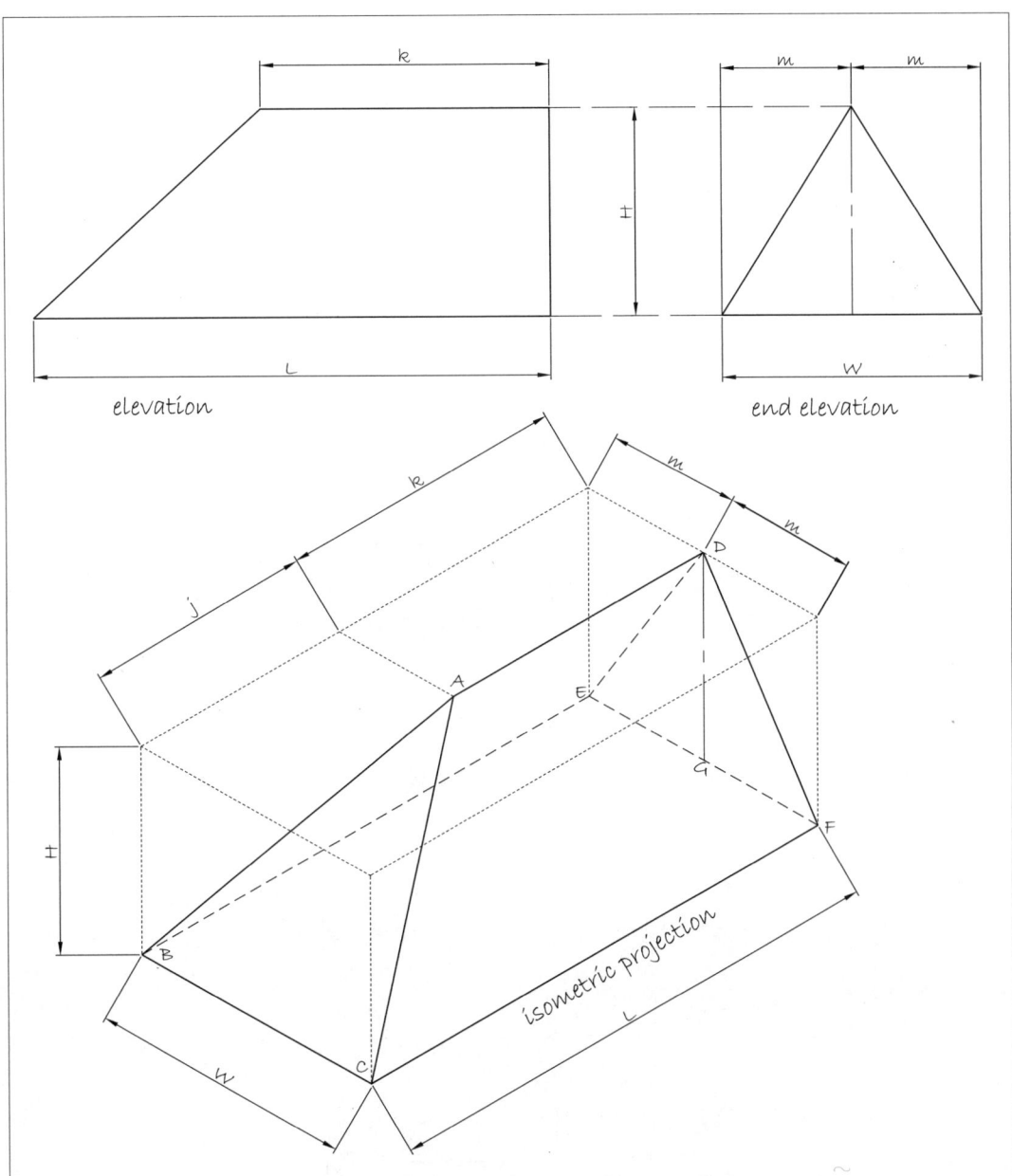

Fig. 1.3

GEOMETRICAL DRAWING FOR CARPENTRY & JOINERY

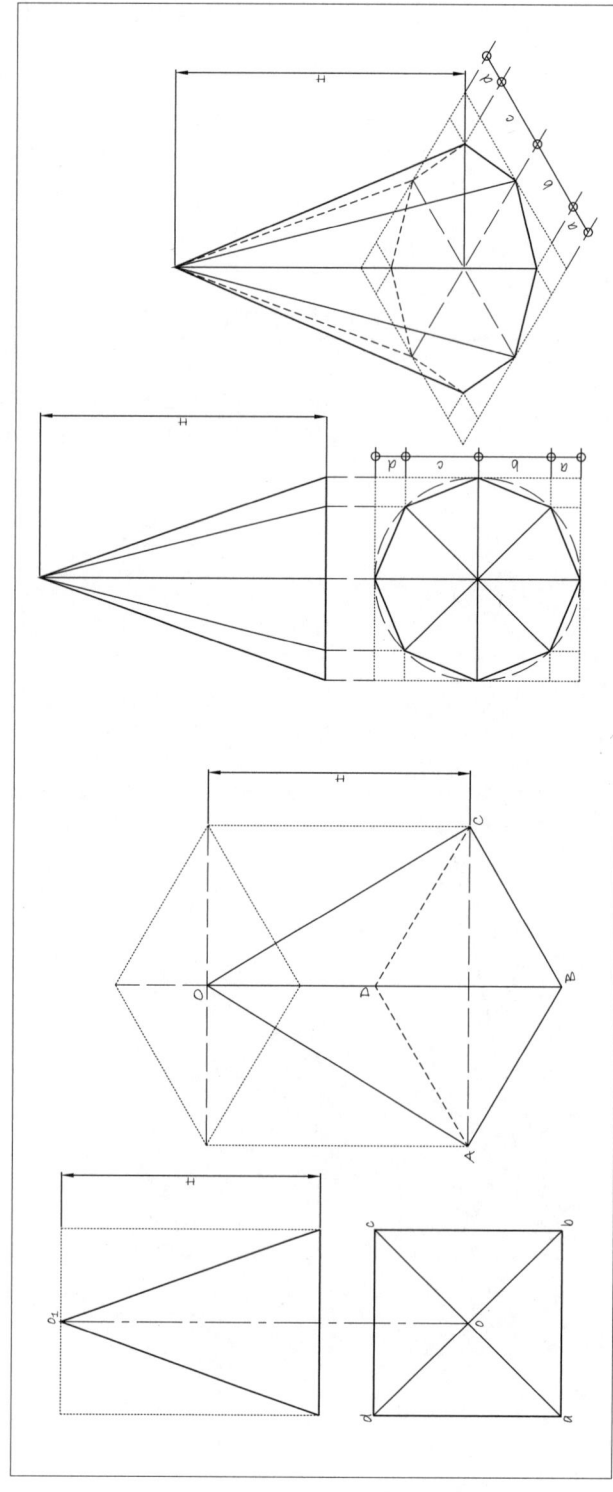

Fig. 1.4
Fig. 1.5

1 ISOMETRIC PROJECTION

By transferring the grid into isometric, the octagon's corners can be easily found.

To complete:

- Mark-off the height **H** from the centre of the isometric grid to find the apex and join to the corners as shown.

SECTION II: CURVES IN ISOMETRIC PROJECTION

When drawing curved objects in isometric, it is necessary to make a grid for the curve, similar to the method for **Fig. 1.5**. In **Figs. 1.6** and **1.7** a cylinder and cone are used to illustrate the procedure.

Method

- Construct the circle of the cylinder in orthographic and describe a square around it as shown in **Fig. 1.6a**, then divide the top side of the square into a number of parts, eight in this case.
- Complete the grid by extending the divisions to the bottom of the square and mark where they intersect with the circle. These points will give measurements **1,2,3,4,3,2,1** to the centre line of the circle.
- Construct the square in isometric (as in **Fig. 1.6b**) and locate on it the division lines, then plot the measurements **1,2,3,4,3,2,1**, to find the points on the curve.
- This curve, which is an ellipse, will be the base of the cylinder; a similar construction gives the top of the cylinder. These two will be set apart by the height as shown.

In **Fig. 1.7** the apex of the cone is found by measuring the height above the centre point of the isometric grid. In the above two figures, **Figs. 1.6** and **1.7**, the isometric grid is lying flat, and consequently the ellipse, because the objects are standing. In **Fig. 1.8** the cylinder is lying on its side.

In **Fig. 1.8** a cylinder is shown lying on its side in isometric projection where a different method is employed to transfer the circle into isometric. A compass is used to draw the arcs that make up the ellipse (a false ellipse in this case).

Method

- Construct the square in isometric in the usual way (but this time standing vertically).
- Draw diagonals **a-c**, **b-d**, and shorter diagonals **e-d** and **b-f**.
- Where **e-d** intersects with **a-c** the centre **j** for arc **e-g** is found.
- Where **b-f** intersects with **a-c** the centre **k** for arc **h-f** is found.
- Point **b** is the centre for arc **g-f**, and **d** is the centre for arc **e-h**.

In **Fig. 1.9** some common items in carpentry and joinery are shown drawn in what is called *exploded isometric*, that is, the items are pulled apart to show what is inside. This device is useful for giving pictorial details of the make-up, or working parts, of the item in question. This form of isometric often accompanies the instructions for flat-pack furniture and appliances, where a certain amount of assembly is required.

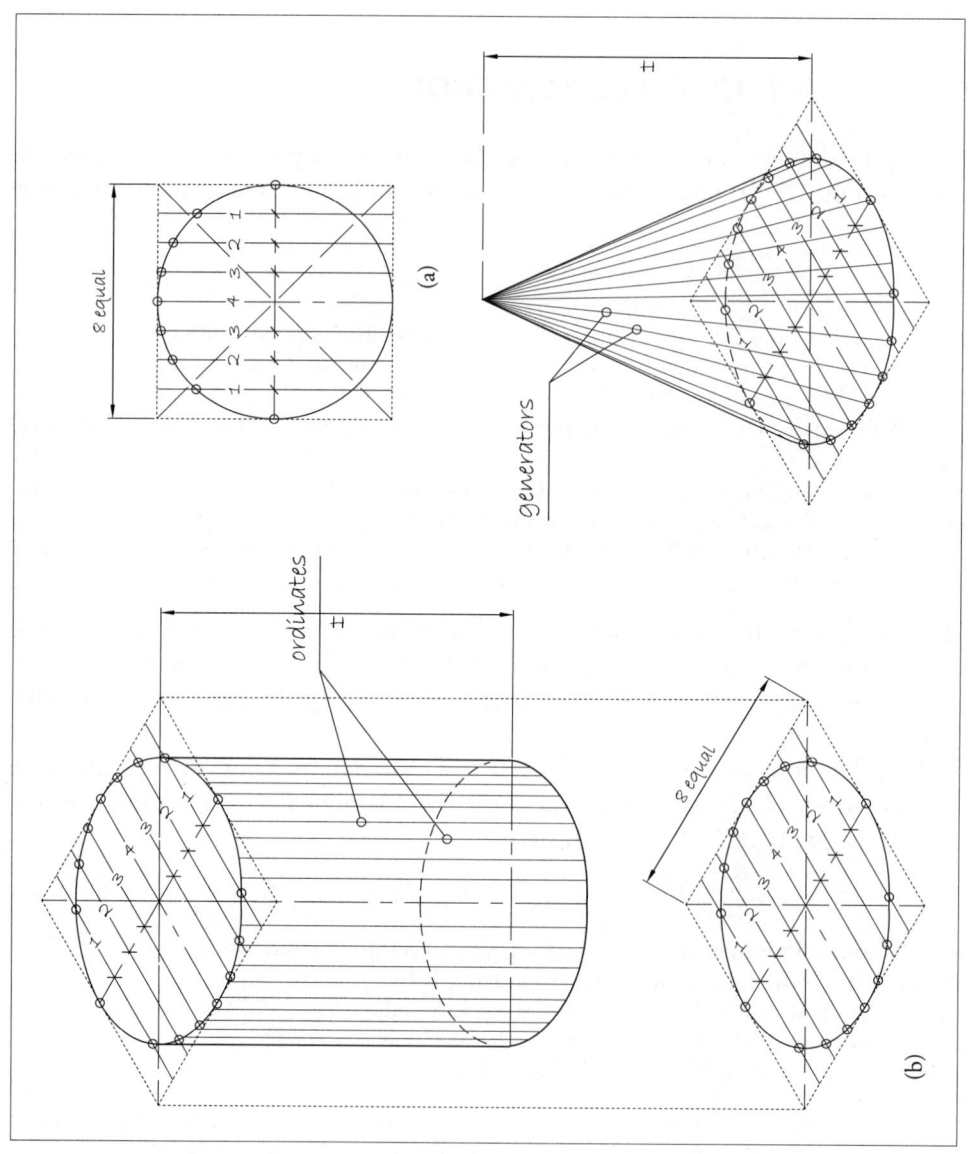

Fig. 1.6a
Fig. 1.6b
Fig. 1.7

1 ISOMETRIC PROJECTION

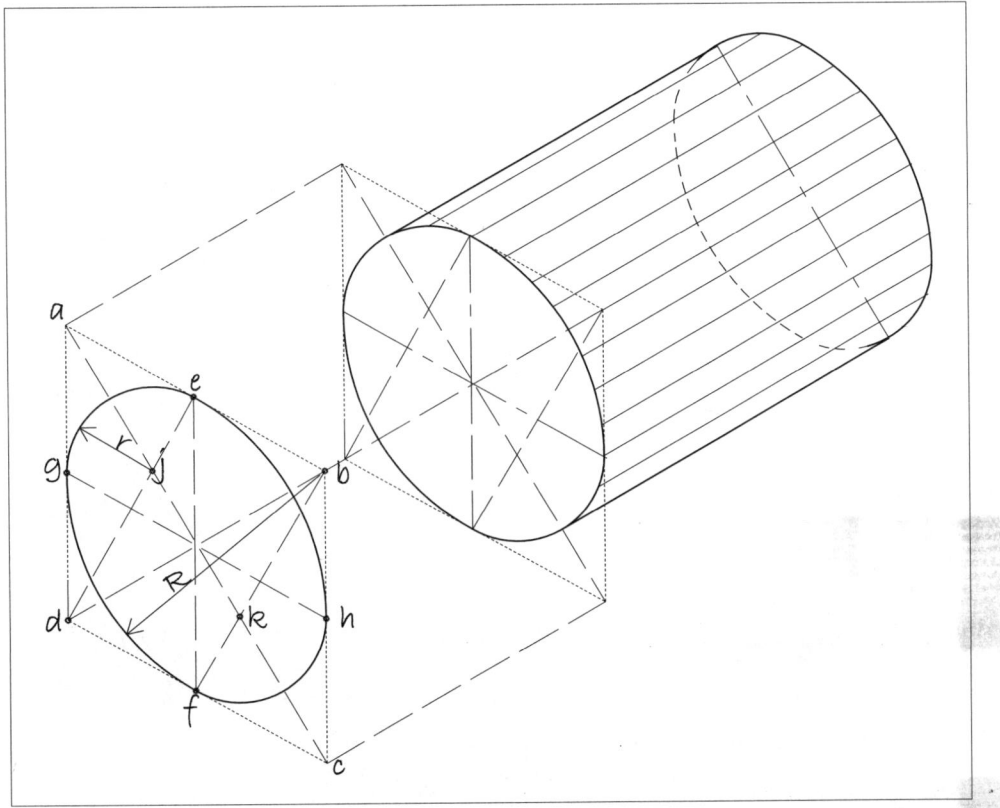

Fig. 1.8

GEOMETRICAL DRAWING FOR CARPENTRY & JOINERY

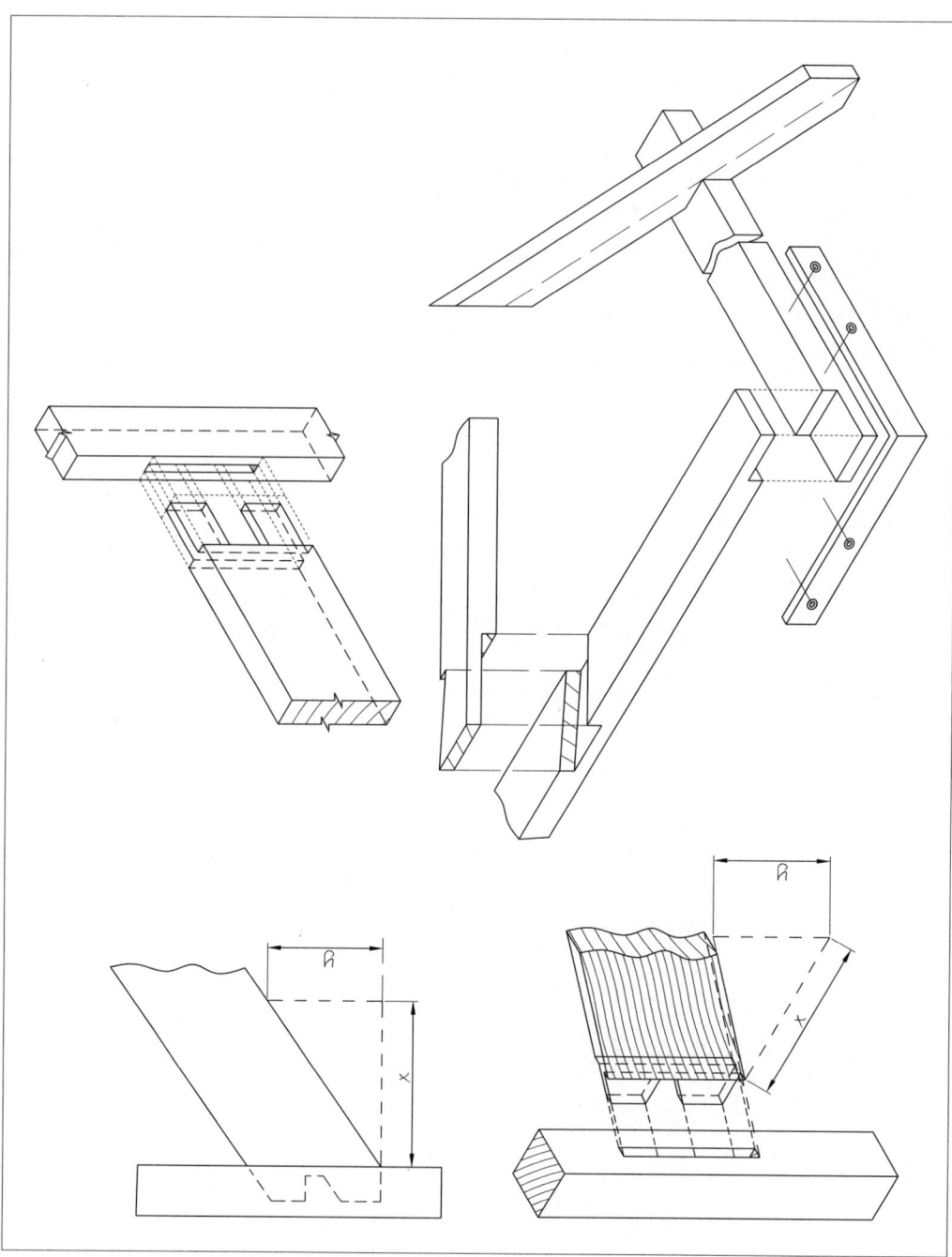

Fig. 1.9

1 ISOMETRIC PROJECTION

Exercise

The orthographic projection of a roof is shown in **Fig. 1.10**. Reproduce **Fig. 1.10** and complete the exercise.

- Draw an isometric projection of the roof. Let the semi-circular part of the roof be the lowest end of the projection.

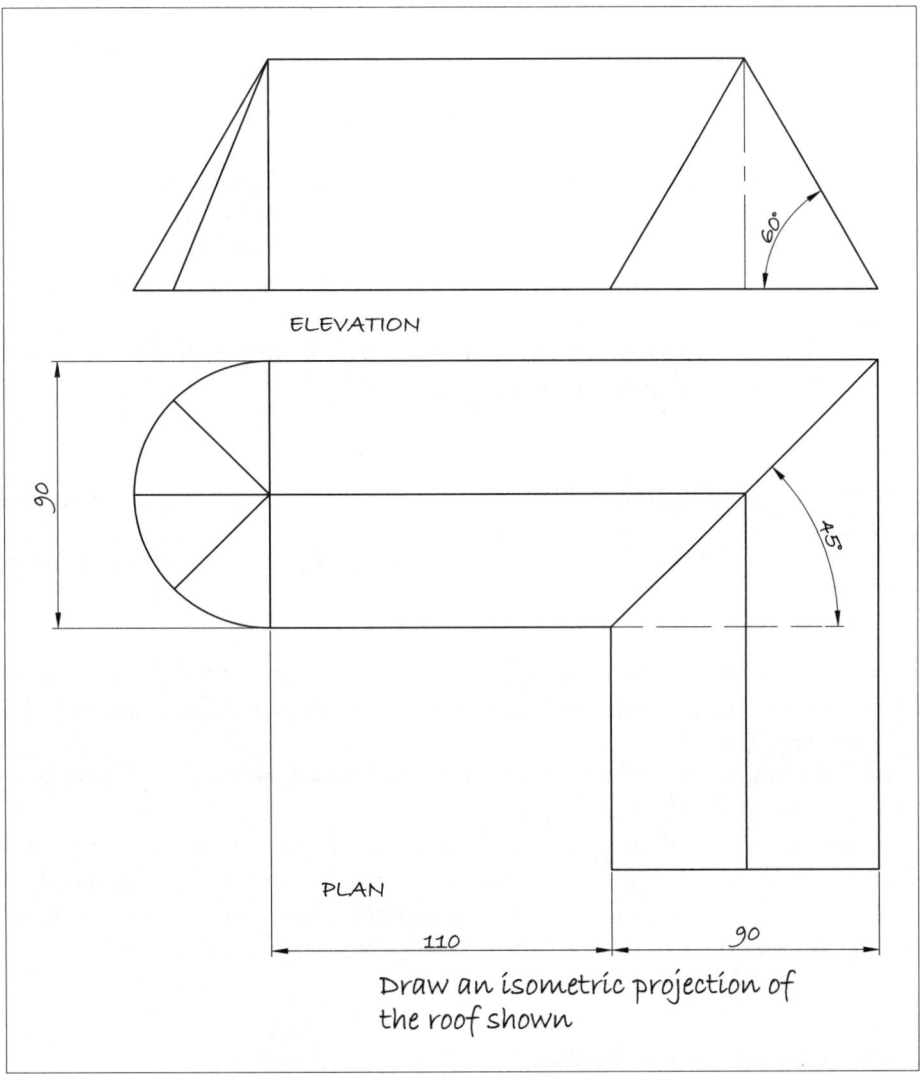

Fig. 1.10

2 PROJECTIONS OF LINES AND TRUE LENGTHS

Out of all the aspects of plain and solid geometry covered in this book, understanding and visualising lines and their true lengths is one of the most important. Once a learner grasps the concept of finding true lengths of lines, it removes a barrier to many problem-solving procedures relating to the content of this book. This chapter deals with lines and their true lengths; for this purpose, lines may be straight or curved.

Straight lines are dealt with as follows:

In **Fig. 2.1** line **a-b** is horizontal in plan and therefore shows a true length in elevation. This always applies and is worth remembering: a line that is horizontal in plan will always show a true length in elevation. The rule also applies to lines that are horizontal in elevation. A line that is horizontal in elevation will always show a true length in plan. In **Fig. 2.1**, therefore, line **c-d** shows a true length in plan.

Line **e-f** is neither horizontal in plan or elevation and does not therefore show a true length in either view. To obtain a true length in this case:

Method

- Pivot **e-f** to a horizontal position in plan, using **e** as the pivot point and dropping **f** to a horizontal position to **fh**.
- In elevation **e1** does not move (being the pivot point) but **f1** in elevation moves horizontally to become **fh1**.
- In elevation, **e1-fh1** now shows the true length (of **e-f**).

In **Fig. 2.2** another approach is used whereby the lines in question (**a-b**, **c-d**, and **e-f**) are looked at as part of a right-angled triangle (e.g. **cdk**), where the hypotenuse in each case is the true length required.

Each line has a run, a rise and a true length. Using line **a-b** as an example, **a-k** is the run; **k-b1** is the rise and **a1-b1** the true length.

Fig. 2.3 shows three curved lines. Line **a-b** is horizontal in plan and shows a true length (true shape) in elevation; conversely line **c-d** is horizontal in elevation and shows a true length in plan.

Line **e-f** is neither horizontal in plan or elevation and consequently does not show a true length. To obtain the true length of **e-f** the following procedure is followed:

Method

- Divide line **e-f** in plan into parts **f1234e**.
- Project to elevation.
- In the elevation, construct the baseline at **f1** and determine height lines **h1**, **h2**, **h3**, and **h4**.
- Take height lines from elevation and project them at 90° to their relevant points in plan, resulting in **E-f**, which is the true shape of curved line **e-f**.

2 PROJECTIONS OF LINES AND TRUE LENGTHS

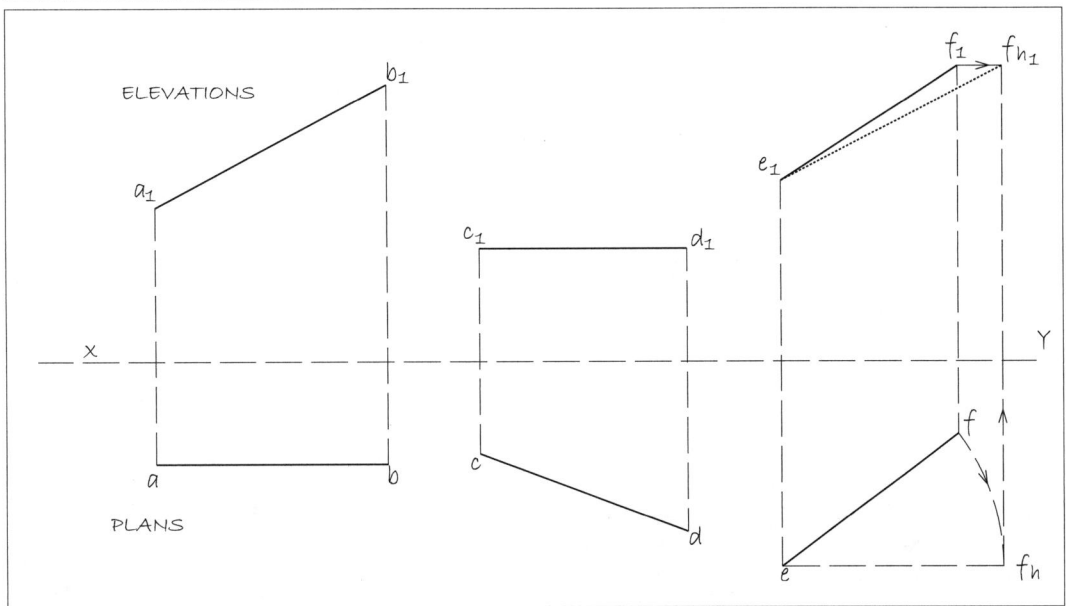

Fig. 2.1

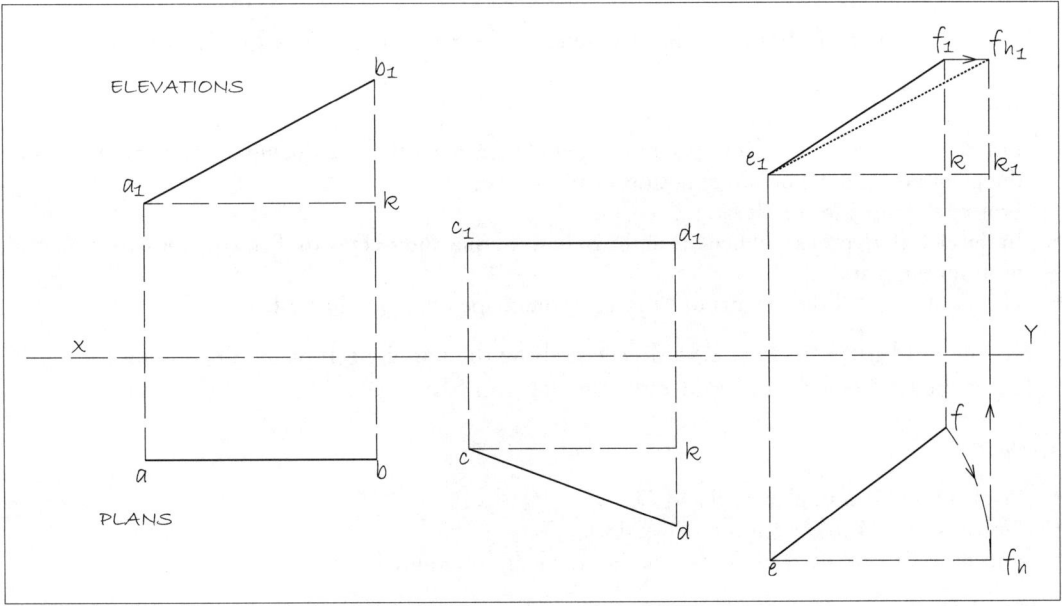

Fig. 2.2

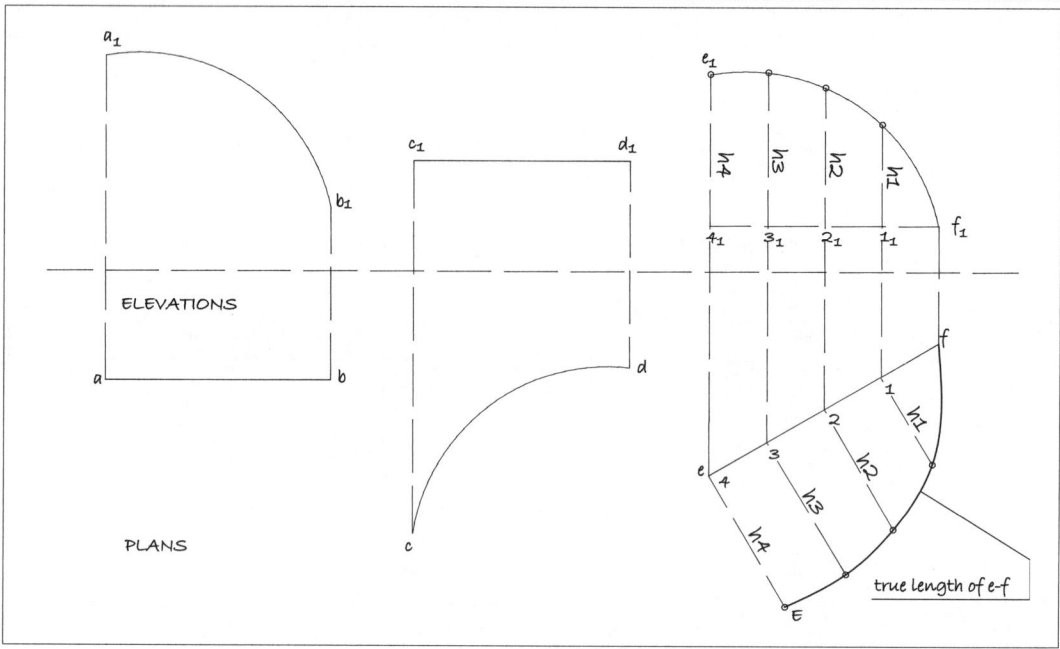

Fig. 2.3

Another method of obtaining the true length of **e-f** is employed in **Fig. 2.4**.

Method

- Divide (as before) line **e-f** into parts in plan and then pivot the line (including its parts), using **f** as the pivot point, to a horizontal position ending in **E**.
- Project all points to the elevation.
- In the elevation, project all heights to the right until they intersect with their counterparts projected up from the plan.
- Join f_1 through all the heights to E_1 to give true shape of curved line **e-f**.

The pyramid turret in **Fig. 2.5a** does not show the true length of its hips in elevation. To get the true lengths of hip **o-f** and smaller hip **f-b**:

Method

- Pivot **b-f** and **b-o** in plan to **F** and **O**.
- Project **F** and **O** to elevation to **F1** and **O**.
- Join b_1 to F_1, and F_1 to **O**. These are the true lengths required.

Fig. 2.5b shows one half of the turret reproduced, but this time the hips are horizontal in the plan view, resulting in the true lengths being revealed in the elevation.

2 PROJECTIONS OF LINES AND TRUE LENGTHS

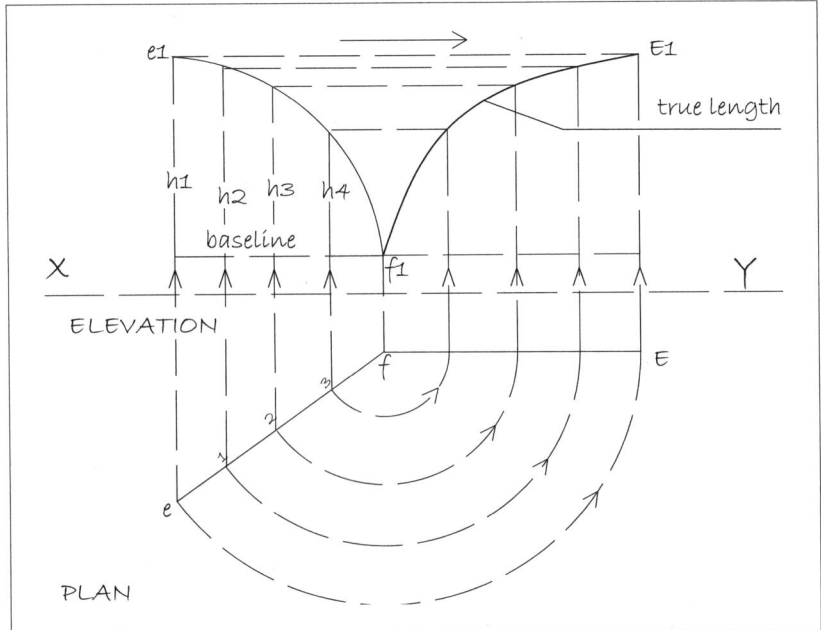

Fig. 2.4

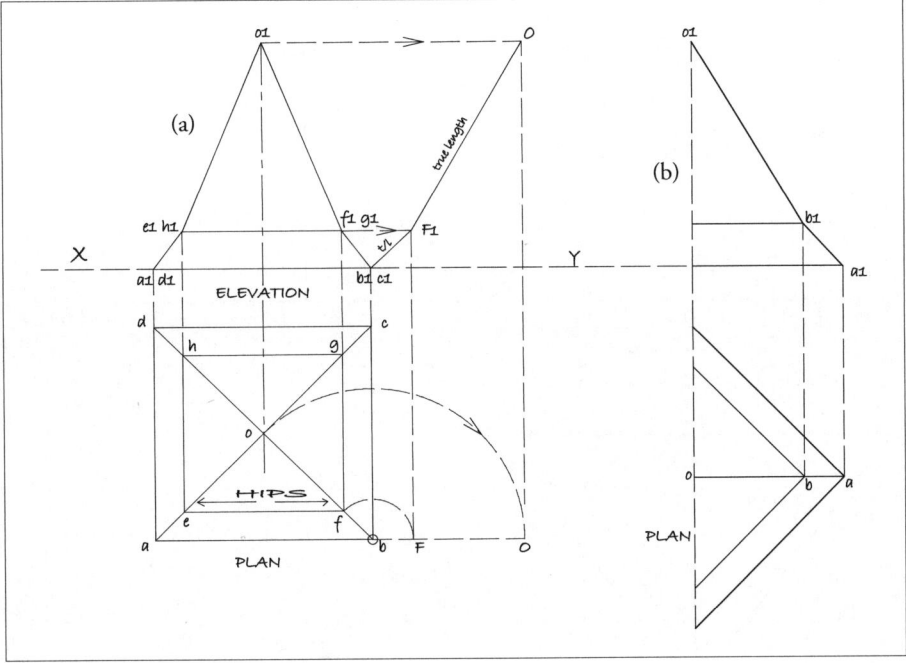

Fig. 2.5a
Fig. 2.5b

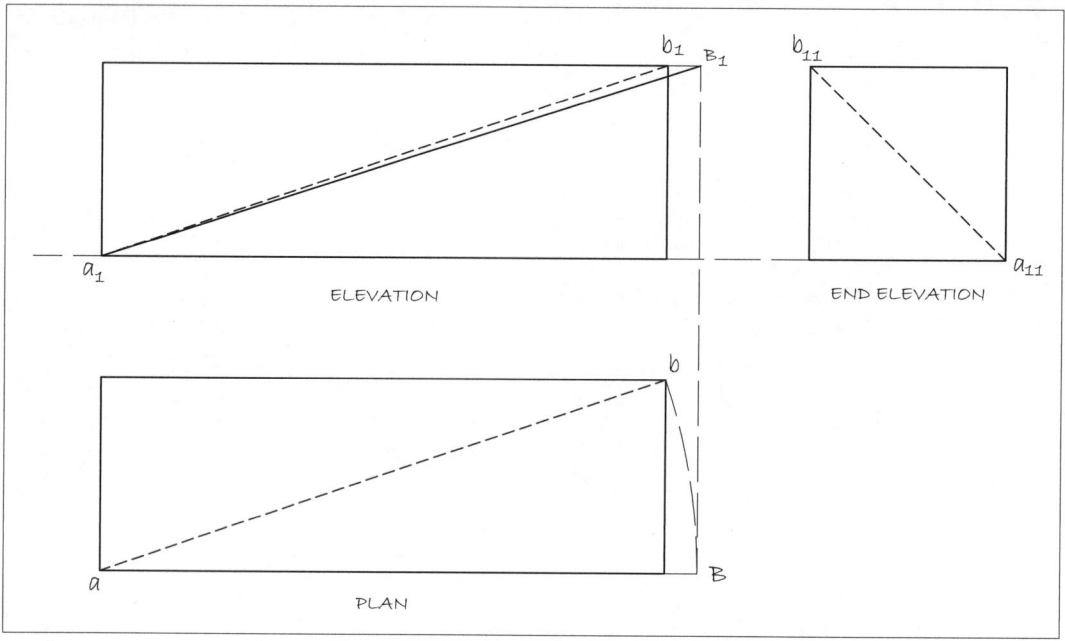

Fig. 2.6

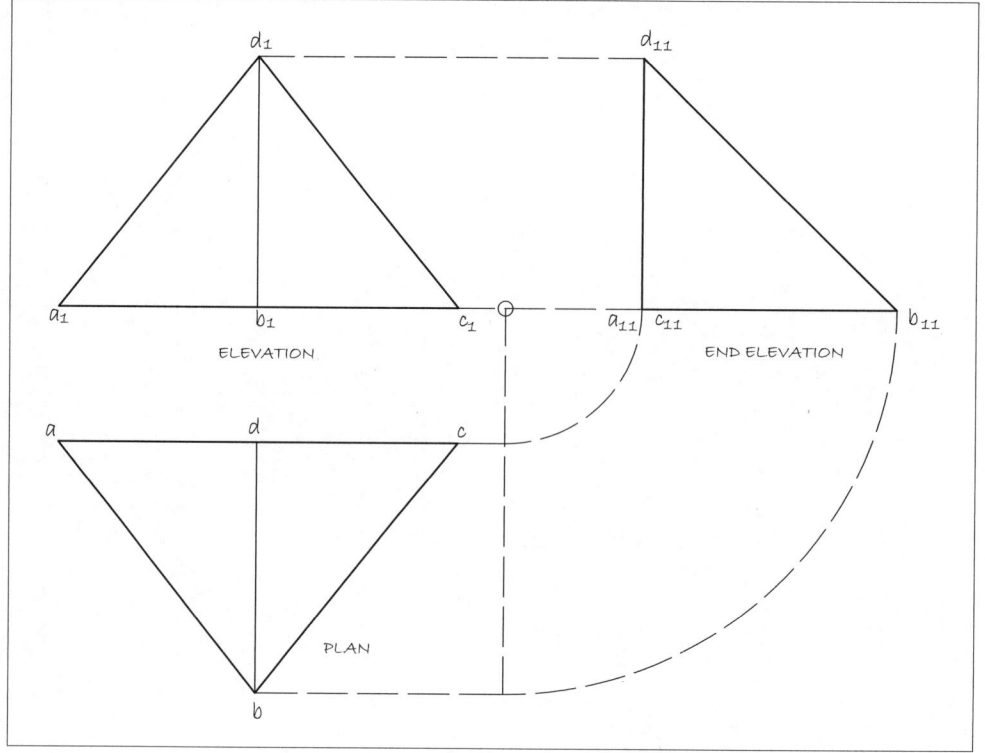

Fig. 2.7

2 PROJECTIONS OF LINES AND TRUE LENGTHS

The plan, elevation and end elevation of a prism is given in **Fig. 2.6**. A line **a-b** representing the longest diagonal of the prism is also shown. The true length of the said diagonal (which is **a1–B1**) is found using the principles employed in the figures above.

A triangular bay window is represented in **Fig. 2.7**. The lines that show true lengths are as follows:

- In plan: **a-b** and **b-c**
- In elevation: **a1-d1** and **d1-c1**
- In end elevation: **d11-b11**

Exercise

1. In **Fig. 2.8 (1)** the plan and elevation of a roof to an angled bay window is given.
   - Identify the lines which are true lengths in both views.

2. In **Fig. 2.8 (2)** the plan and elevation of a square prism is shown.
   - Find the true length of the lines **a**, **b**, **c** and **d**, which are within the prism.

3. In **Fig. 2.8 (3)** the plan and elevation of a semi-elliptical disc is given.
   - Find the true shape of the line **a-b-c**.

4. In **Fig. 2.8 (4)** the plan and elevation of a plinth are given.
   - Find the length of line **a-b**.

5. In **Fig. 2.8 (5)**:
   - Find the true length of the lines **e-b** and **f-c**.

6. In **Fig. 2.8 (6)**:
   - Find the true lengths of the inner generators of the cone.

GEOMETRICAL DRAWING FOR CARPENTRY & JOINERY

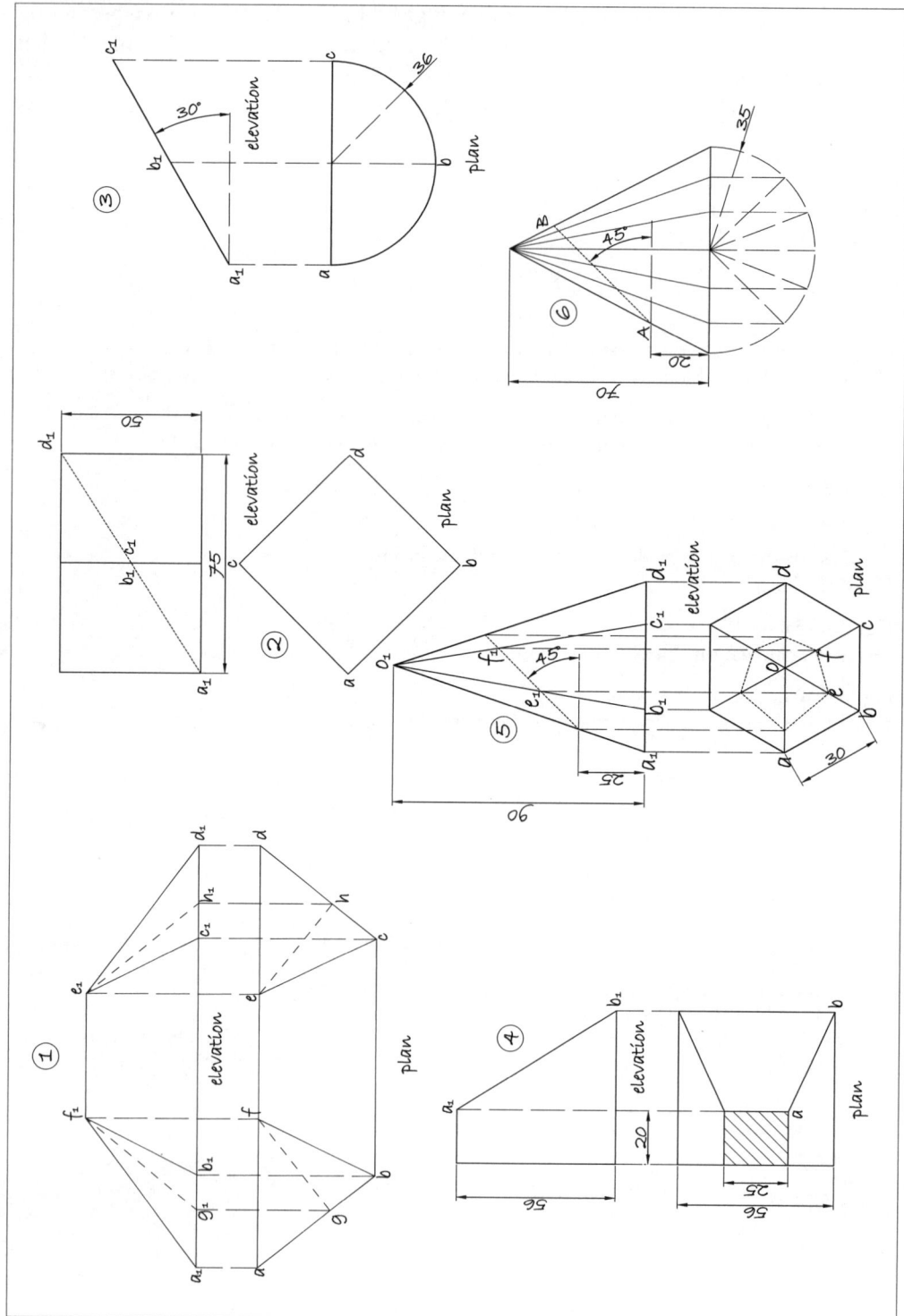

Figs. 2.8

3 AUXILIARY ELEVATIONS AND PLANS

SECTION I: AUXILIARY ELEVATIONS

While Chapter 17 on constructions covers planes of projection including an explanation of orthographic projection, this chapter explores further the vertical planes of projection onto which auxiliary elevations are projected. While a more rounded knowledge of orthographic projection is a prerequisite for the course of study covered in this book, auxiliary elevation deserves some coverage here, in view of its importance in solving many of the problems encountered with developments, true lengths, and so on.

What is an auxiliary elevation? Imagine a building in the centre of a town and you are walking around that building. As you come round the building, its shape changes with every step, and you see a new elevation (auxiliary elevation) every time.

In **Fig. 3.1**, assume **A**, **B** and **C** to be different positions of a spectator viewing the building shown. At position **B**, you see the building with its front square-off to your line of vision. At position **A**, you see the front of the building as well as some of the sidewall on the left; at **C**, the front wall and side wall to the right are visible.

That is what auxiliary elevations are all about: the changed appearance of things as you view them from different angles. Note: It is important to understand that as you view the building from different positions on the ground, the heights do not change (other than in perspective projection, which is not relevant here).

Method

- When the line of vision is determined, construct an **X1-Y1** line square-off (the line of vision), preferably clear of the existing elevation.
- Project all relevant points from the plan, parallel with the line of vision.
- Construct heights **h1** and **h2** square-off **X1-Y1**.
- Join all relevant points, at the relevant heights, to reveal the auxiliary elevation.

Note: There are many auxiliary elevations possible from point **A**. The auxiliary elevation shown is projected parallel with the 'line of vision' indicated by the arrow (more about line of vision to follow).

In **Fig. 3.2** the plan and elevation of a square-based pyramid is shown. Projected from the plan are two auxiliary elevations; the line of vision from point **A** is at 45° and from point **B** the angle is random (in this case, somewhere less than 45°).

GEOMETRICAL DRAWING FOR CARPENTRY & JOINERY

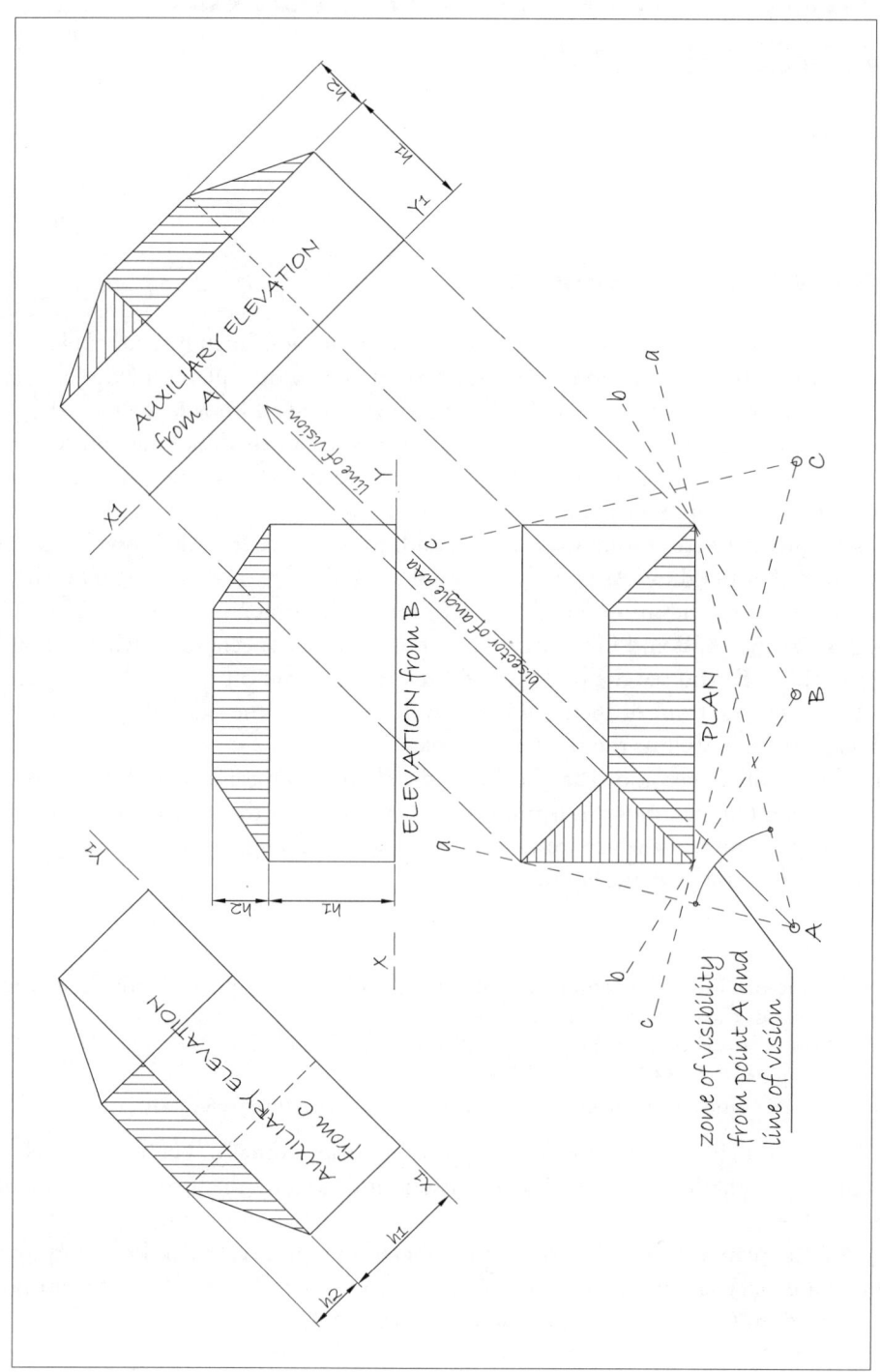

Fig. 3.1

3 AUXILIARY ELEVATIONS AND PLANS

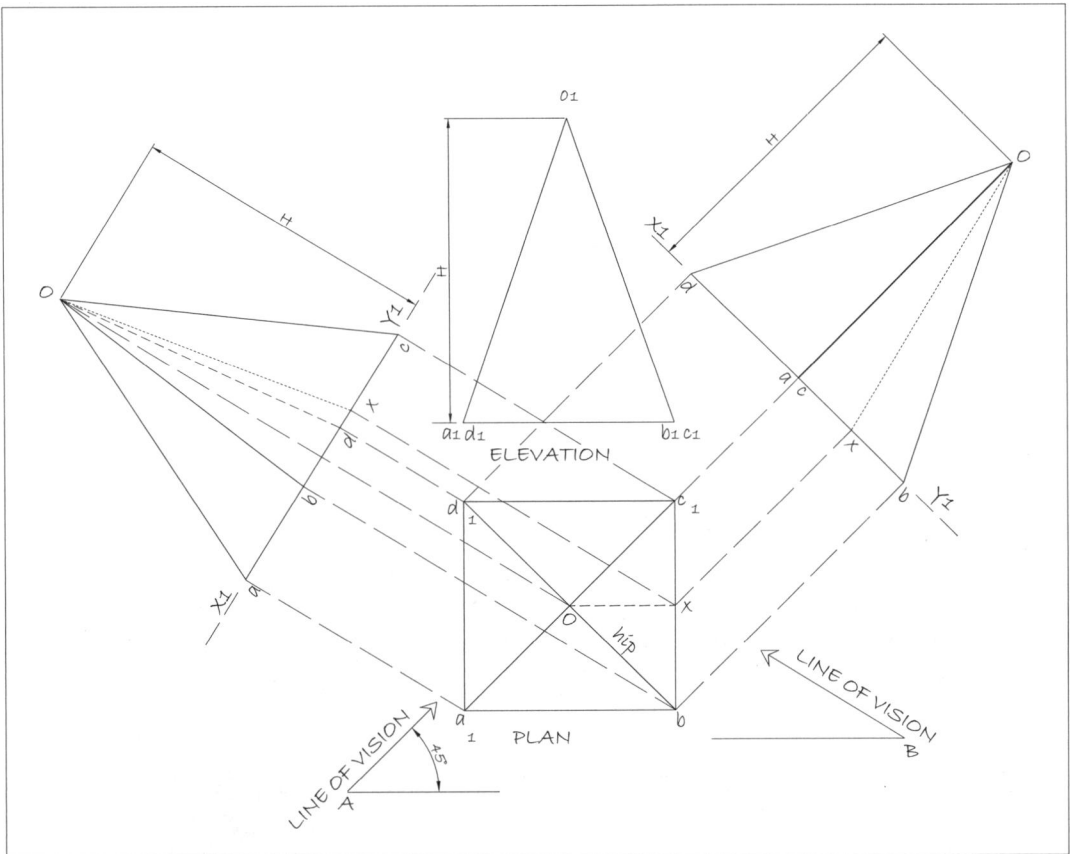

Fig. 3.2

Method

- Construct new **X-Y** line (**X1-Y1**) at right angles to line of vision. This line may be constructed anywhere but preferably clear of existing elevation.
- Construct height line(s), in this case, **H**, at right angles to **X1-Y1** line.
- Project all relevant points from plan onto auxiliary elevation and join them up.

The hip **o-b** of the pyramid is at 45° in the plan and therefore at right angles to the line of vision from **A**; in other words, the hip is square-off the line of vision. When something is being viewed square-off in this way, a true length is visible in the projected view. Hip **o-b**, therefore, shows a true length in auxiliary elevation, right-hand side. In the auxiliary elevation projected from **B**, no true lengths are visible (except the height H).

Note: The line **o-x** in the plan shows a true length in the elevation (as **o1-b1**).

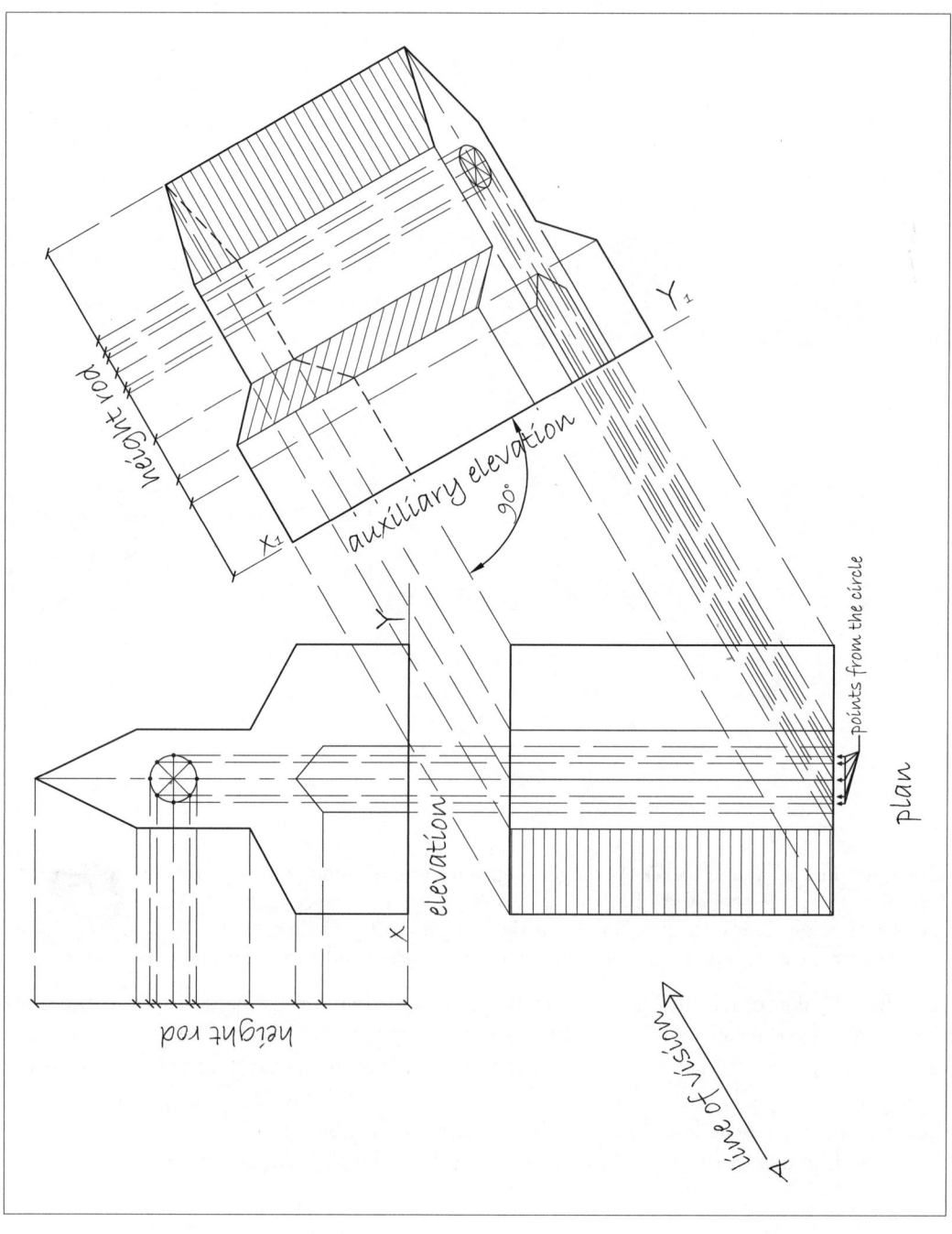

Fig. 3.3

3 AUXILIARY ELEVATIONS AND PLANS

Fig. 3.3 shows the plan and elevation of a building with an auxiliary elevation projected from point **A**. In this case, a circular window is shown in the elevation. The circumference is divided into parts (in this case, eight), which are projected to the height rod; thus enabling the points on the curve to be found in the auxiliary elevation. In the auxiliary elevation the circle appears as an ellipse.

Note: A height rod is a convenient device for transferring heights from the elevation to the auxiliary elevation.

SECTION II: AUXILIARY PLANS

Fig. 3.4 shows the plan, elevation and auxiliary plan of a truncated square-based pyramid.

The auxiliary plan is projected from point **A1** at the given angle of projection, which in this case is at 90°, to the sloping surface of the pyramid.

Method

- Construct an **X1-Y1** at 90° to the line of vision and set off the widths, taken square-off from the **X-Y** line in plan as shown.
- Project all points of the pyramid from the elevation to intersect at the relevant projected widths.

In this situation, imagine yourself as a spectator at point **A1** looking at the pyramid at the given angle of projection, which in this case is at 90° to the sloping surface of the pyramid. You see the true shape of surface **ABCD**, projected in the auxiliary plan. As you view the pyramid from **A1**, which is **A** in the plan, the widths from the **X-Y** line do not change.

A cylinder truncated at an angle of 45° is shown in **Fig. 3.5**. The line of vision for the auxiliary plan is at 45°; consequently, the true shape of the sloping surface of the cylinder will be revealed in the auxiliary plan.

Method

- In the plan, the circumference of the cylinder is divided into a number of parts (in this case, eight). These points are found on the sloping surface in the elevation and projected square-off (or parallel with the line of vision).
- **X1-Y1** is set up square to the line of vision along with the widths **a** to **e**. Where the relevant projection lines intersect, points on the true shape are found.

The true shape is an ellipse, the major axis of which is the length of the sloping cut, and the minor axis is the diameter of the cylinder.

In **Fig. 3.6** a pitched roof with a hipped end is shown in plan and elevation. Projected from the elevation is an auxiliary plan of the hipped end. Again, the spectator at **A1** (top left) is looking at the hipped end of the roof square-on and, therefore, sees all parts in true shape, including the roof light opening and all rafters.

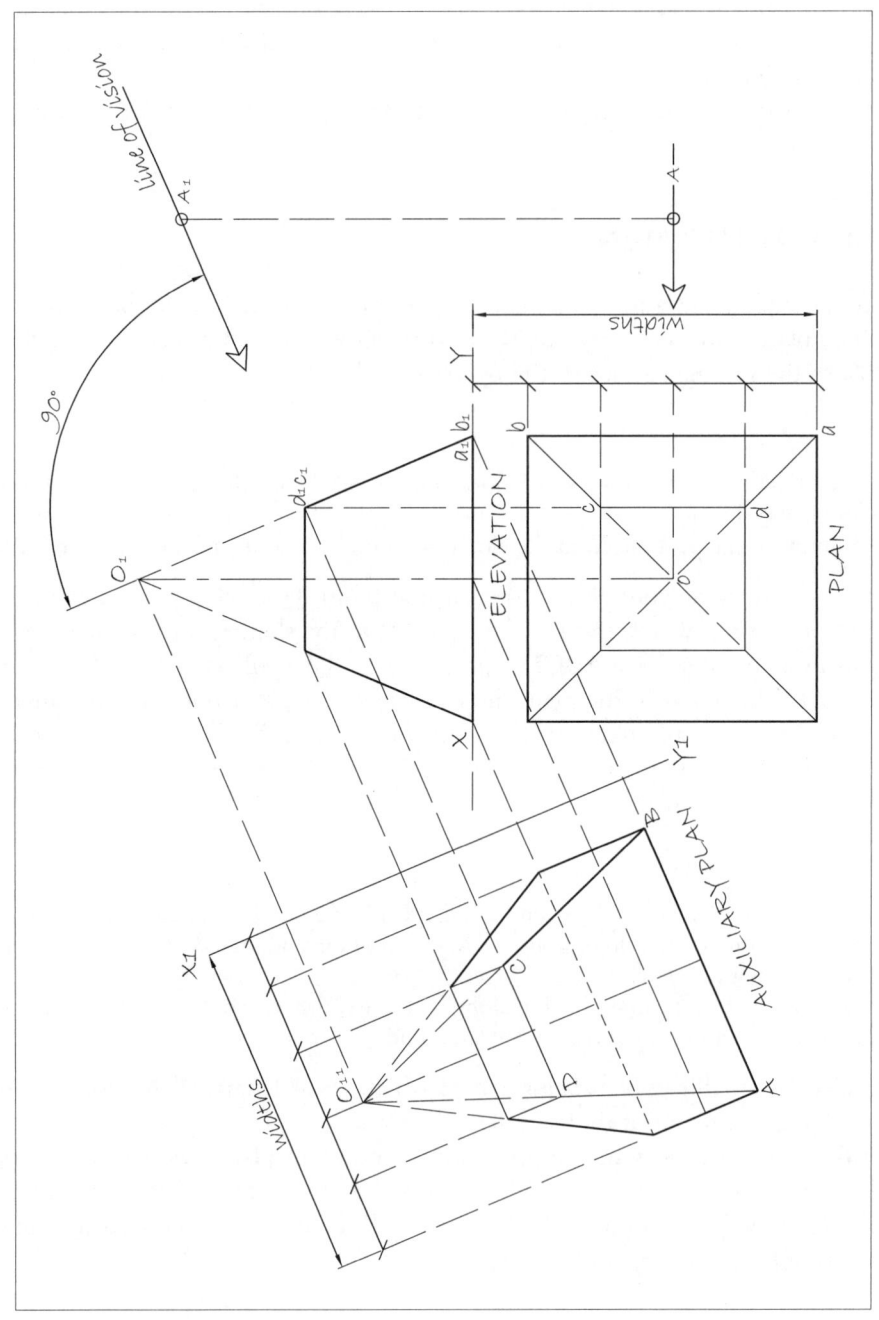

Fig. 3.4

3 AUXILIARY ELEVATIONS AND PLANS

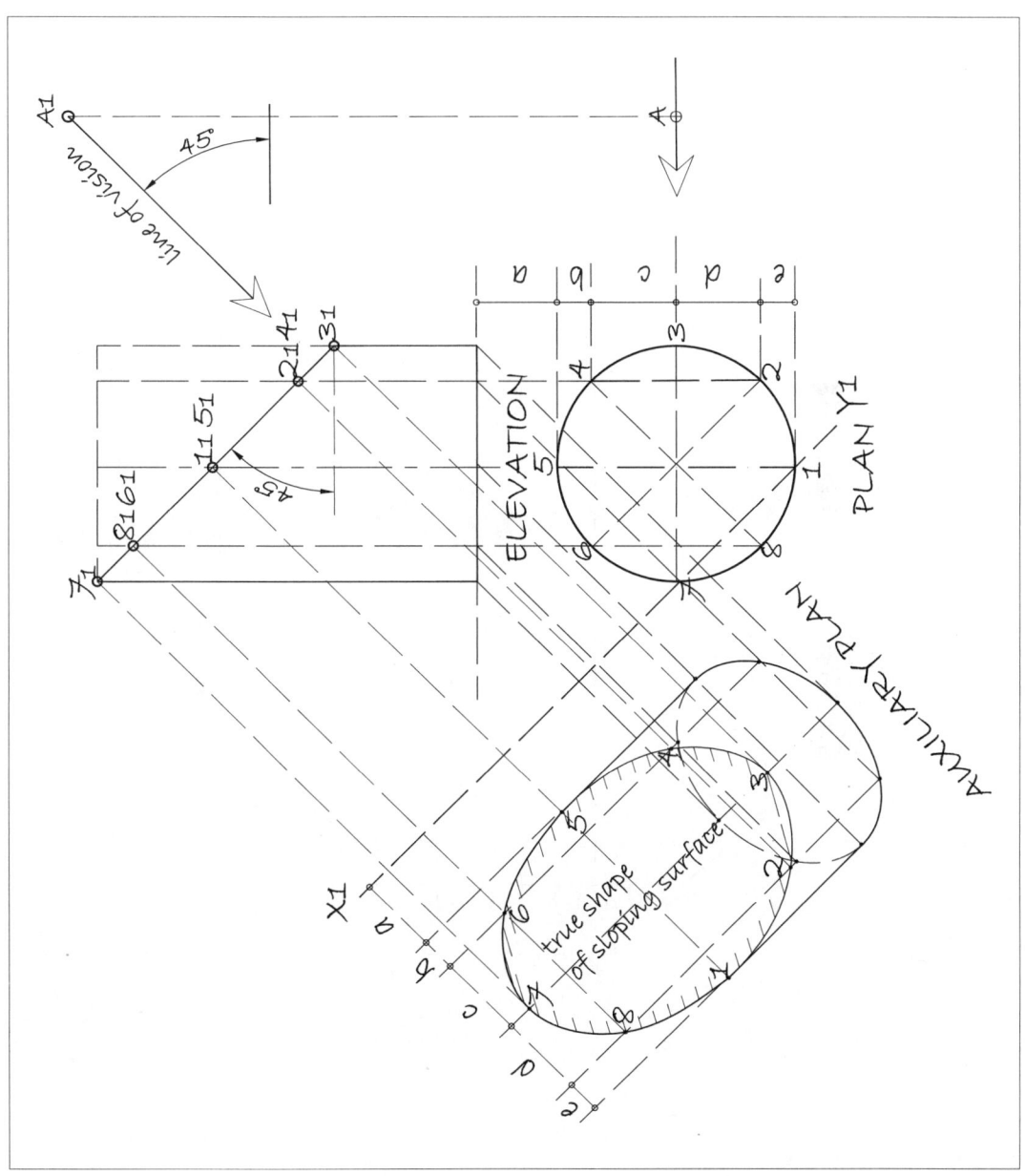

Fig. 3.5

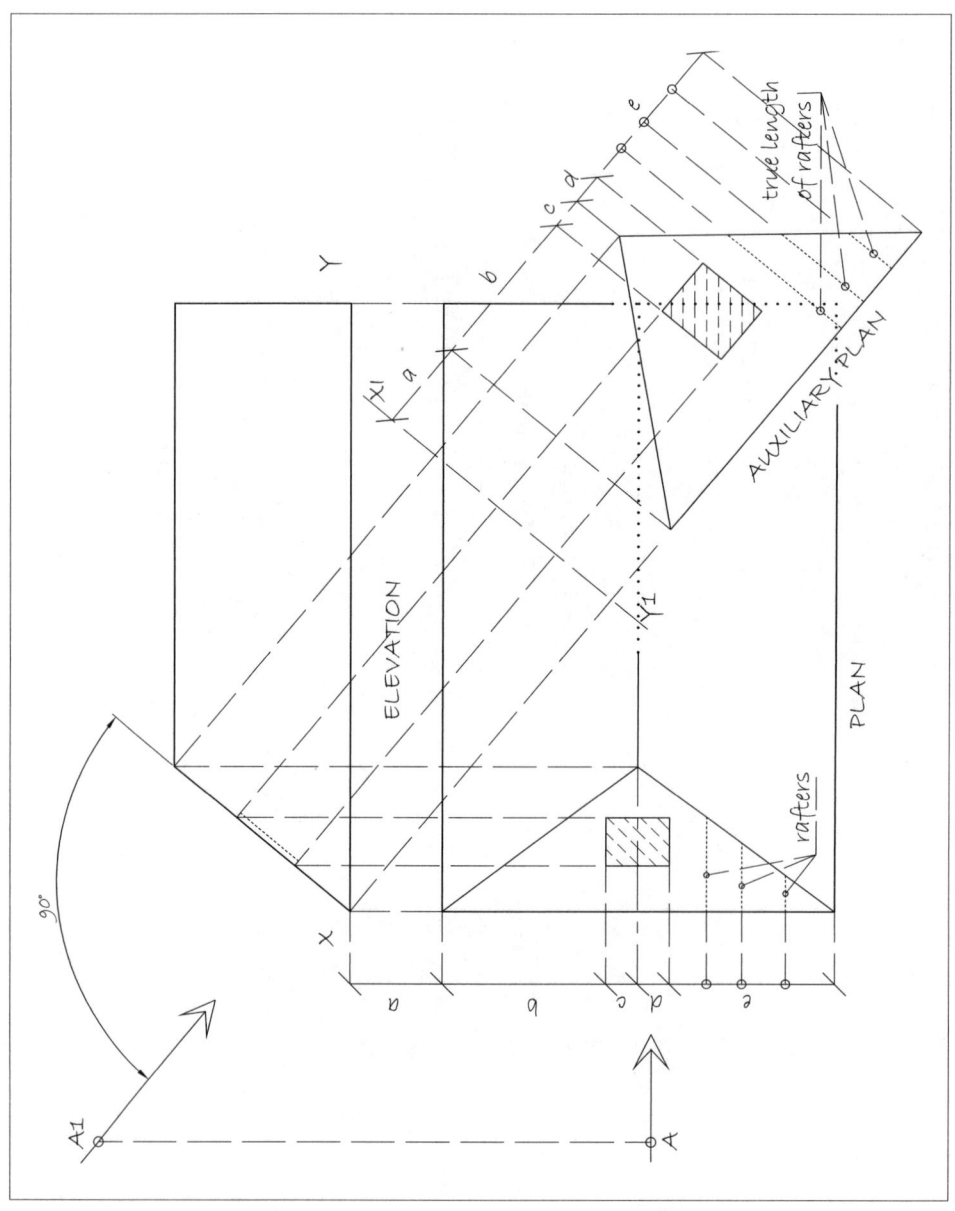

Fig. 3.6

3 AUXILIARY ELEVATIONS AND PLANS

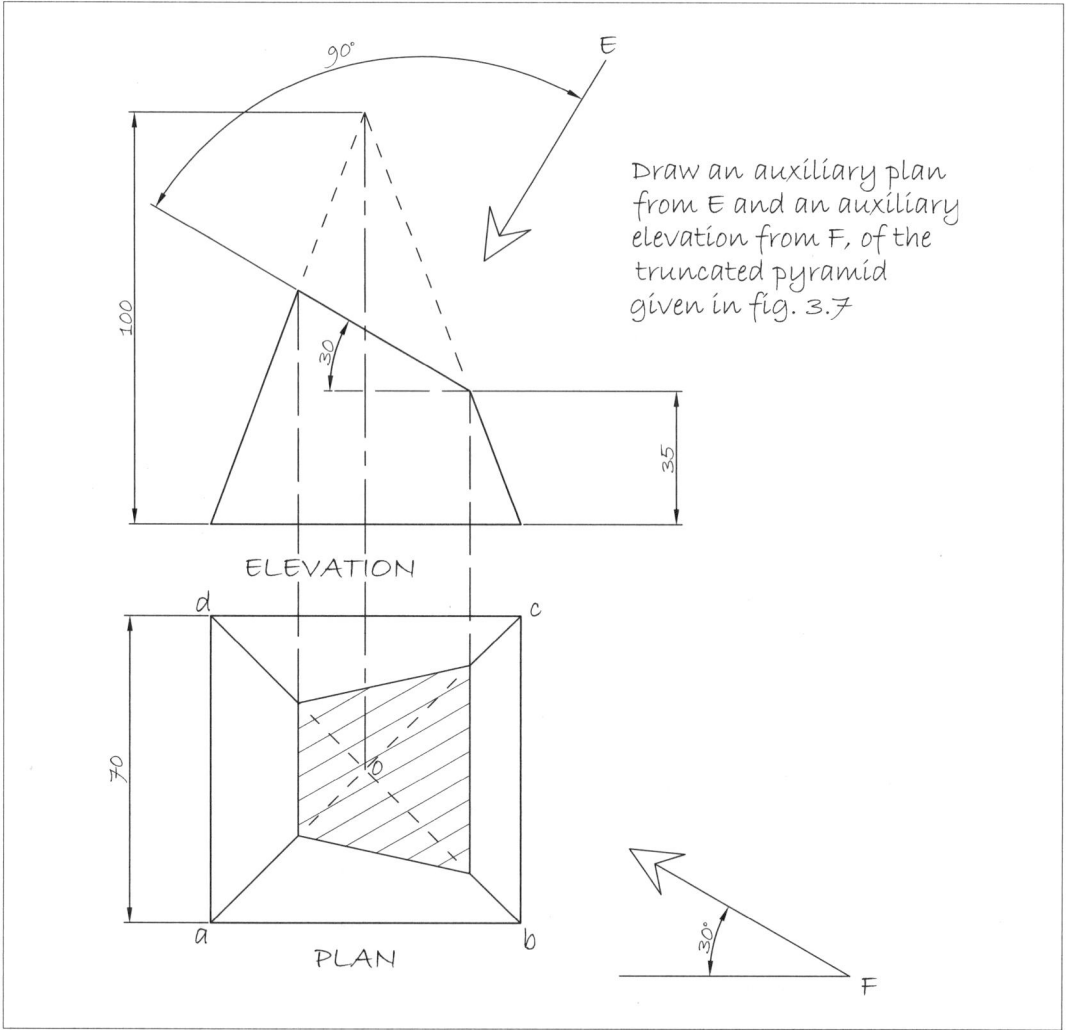

Fig. 3.7

Exercise

1. **Fig. 3.7** shows the plan and elevation of a truncated pyramid. Reproduce **Fig. 3.7** and complete the exercise.
   - Draw an auxiliary plan from **E**.
   - Draw an auxiliary elevation from **F**.

2. **Fig. 3.8** shows the plan and elevation of a pitched roof with a dormer window and a pyramid roof intersecting. Reproduce **Fig. 3.8** and complete the exercise.
   - Draw an auxiliary elevation of the roofs, where the line of vision is at 45° to the plan as shown.

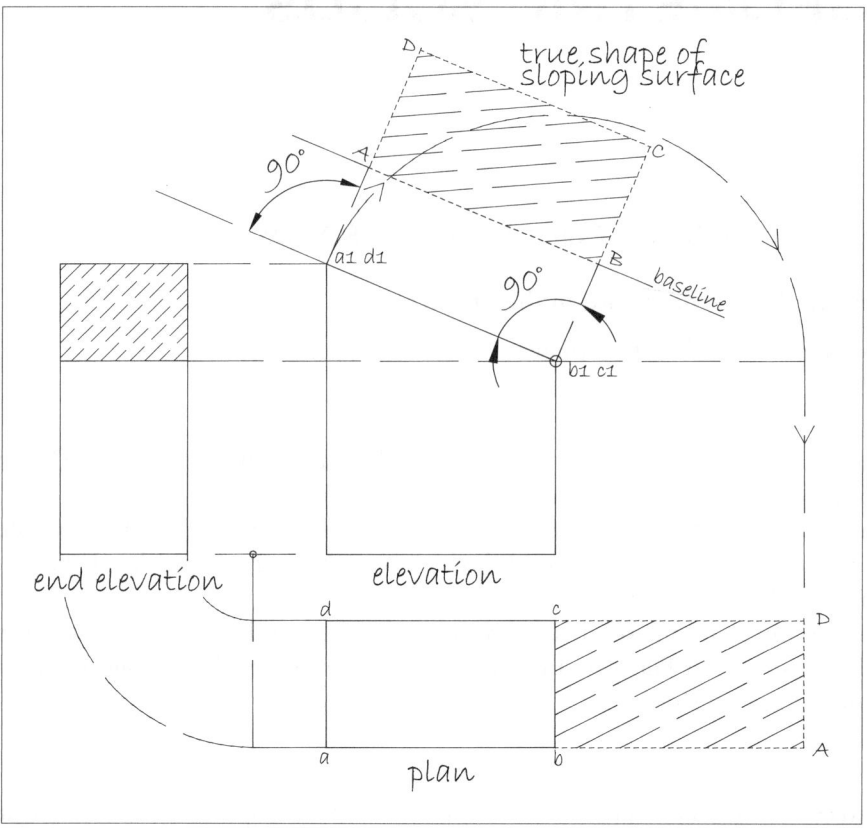

Fig. 4.1

Method 1

- This time in elevation a centre line (**c/l**) is drawn rather than a baseline, at a convenient distance outside the elevation. This **c/l** must be parallel with the sloping surface.
- The half-widths are then taken from the **c/l** in plan to **e**, **f**, **g** and **h** and plotted from the **c/l** in elevation on their relevant projection lines, ending in **E**, **F**, **G**, and **H**. This is the true shape of the cut surface.

Method 2

- As with Method 2 above, the cut surface is made horizontal in the elevation; this time using **e1 f1** as the pivot point (and swinging **h1 g1** to a horizontal position) and projecting to the plan.
- Project **g** and **h** to become **G** and **H**; then join back to **e** and **f** to give the true shape.

4 GEOMETRIC SOLIDS

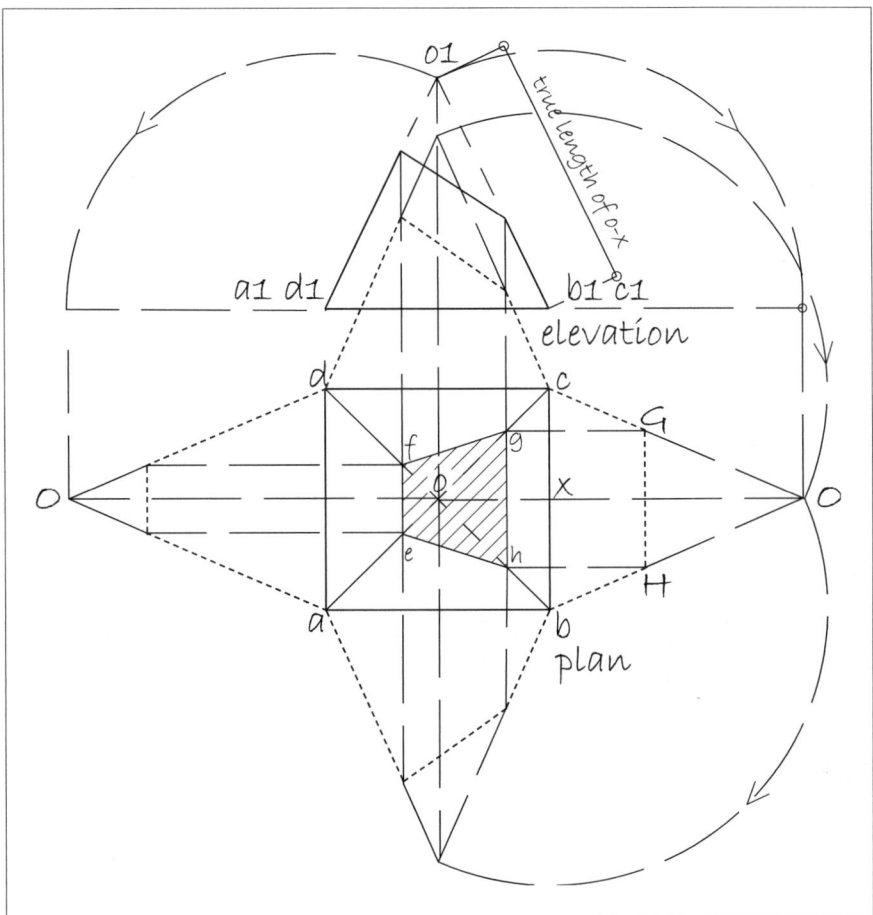

Fig. 4.4

Note: In this development the true length of **o-x** (in roofing terminology, the crown rafter) is used to obtain the desired result. In the **Fig. 4.3**, the true length of the sloping edge (hip) was used.

Fig. 4.5 shows the orthographic projection of truncated cylinders. On the left, the truncation is at an angle of 45° resulting in an end elevation, which shows the cut as a circle. The example on the right shows an ellipse in end elevation.

In both examples in **Fig. 4.5**, the true shape of the cut surface is an ellipse. The method of acquiring the true shape is outlined in both examples.

In **Fig. 4.6** a doubly truncated cylinder is given, and one of the inclined (sloping) surfaces is developed as well as the curved surface of the entire cylinder. The cylinder development, minus its truncated part is also shown.

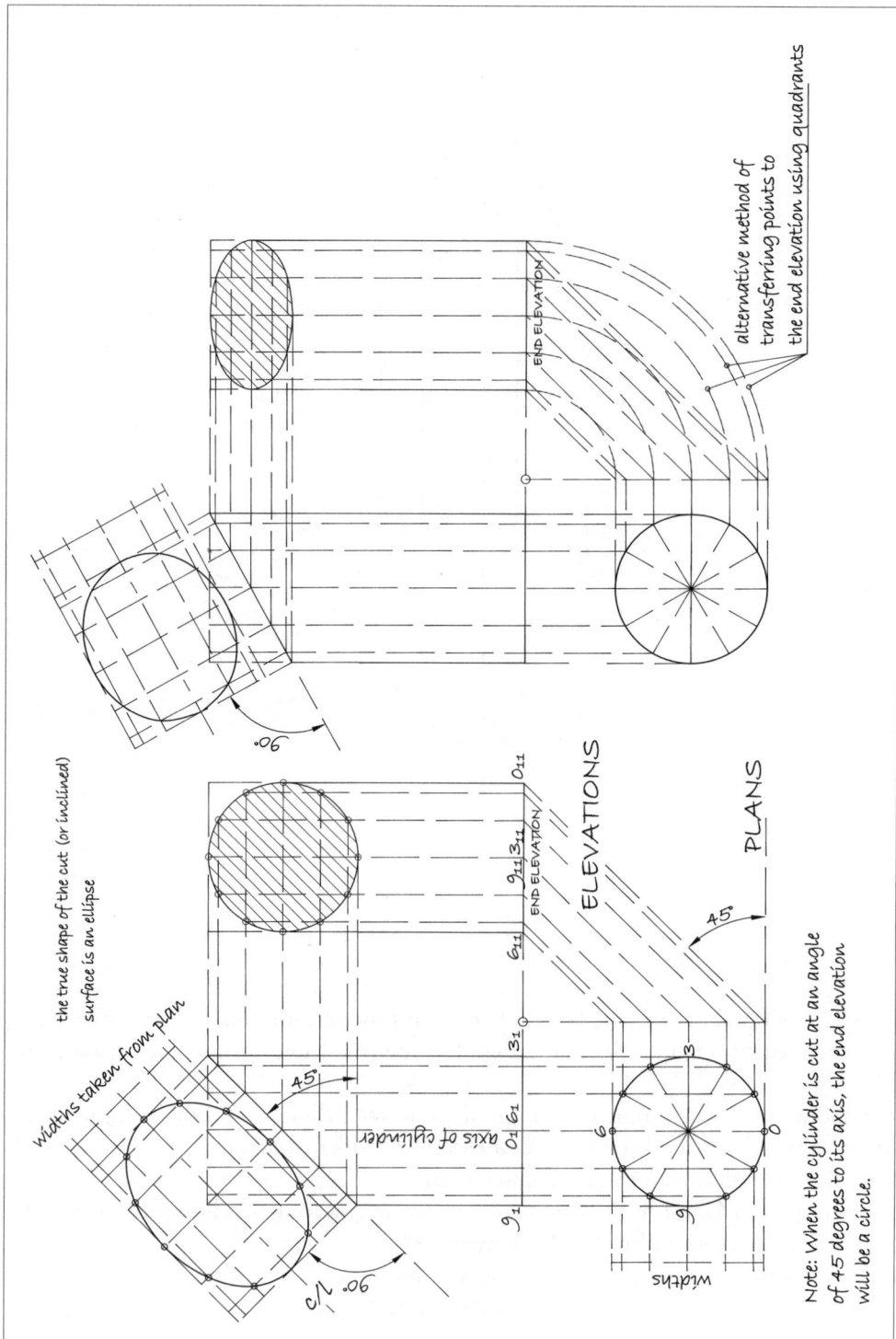

Fig. 4.5

4 GEOMETRIC SOLIDS

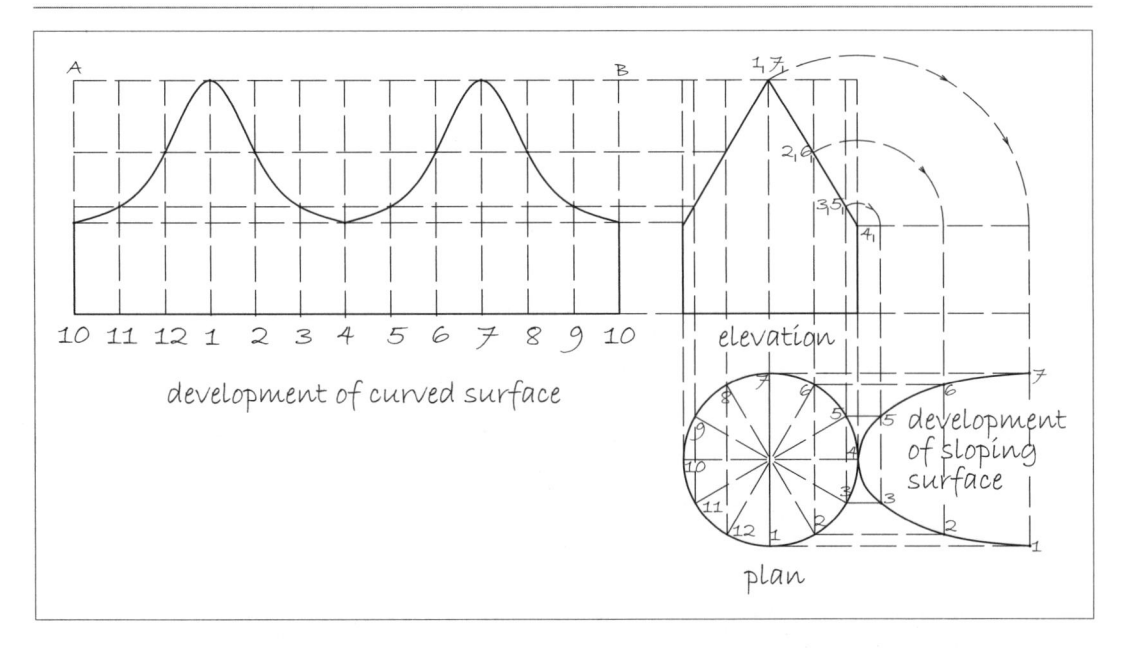

Fig. 4.6

Method

- In the plan, the circumference is divided into the usual 12 parts and the points are projected to the elevation, becoming ordinates on the curved surface of the cylinder.
- To develop the sloping surface, pivot the surface and all points on it to a horizontal position in the elevation and project to the plan.
- Then project points **1**, **2**, **3**, **5**, **6** and **7** horizontally to the right to intersect with the projected points from the elevation.
- Draw a curved line through the relevant intersections to obtain the true shape.

In this development, the surface hinges on point **4** (**4₁** being the pivot point in the elevation).

To develop the curved surface of the cylinder: step-off the 12 equal distances from the circumference in the plan to the left of the elevation as shown and draw construction lines through the same; then project all relevant points from the sloping surface to the left; the grid in which the development lies is now constructed. Join up all relevant points to reveal the development.

Note: **10-A-B-10** represents the development of the complete cylinder.

Fig. 4.7 shows a practical application of developing a truncated cylinder.

A line drawing of a pitched roof with a dormer window that has a segmental roof is given. The exercise is to develop the true shape of the dormer roof.

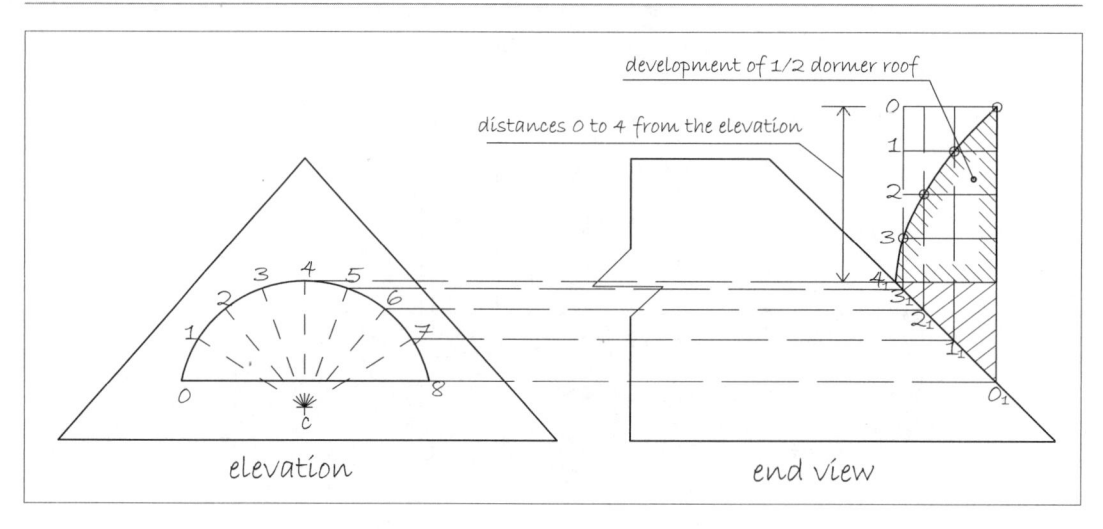

Fig. 4.7

Method

- Divide the curve in the elevation into an equal number of parts, index, and project to the end view.
- In the end view step-off distances **0** to **4** (taken from the elevation) as shown and project construction lines vertically from **0₁**, **1₁**, **2₁**, **3₁** and **4₁**. The grid into which the development fits is now drawn,
- Draw a curved line through the relevant points to complete the development.

In **Fig. 4.8a**, a truncated prism with an octagonal base is drawn in plan, elevation and end elevation. The developed vertical surfaces of the prism are shown on the right.

In **Fig. 4.8b**, the same prism is reproduced with the sloping surface developed, using two methods already covered in this chapter.

Fig. 4.9 shows the projections of a hexagonal-based prism, which is truncated at the given angle. The inclined surface is developed using three methods: two of these have been covered above; the third shows the inclined surface 'flipped' vertically in the end elevation. Part of the elevation is reproduced to facilitate the third method in order to avoid 'crowding' of the given elevation.

As a geometric solid, the cone is possibly the most interesting and intricate of all. It will be dealt with more comprehensively in Chapter 7 on conic sections. **Fig. 4.10** is included here to illustrate how the curved surface is developed.

A cone is generated by a sloping line, which revolves round a vertical axis. Imagine spinning a right-angled set square on one of its vertical sides; the hypotenuse will generate a cone.

A cone and its practical application as a traffic cone is represented in **Fig. 4.10**, the outside line in the elevation shows the true length of the cone's generator.

4 GEOMETRIC SOLIDS

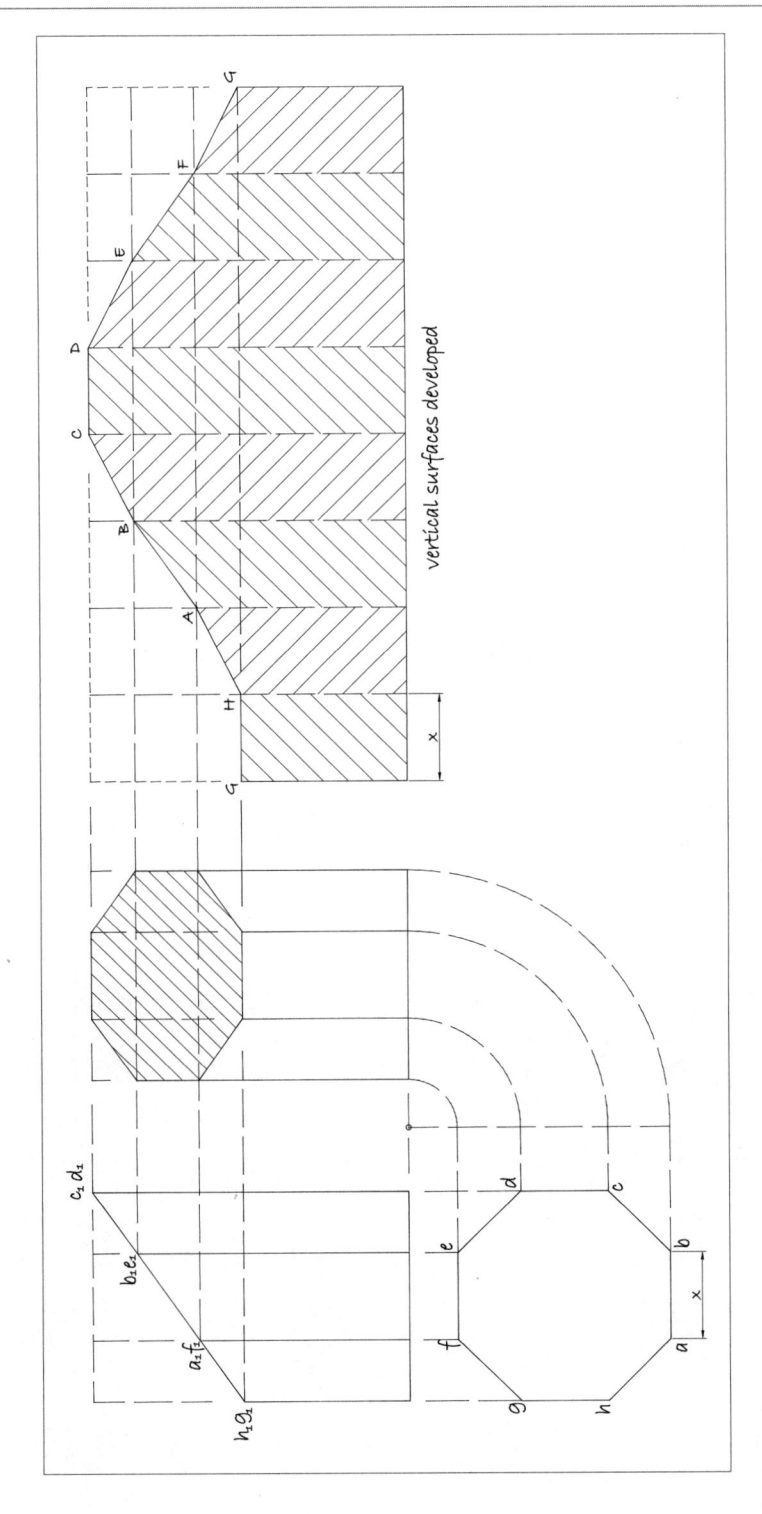

Fig. 4.8a

The curved surface of a cone can only be developed from its apex. This is done as follows:

Method

- From the apex, swing the true length of the generator as an arc.
- On the arc set off the length of the circumference, taken from the plan and join back to the apex. This gives the full curved surface developed.
- To develop the shaded part, swing the relevant arcs again from the apex to complete.

Exercise

1. **Fig. 4.11** shows the plan and elevation of a pentagonal-based prism, which is truncated by the cutting planes as shown. Reproduce **Fig. 4.11** and complete the exercise.
   - Draw an end elevation on both sides.
   - Develop the vertical surfaces of the prism.
   - Develop the sloping surfaces of the prism.

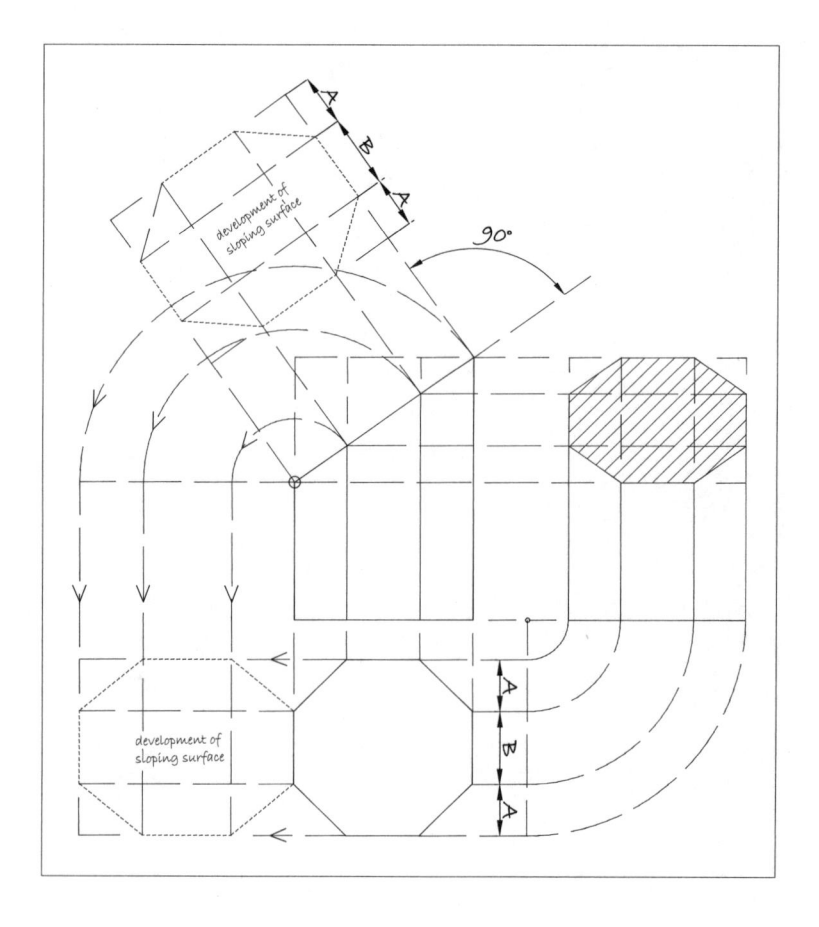

Fig. 4.8b

4 GEOMETRIC SOLIDS

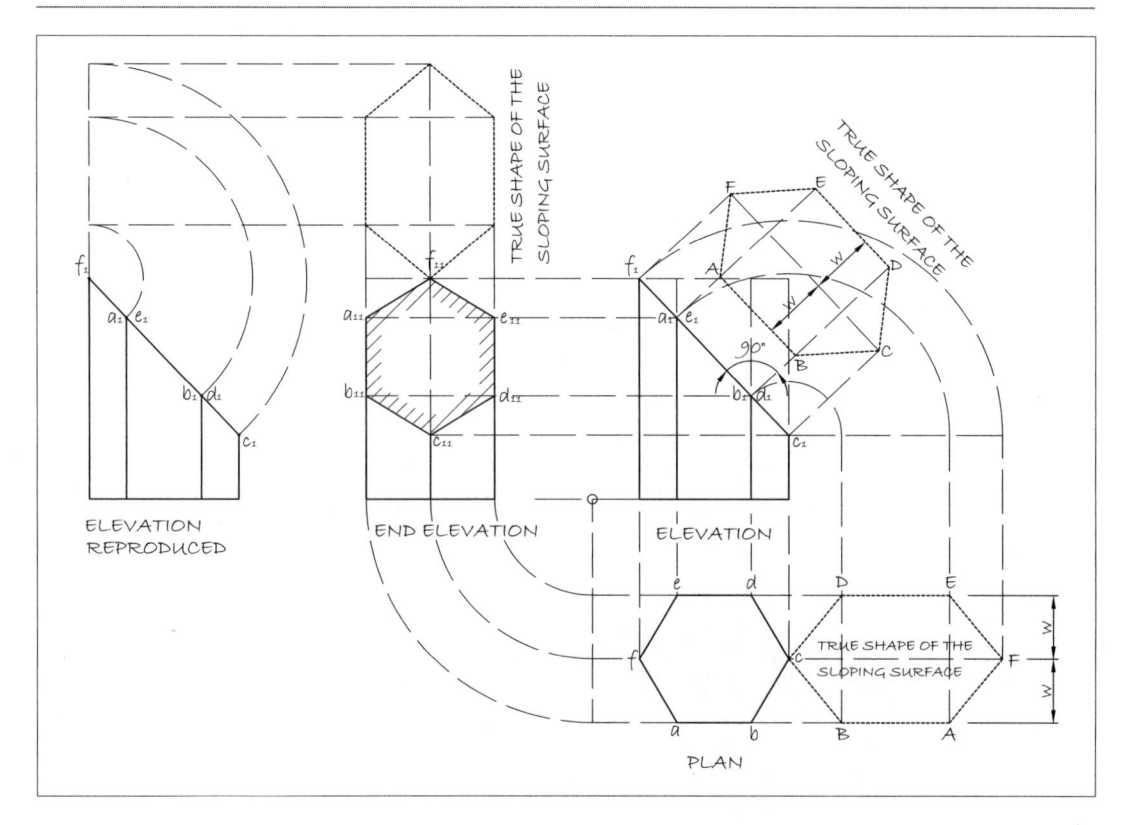

Fig. 4.9

2. **Fig. 4.12** shows the incomplete orthographic projection of a truncated, hexagonal pyramid. Reproduce **Fig. 4.12** and complete the exercise.
    - Complete the plan and end elevation.
    - Develop the truncated surface.

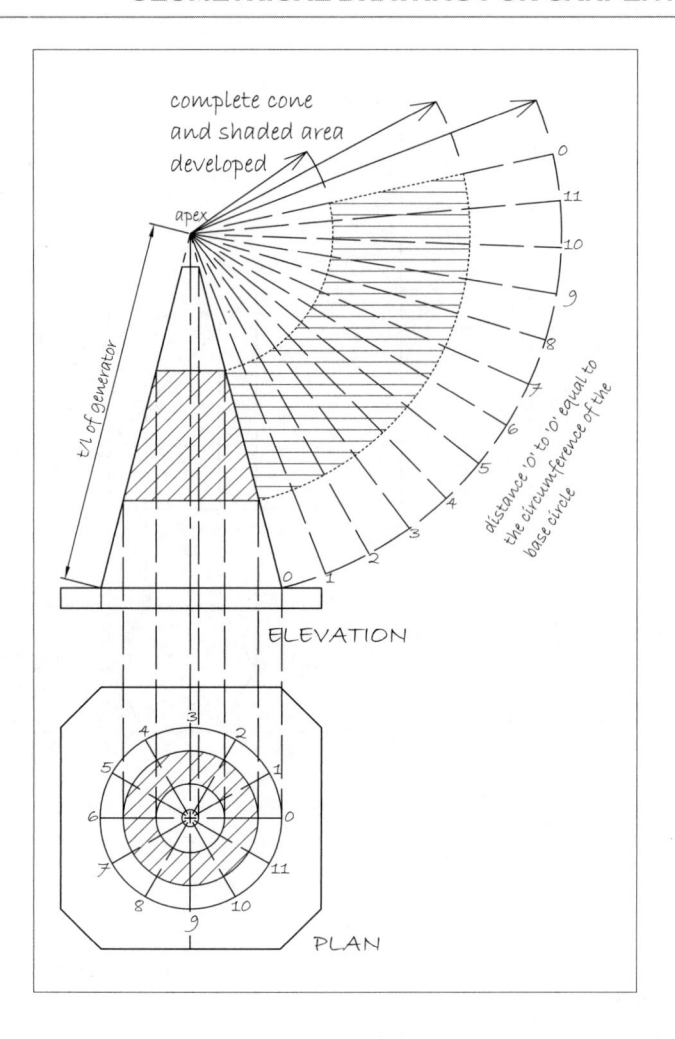

Fig. 4.10

4 GEOMETRIC SOLIDS

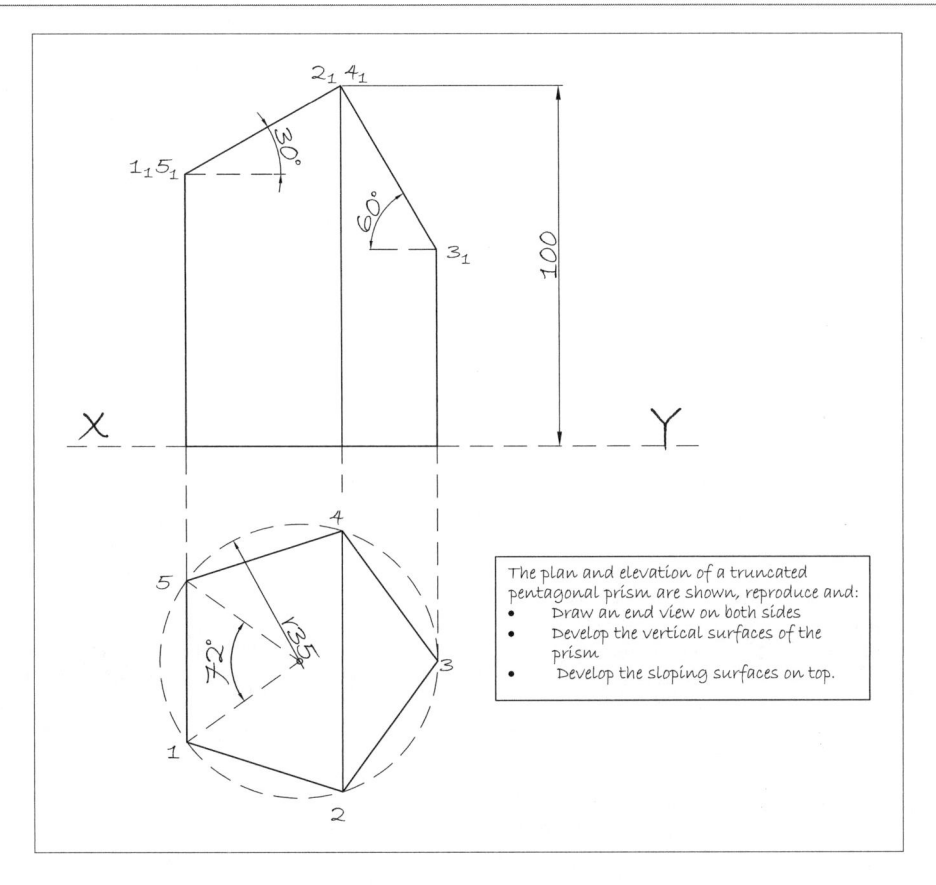

Fig. 4.11

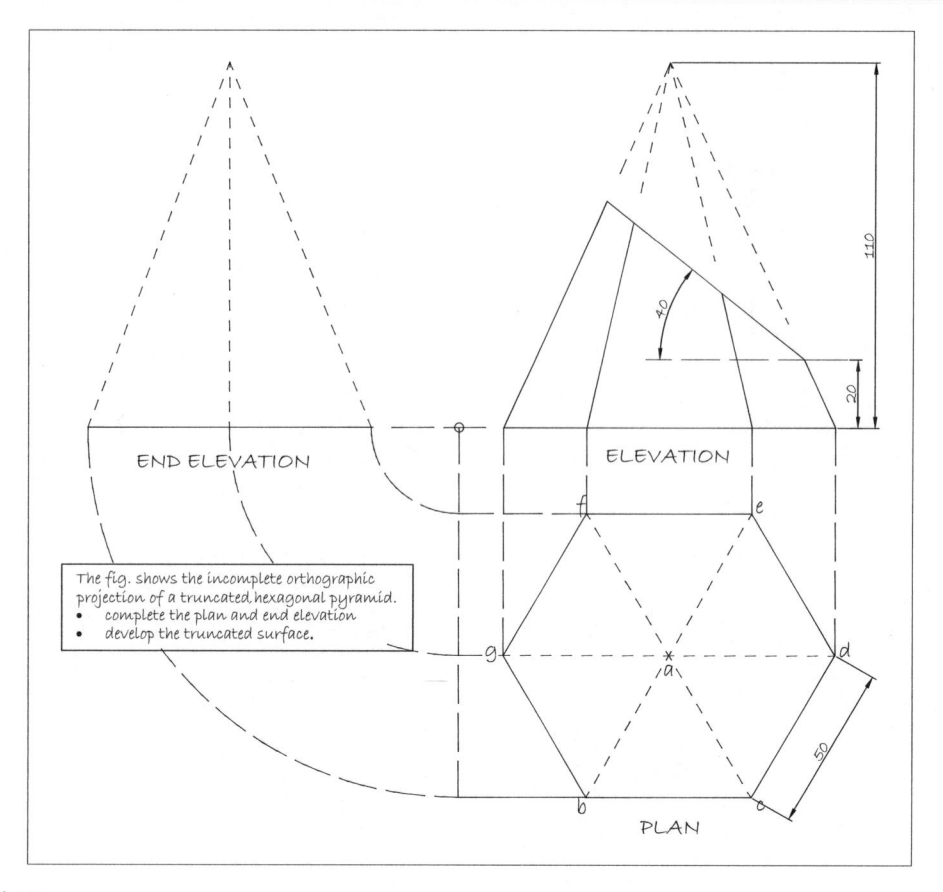

Fig. 4.12

SECTION II: REBATMENT OF SURFACES

When developing a flat (plane) surface of an object, it is often helpful to hinge the surface to the object in question. This procedure is called *rebatment*. The core principle of this procedure is that all points on the surface being developed will move at an angle of 90° to the hinged line. The method has been used in this chapter and in several other chapters of this book.

In **Fig. 4.13** a pyramid and the development of some of its surfaces, in orthographic projection (on the right), is illustrated beside a pictorial view of the process. Note that surface **B** is hinged to the base of the pyramid, and all the relevant points on its surface move at 90° to the hinged line. This is the same for surfaces **A** and **C**.

Using surface **C** as an example, the hinged line in plan is a 'line' and in elevation it is a 'point'. It does not move in either view. Figures **4.14** and **4.15** show the principle of rebatement applied to various roof surfaces.

4 GEOMETRIC SOLIDS

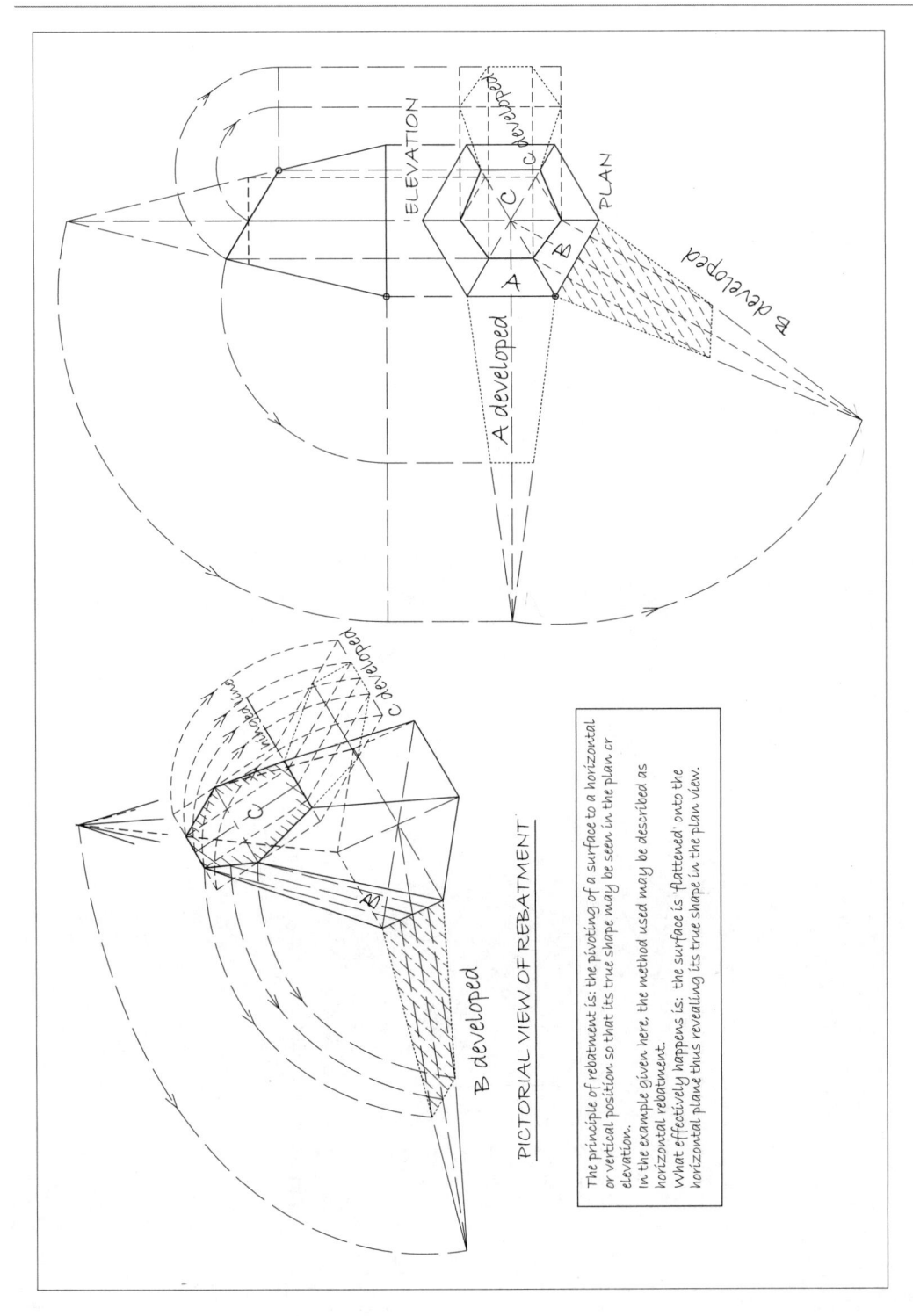

Fig. 4.13

GEOMETRICAL DRAWING FOR CARPENTRY & JOINERY

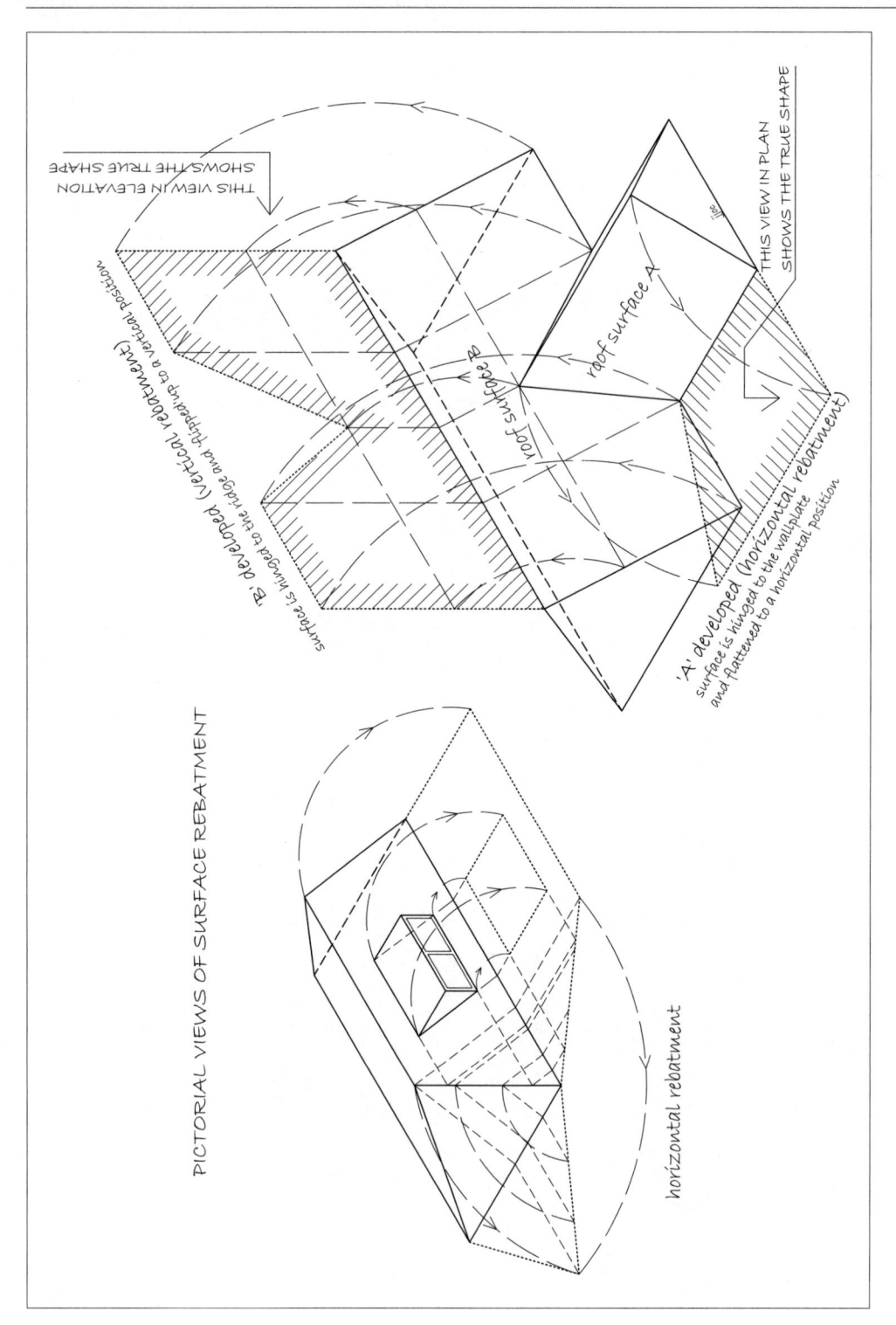

Fig. 4.14
Fig. 4.15

5 SPLAYED WORK

Fig. 5.1 shows the plan and elevation of a hopper (a funnel-type object), with two of its surfaces developed. Two methods are employed, using the same principle of rebatement.

Method 1

- Both surfaces are pivoted to a horizontal position in the elevation.
- Surface **J** is pivoted upwards and surface **K** downwards.
- As **a1** and **d1**, pivot upwards in the elevation, their counterparts **a** and **d** travel horizontally to the left in the plan to become **A** and **D**.
- To complete the development, join **e A D h**, and similarly with surface **K**.

In **Fig. 5.2**, a similar hopper is shown with thickness of material added.

Method 2

- On the right, surface **abcd** is developed in the same manner as in **Fig. 5.1**, but this time with the edge **bcfe** joined on. The pivot point is **a1** in the elevation.
- On the left, the complete side wall **L** of the hopper is knocked flat in the elevation (pivoting on point **P**) and a plan is projected, which reveals the true shape of the inner and outer surfaces.

In **Fig. 5.3** the part plan and elevation of a doorway entrance with splayed reveals (jambs) is shown. Methods of developing the right-hand reveal, surface **R**, and the top surface **T**, are given.

Method

- Surface **R** is pivoted on point **a** in the plan, bringing **b** to a horizontal position (giving dimension **w**) and projecting to the elevation, where **b1** travels to the right and becomes **B**.
- To obtain developed surface **T**, swing **a1-B** to intersect with **b1** projected vertically.

Fig. 5.4a shows a line drawing of a name board which goes round a corner to a shop front. Surfaces **A** and **B** are developed using methods shown in the two previous examples.

Method

- Surface **B** is pivoted on point **P** to a horizontal position in the elevation and projected to the plan, where **d** and **c** move horizontally to become **D** and **C**.
- Surface **A** is found by swinging **C**, to **c** extended in the plan.

Fig. 5.4b shows an auxiliary elevation of the mitre line between the two surfaces **A** and **B**. This results in the true length of the mitre in the projected view. The dihedral angle between the surfaces is found as follows:

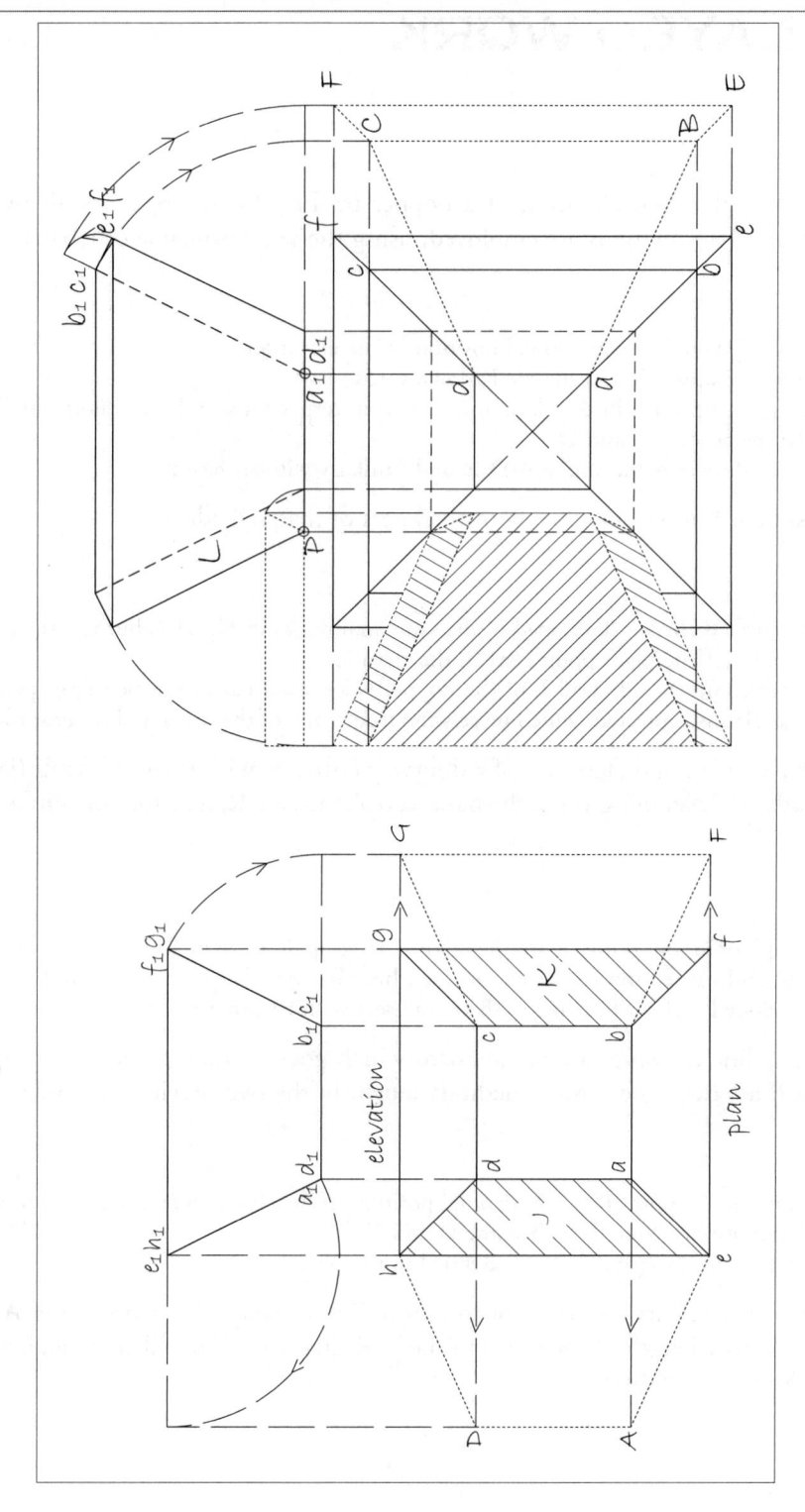

Fig. 5.1
Fig. 5.2

5 SPLAYED WORK

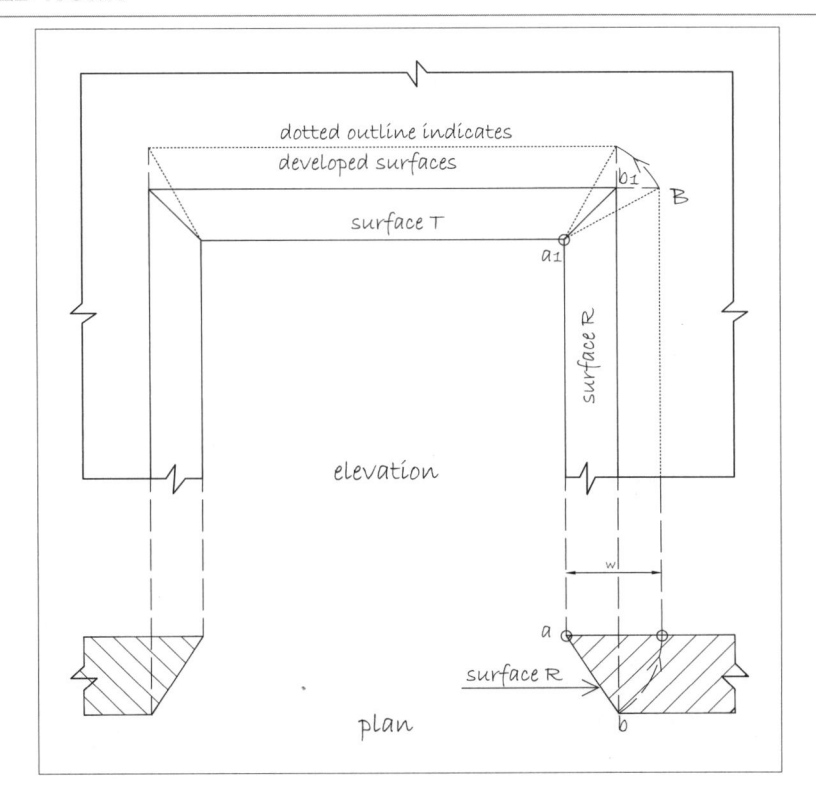

Fig. 5.3

Method

- Draw a line square-off the mitre line to **e1 f1** and pivot to **g1**.
- Project from **g1** to the plan of the mitre and join to **e** and **f**.
- The dihedral angle is contained as shown.

Fig. 5.5 shows the plan and elevation of a reception counter with an angled corner and splayed front. The splayed surfaces **A** and **C** are shaded in the plan, with surface **B** shaded in the elevation.

The developments of the three surfaces are shown using dotted outline.

The first surface developed is **C**, which is knocked flat in the elevation, **f1 e1** being the pivot, and projected to the plan.

On all of the developed surfaces the length of the mitres **f-C** and **g-B** are constant.

Fig. 5.6 shows the plan and elevation of a shaped post, with two struts of different sections abutting. On the left, the strut is square-sectioned and, on the right, the section is triangular and equilateral. In both cases, the struts are cut to fit snugly around the post.

To achieve the correct cuts the bevels must be found by developing the surfaces of the struts. This is done by 'flattening out' the true widths of the struts as shown.

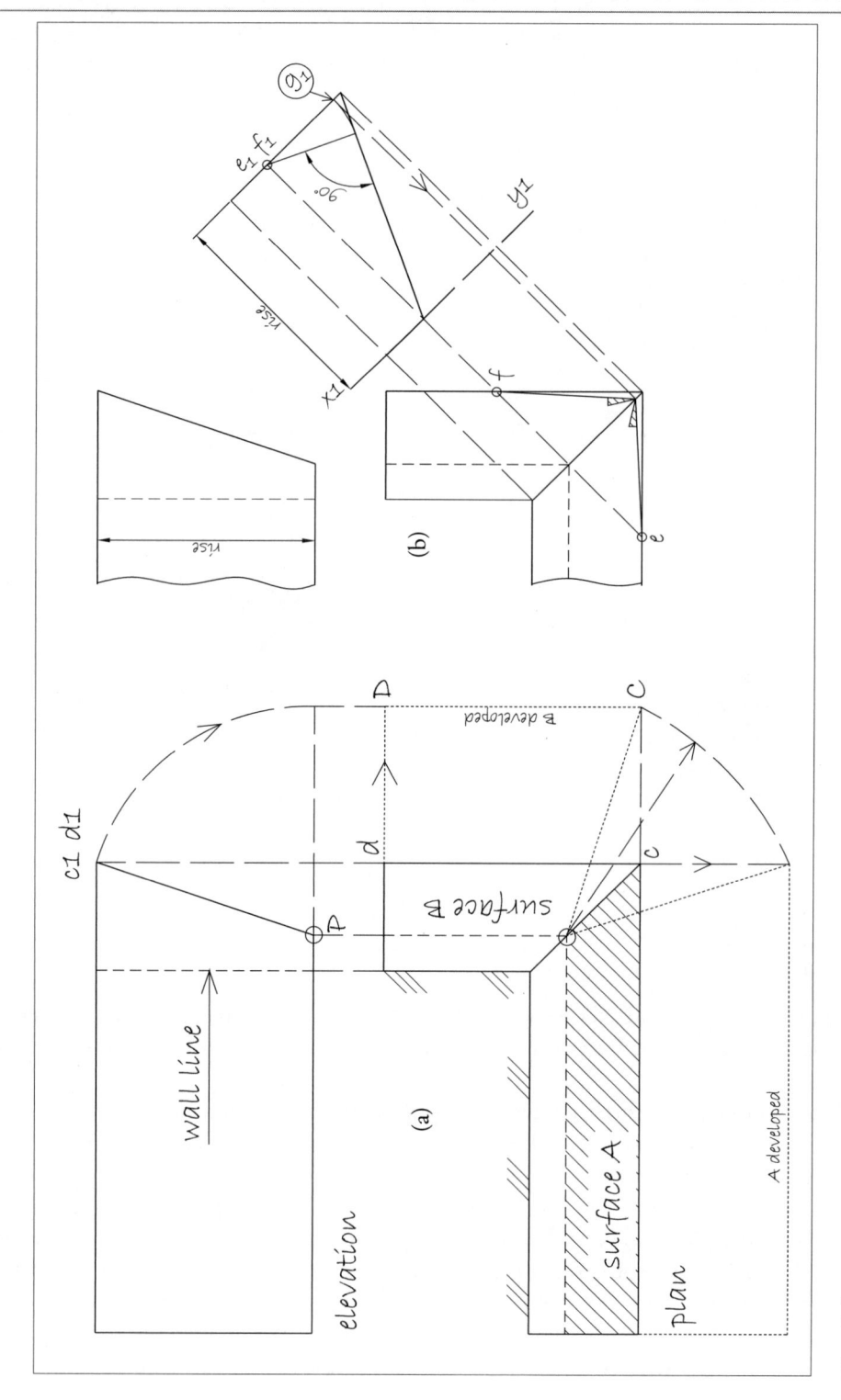

Fig. 5.4a
Fig. 5.4b

5 SPLAYED WORK

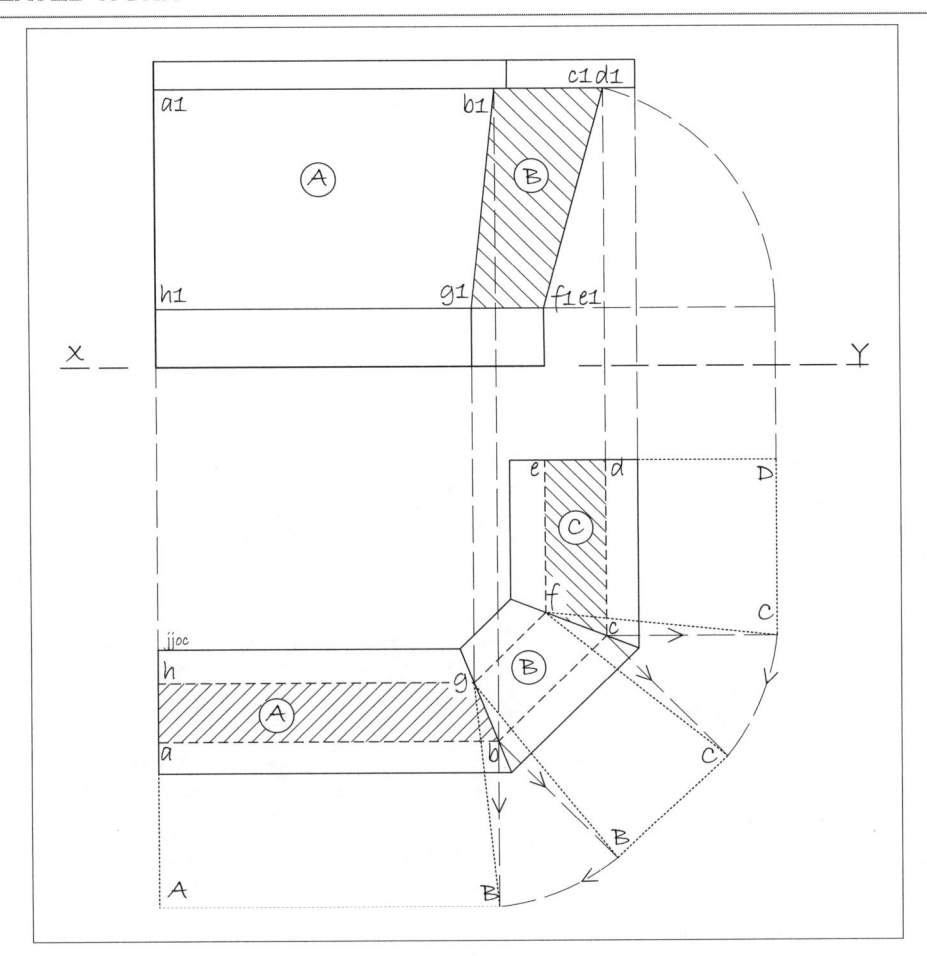

Fig. 5.5

The developments are hinged onto the upper edge of each strut, and all projection lines travel at 90° to it (the hinged edge).

Note: An important rule to remember is that when hinging developments in this manner, all points on the developing surface will move at 90° to the hinged line.

Exercise

Fig. 5.7 shows a line drawing in plan and elevation of a hexagonal hopper. Reproduce **Fig. 5.7** and complete the exercise.

- Determine the true shape of the shaded area.
- Find the dihedral angle between surfaces **A** and **B**.

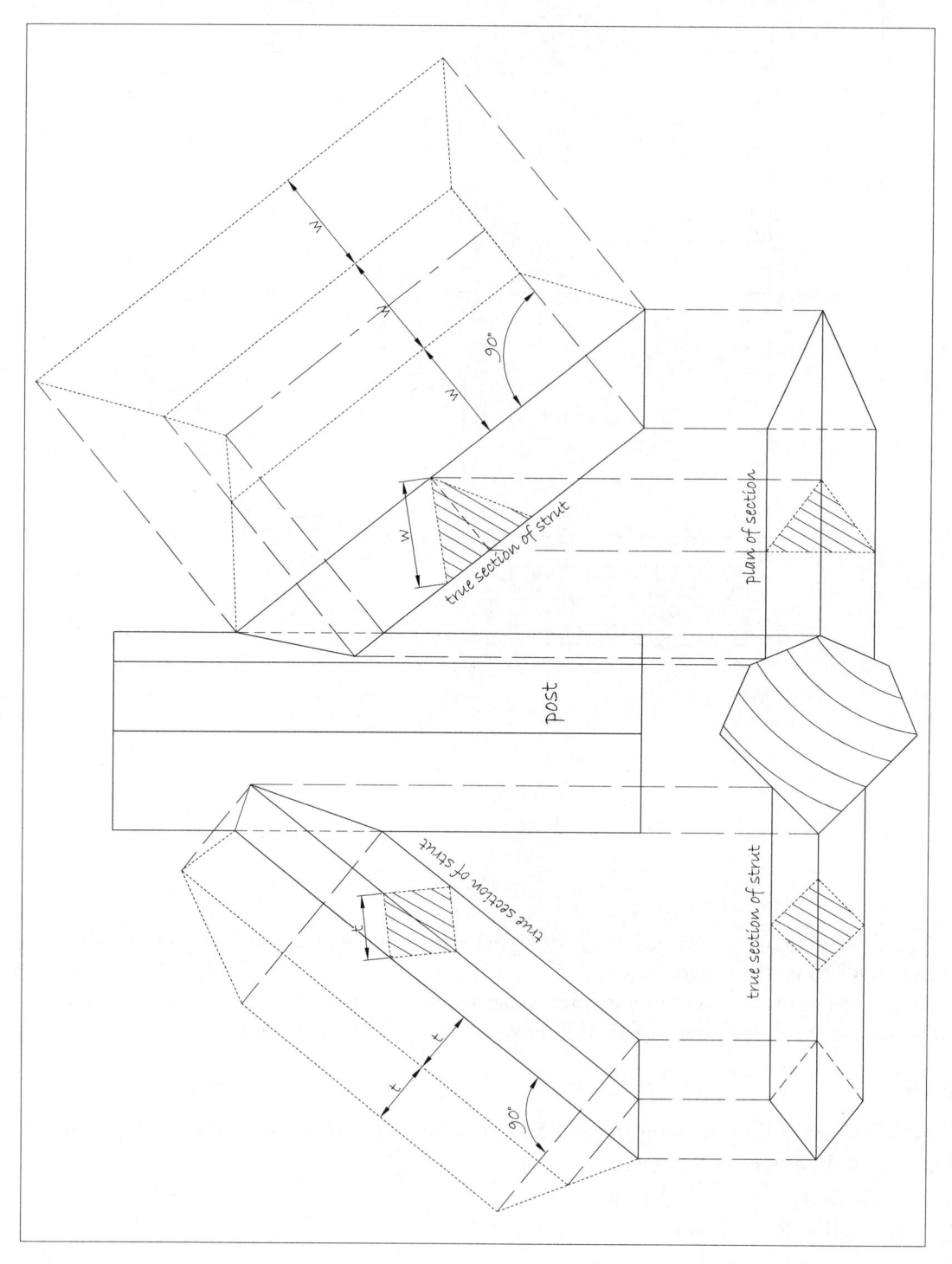

Fig. 5.6

5 SPLAYED WORK

ELEVATION

PLAN

Fig. 5.7 shows the plan and elevation of a hexagonal hopper.
- develop the shaded area
- find the dihedral angle between surfaces A and B.

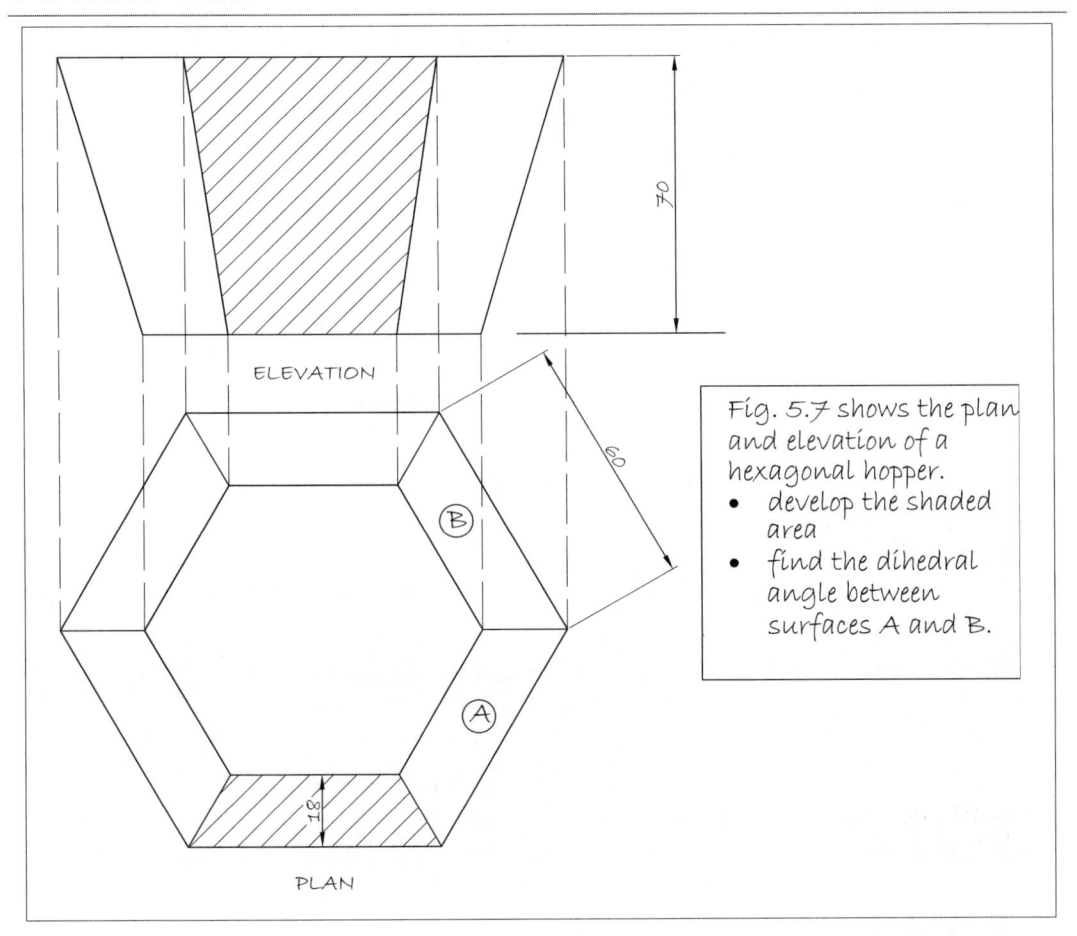

Fig. 5.7

6 GEOMETRIC CONSTRUCTION OF ARCHES

SECTION I: ARCH OUTLINES

Arches and their geometrical construction will be dealt with in this chapter, covering various arch outlines. Arches have:

- a *springing line*, the line from where the arch starts to curve
- a *span*, the width of the arch
- a *rise*, the height of the arch.

Fig. 6.1 shows a semi-circular arch, and **Fig. 6.2** is a segmental arch, meaning its curve is part of a circle. On the figures are shown:

- a *normal*, which is a line originating at the centre point for the curve and extending beyond it (the curve)
- a *tangent*, a line which is at right angles to the normal at the same point on the curve.

When an arch has brickwork or cut stone, the joint lines are *normal* to the curve. A *normal* may be described as a line that is square-off a curve.

Fig. 6.3a shows a semi-elliptical arch, which is constructed using concentric circles.

Method

- Having determined a span and rise for the arch, construct semi-circles on both, the span is the major axis of the ellipse and the rise is half the minor axis
- Draw a series of lines as **c-g** (these can be random).
- Drop a vertical from **g** and draw horizontal from **h** to give point **p**, which will be a point on the curve.

To draw a normal to an ellipse, it is necessary to find points **f1** and **f2**, which are the *focal points* of an ellipse, and are important for other methods of construction (see also **Fig. 6.3c**).

Method

- To find **f1** and **f2**, take length **a-c** (which is half the major axis) and strike it from **d** to intersect with the major axis at **f1** (do similar for **f2**).
- To construct a normal, join **f1** to any point on the curve, **p** and **p2** in this case, and continue to **f2**. This gives angle **f1 p f2**.

The bisector of this angle establishes the normal at **p**, and a line at right angles gives the tangent.

6 GEOMETRIC CONSTRUCTION OF ARCHES

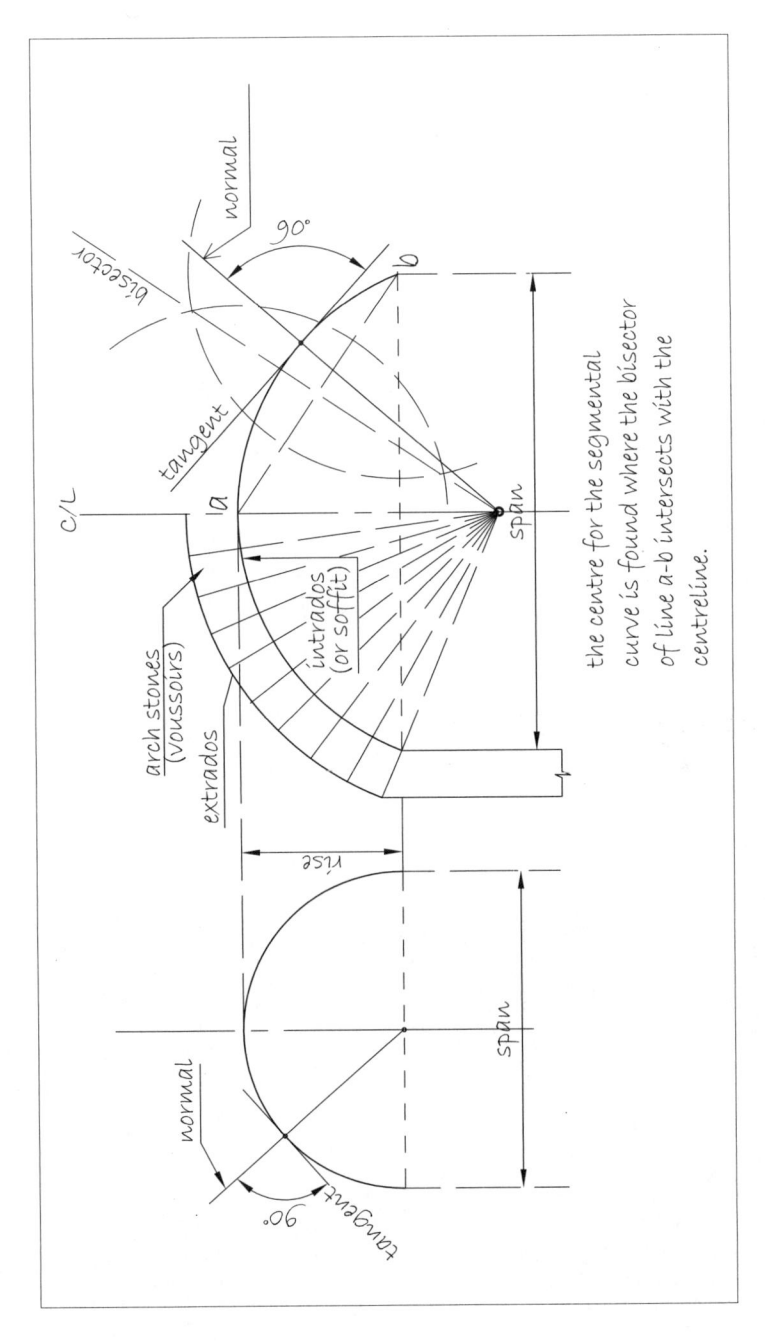

Fig. 6.1
Fig. 6.2

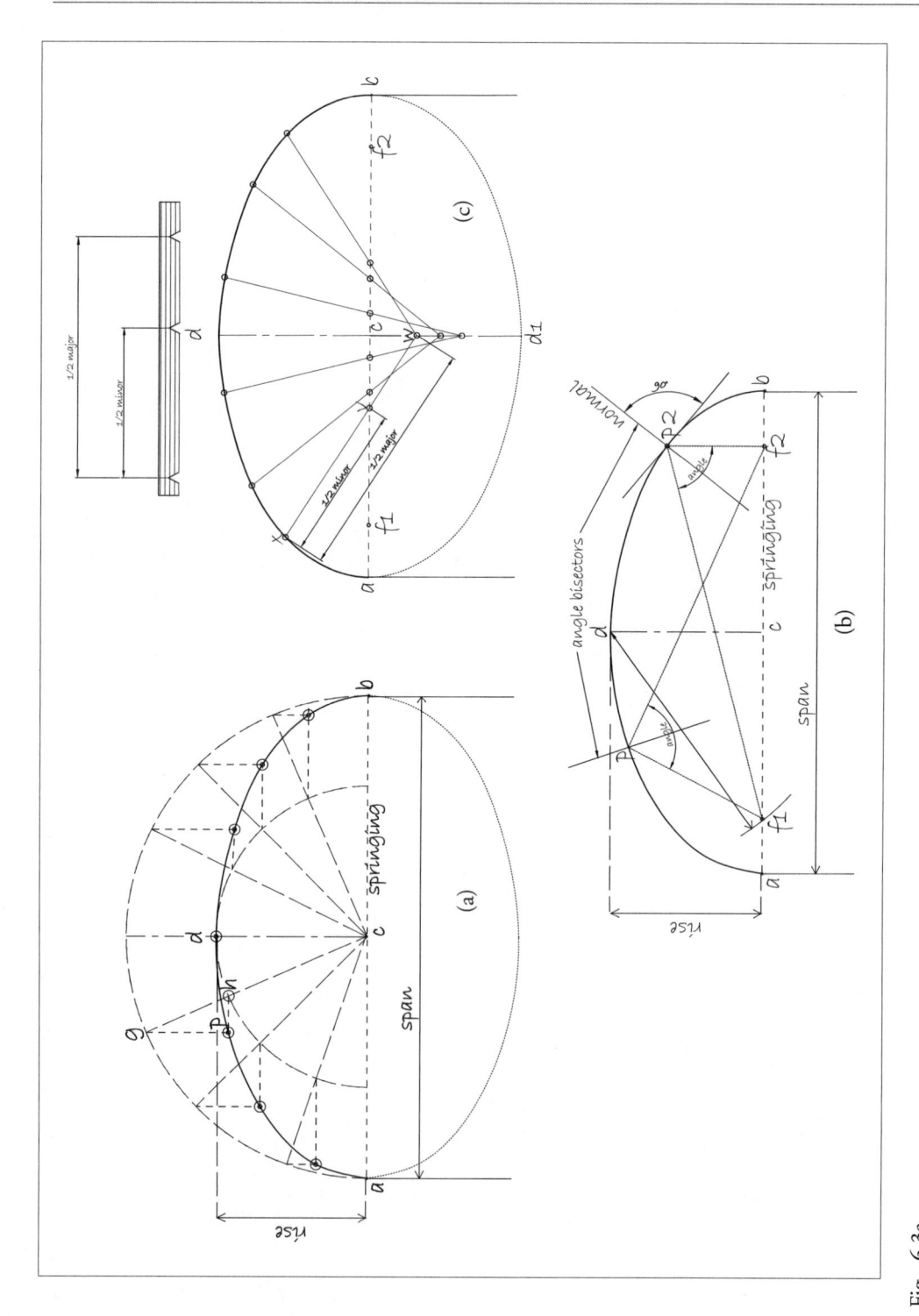

Fig. 6.3a
Fig. 6.3b
Fig. 6.3c

6 GEOMETRIC CONSTRUCTION OF ARCHES 53

Note: A point to remember at all times is that the cumulative lengths **f1-p** and **p-f2p** will always be equal to the length of the major axis.

A second method of drawing an ellipse is given in **Fig. 6.3c**. A *trammel,* which may be a rod or a lath of wood or paper, has the half-lengths of the major and minor axes marked on it. It is then placed on the axes, with the marks lying on the axes, and making the trammel-end a point on the curve.

A *three-centred* arch is shown in **Fig. 6.4a** this may also be called a *false,* or *pseudo,* ellipse. This arch is made up of arcs of different radii.

Method

- After deciding on the span and rise, make **a-b** and **b-d** equal.
- Join **o** to **a** and swing **o-d** to **e**.
- Bisect line **a-e** and extend the bisector to **c1** (the centre line). Point **c2** is established where the bisector crosses **a-c** (**c3** is equidistant from **b**).
- Scribe an arc from **c1** with radius **c1-o** to stop at **p** and **p2**. Scribe an arc from **c2** to **p** and **a**, and work similarly with the right-hand side.
- **p** is the point of contact between both curves, and **c1-p** (the bisector) establishes the normal at point **p**. This is a *common normal,* meaning it is common to both curves.

Fig. 6.4b shows a variation of the construction in **Fig. 6.4a**. In both cases, the end result is the same.

Method

- Complete rectangle **aghd**.
- Pivot **d-a** to **e**, pivot **h-e** to **f**.
- Bisect **a-f** to give centre **c1** and centre **c3** (on extended centre line **h-g**).

Fig. 6.5 also shows a *three-centred* arch, which is constructed using arcs of three different circles, but the difference between this one and one in **Fig. 6.4** is that the rise is 'self-determining'.

Method

- Construct **a-b** as the span.
- Draw two circles on the span as shown.
- Draw a line from **c2** at 45°, as shown, to touch the circle, and construct an angle of 90° off this line to intersect at **c1**.
- Construct two lines from **c1** through **c2** and **c3**, respectively. These lines will become common *normals* beyond the curve.
- Complete the curve by constructing an arc **c1-p** to **p1**.

The arches shown in **Figs. 6.6** and **6.7** are drawn using compass construction and are self-explanatory in method. A four-centred arch is given in **Fig. 6.8**. The construction is evident from the drawing. Note, however, that the figure **c1 c2 c3 c4** is a square. The common normal is shown to conform to the examples above. This arch is also referred to as *Tudor arch.*

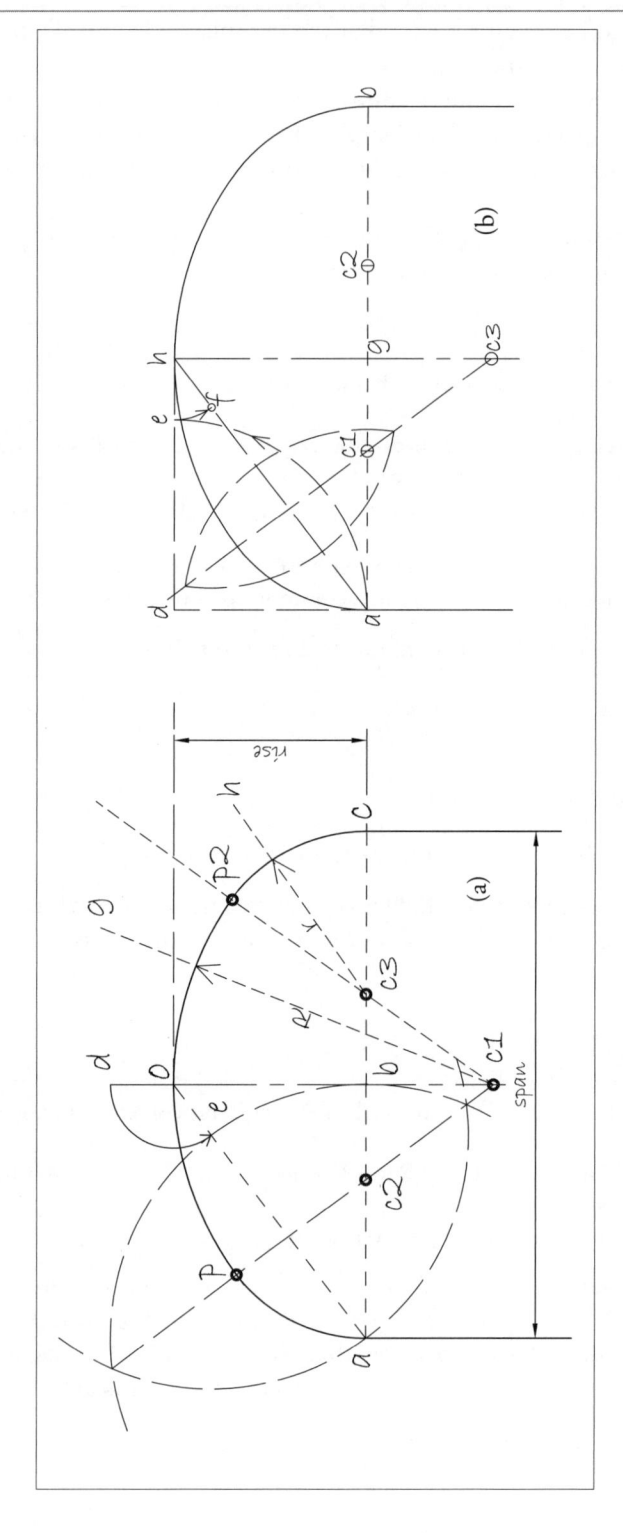

Fig. 6.4a
Fig. 6.4b

6 GEOMETRIC CONSTRUCTION OF ARCHES

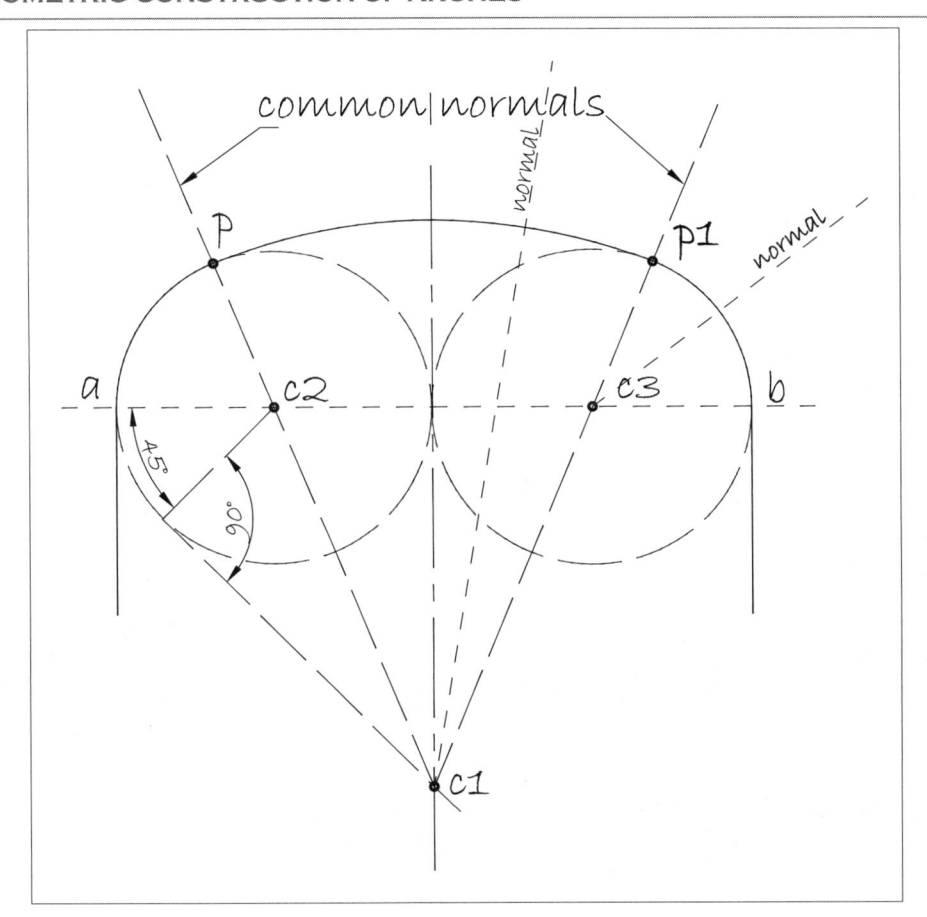

Fig. 6.5

GEOMETRICAL DRAWING FOR CARPENTRY & JOINERY

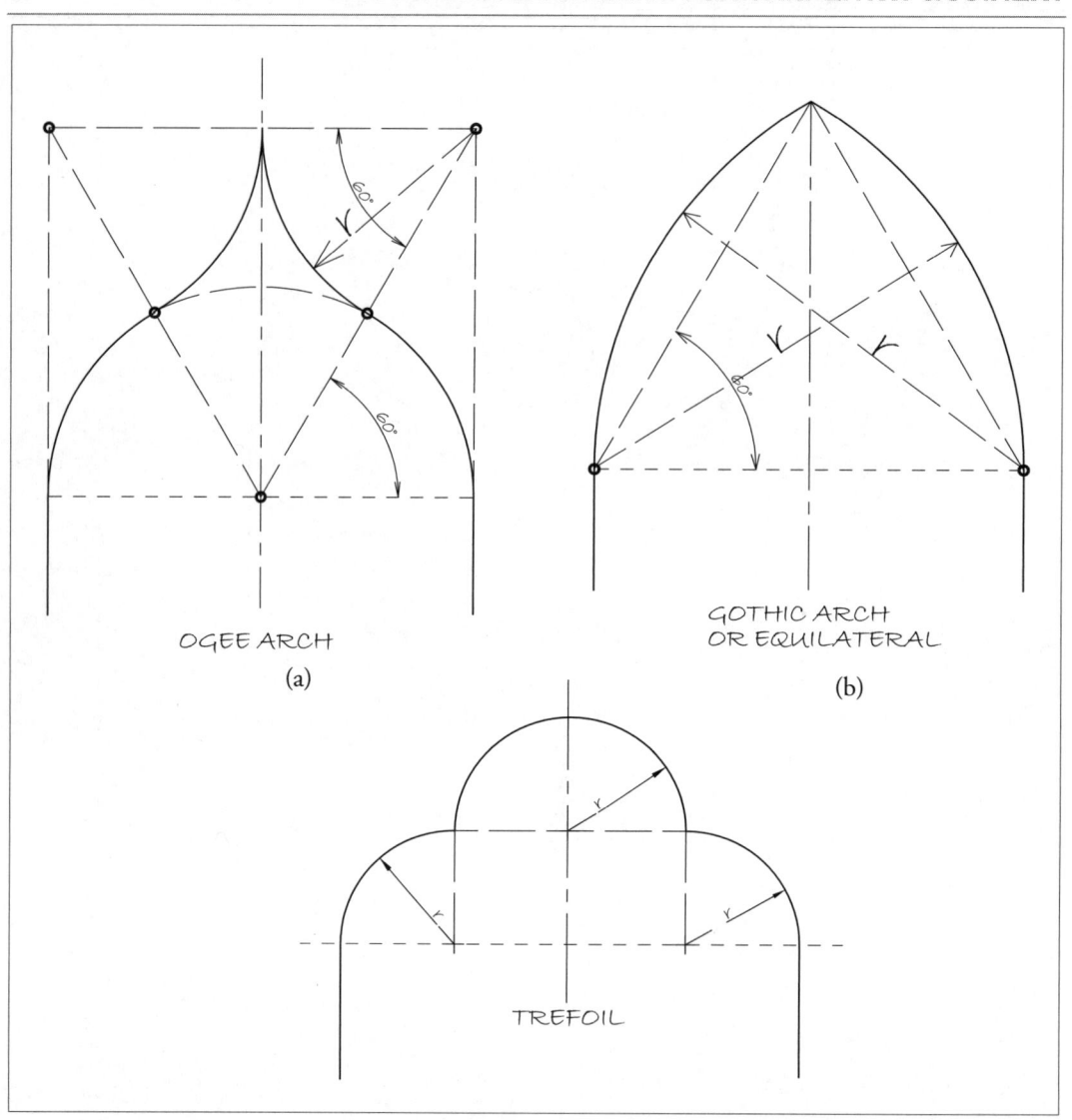

OGEE ARCH
(a)

GOTHIC ARCH
OR EQUILATERAL
(b)

TREFOIL

Fig. 6.6a
Fig. 6.6b
Fig. 6.7

6 GEOMETRIC CONSTRUCTION OF ARCHES

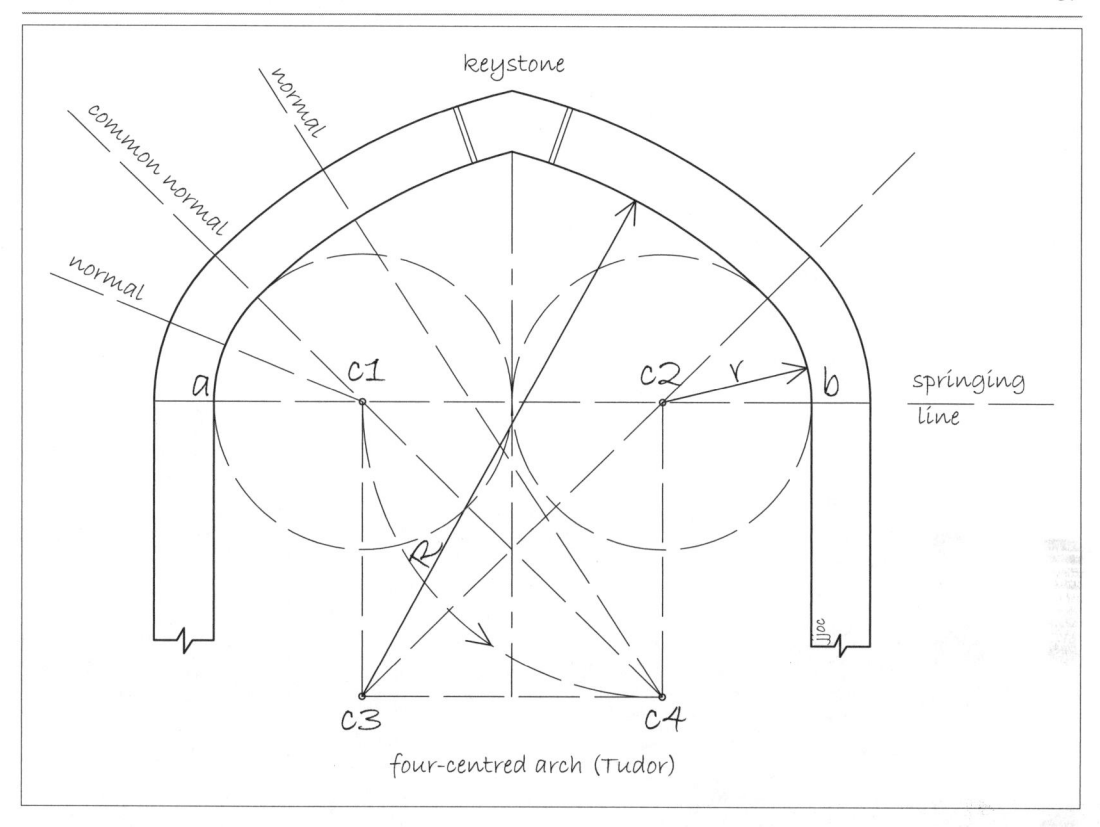

Fig. 6.8

SECTION II: SOFFIT DEVELOPMENT

In Section 1, arches were dealt with as profiles (shapes) without any thickness. Here, we consider the thickness of the wall of which they are part. The examples in this section will deal with arches that are 'square' to the wall. In other words, the cut-out of the wall will be at 90° to the front surface of the wall.

In Section III below, *skew* arches will be covered.

The soffit of an arch is the part that is over your head as you walk underneath. In the examples given in Section I above, and in general, the soffit begins and ends at the springing line. The developed shape of the soffit is dealt with here.

Fig. 6.9 shows the projections of a semi-circular arch and the development of the soffit, using two methods. In the first method, the elevation, the arch outline is 'stretched out', using a method explained in Chapter 17 on constructions. The thickness of the wall (**t**) is added to complete the development. In the plan view, the dimensions **0** to **1**, **1** to **2** etc. are laid out as shown to give the length of the development.

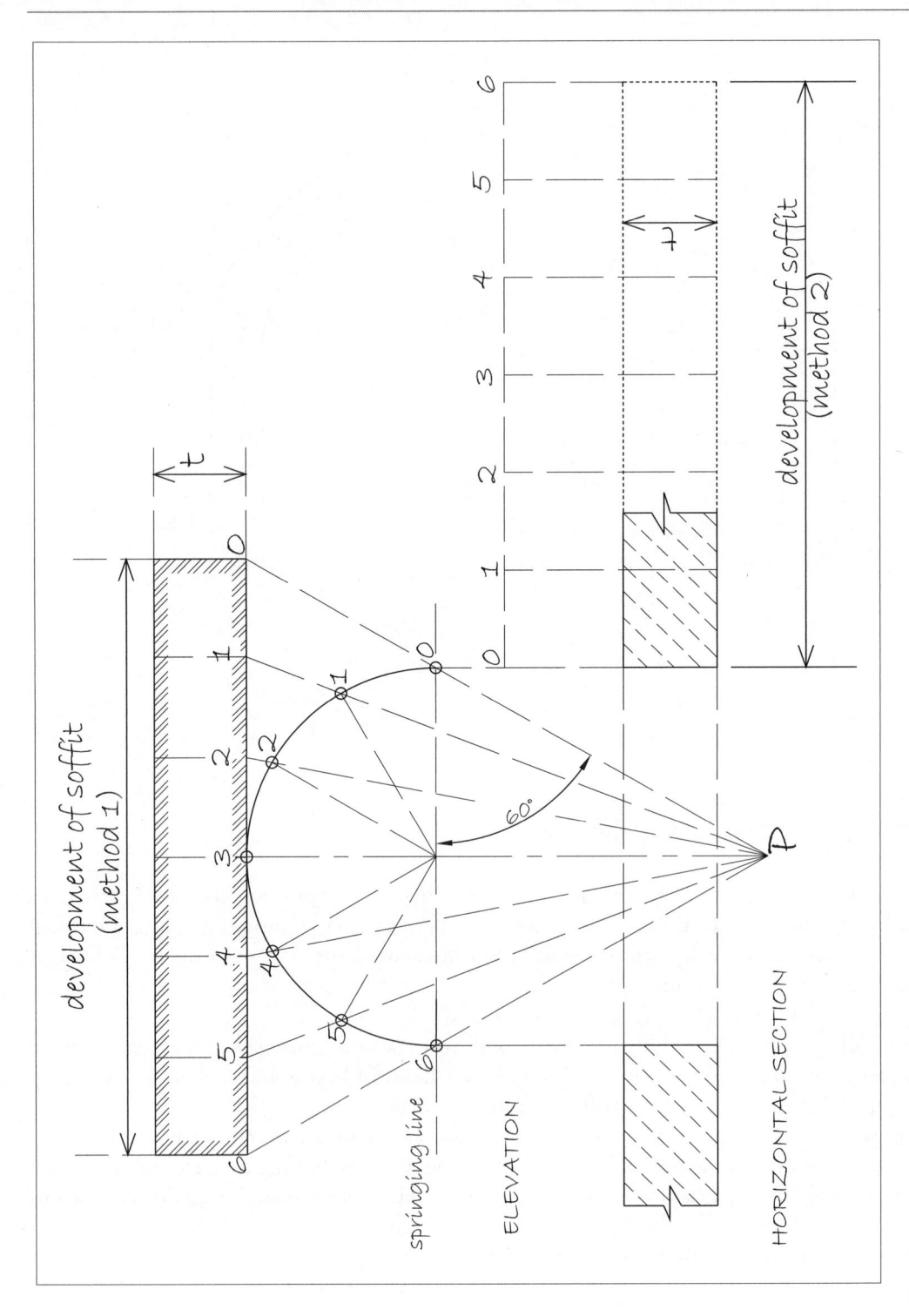

Fig. 6.9

6 GEOMETRIC CONSTRUCTION OF ARCHES

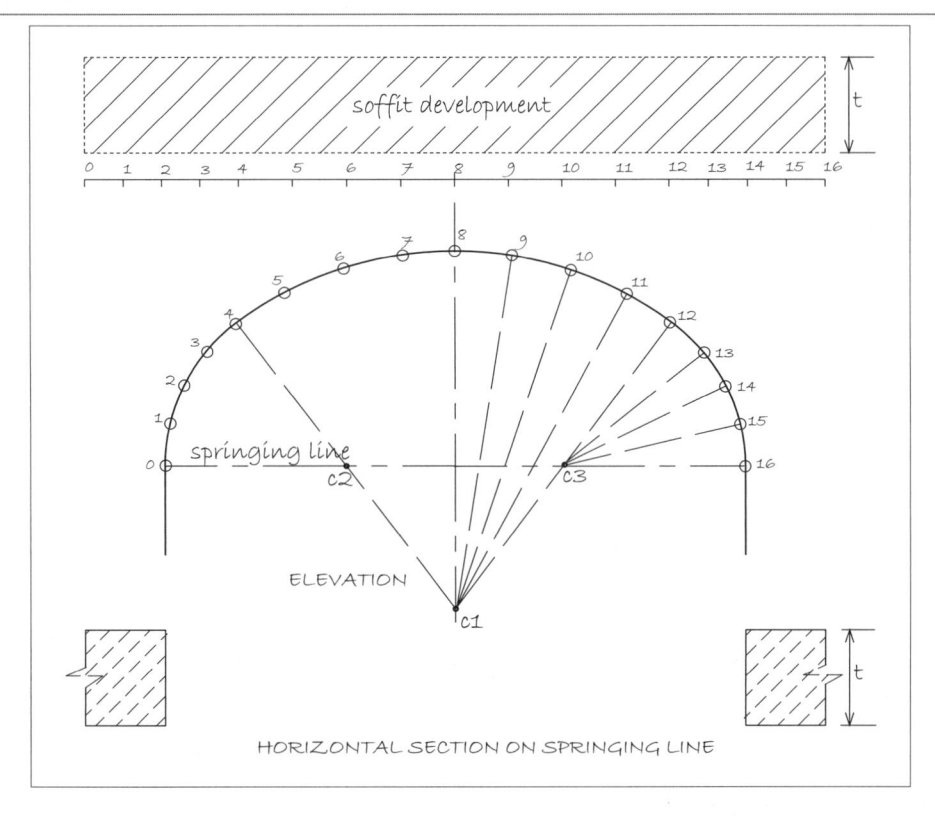

Fig. 6.10

Regardless of which method is used, the ultimate length of the 'stretch-out' must equal the length of the semi-circle in the elevation.

The arch in **Fig. 6.10** is 'three-centred', and in this case the distances have to be taken from the arch outline and plotted on a straight line to give the stretched-out development.

The arches in **Figs. 6.9** and **6.10** had their *jambs* (or linings) and consequently their soffits square-off the wall. In the example in **Fig. 6.11** the cut-out in the wall is splayed at the angle given in the horizontal section, and the soffit will have the same splay. The surface of this soffit is part of a cone and the cone in question is outlined in the end elevation as **A1-31-E**. The same cone, to a smaller scale, is shown to the left of the elevation, right way up.

To develop the soffit:

Method

- Divide the curve in the elevation into a number of parts; in this case, **0** to **6**.
- In the plan, extend **c-d** and **0-b** to **A**, which is the apex of the cone in question.
- Draw arcs **A-0** and **A-b** and mark off the distances **0** to **6** from the elevation, as shown on the outside arc.
- Join the last point, which is **6**, back to the apex **A**, to complete the development.

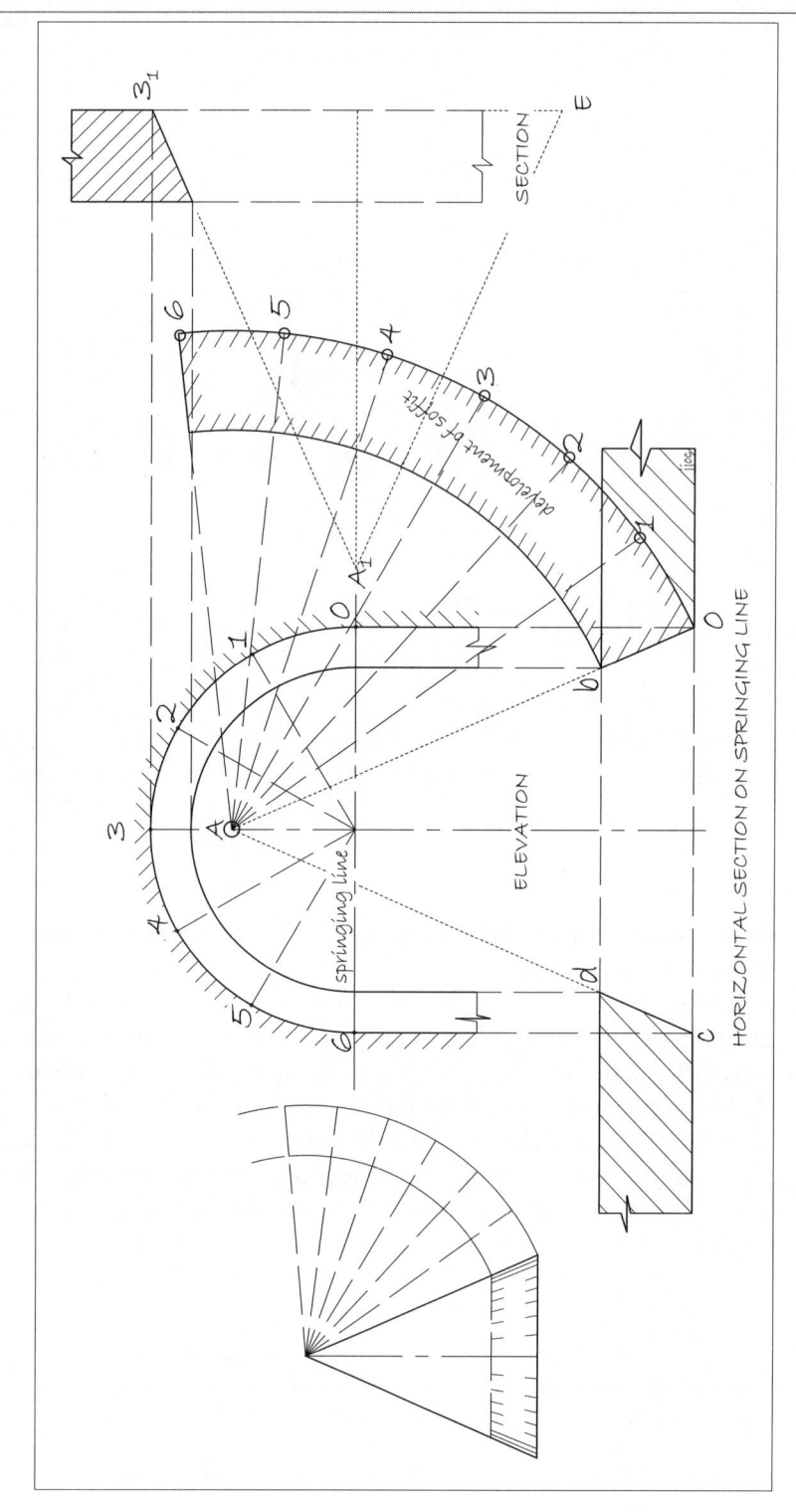

Fig. 6.11

6 GEOMETRIC CONSTRUCTION OF ARCHES

SECTION III: SKEW ARCHES

The following examples are of arches that are oblique to the wall surface, commonly called *skew*, or skewed, arches. Their elevational outlines may be similar to the examples above, but the development of their soffits involves considerably more taxing geometry. The angle of cut through the wall is a variable and the one used in the examples following is expedient.

In **Fig. 6.12** the arch shown is semi-circular when viewed parallel with the angle of skew, as from point **S**, but when the arch is looked at in conventional elevation, square-off the wall, it is elliptical. The arch is, therefore, based on a *right-cylinder,* which is angled to and cut by

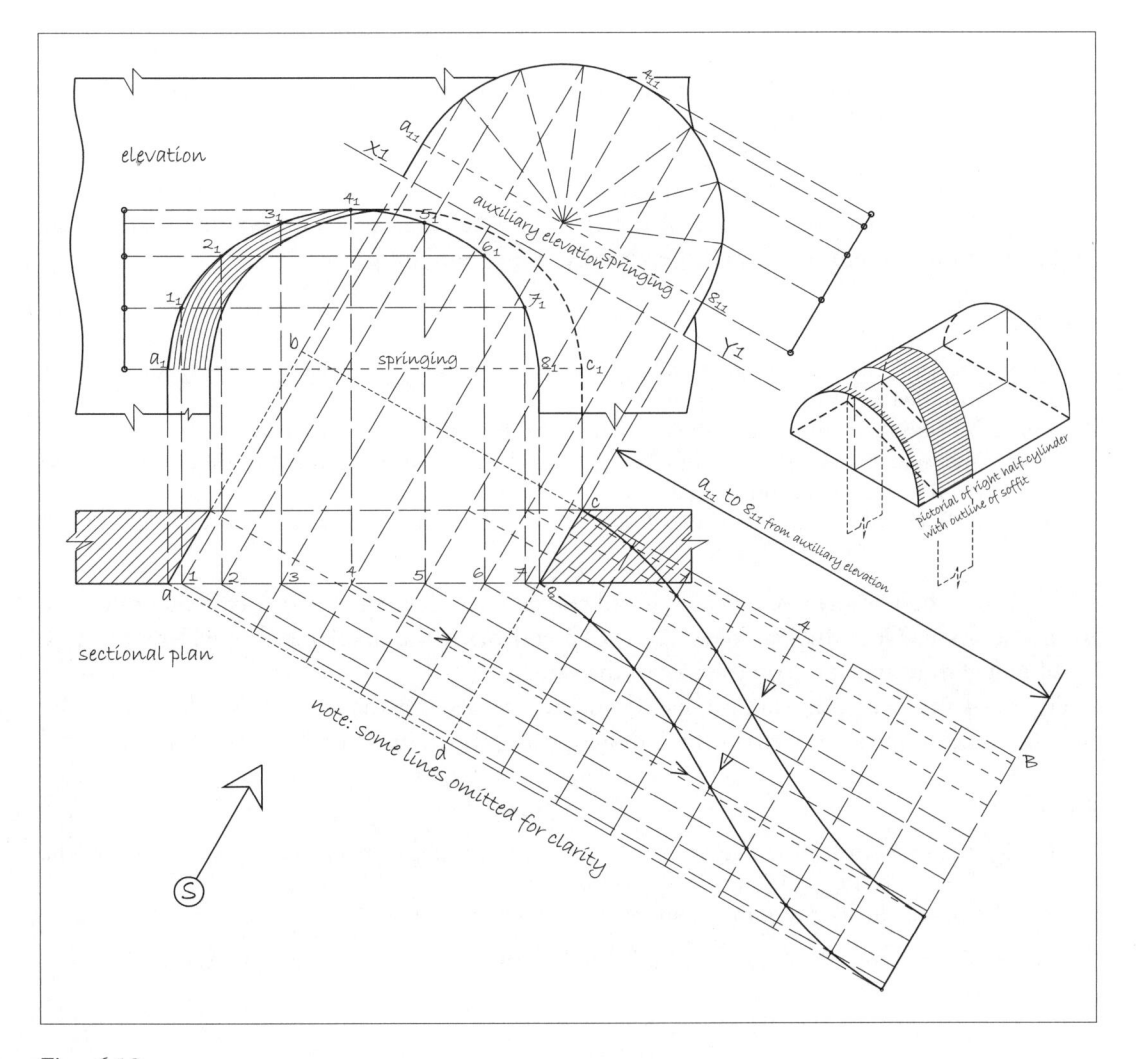

Fig. 6.12

the planes of the wall. The surface of the soffit will consequently be part of a right cylinder, which is indexed **abcd** in plan view, and will be developed as follows:

Method

- Divide the semi-circle in auxiliary elevation into a number of parts; in this case, eight.
- Project them to the face of the wall in plan, from **a** to **8**.
- To complete the elevation, project **a** to **8** vertically and set up a height rod as shown. Join up the points to give the semi-elliptical outline.
- In the plan, set out **c-B** equal to **a11-811** (the length of the semi-circumference from the auxiliary elevation) and the sub-divisions that represent ordinates on the cylinder. This whole represents the roll-out of the half cylinder.
- Using line **c-8** as the hinged line for the development, project the remaining points, **7** to **a**, from the face of the wall at right angles to the skew.
- Where the projection lines intersect with the relevant ordinates, points on the developed soffit are found; as an example, see ordinate **4**.
- The points for the rear of the wall are found in the same way, though for clarity some are omitted in the drawing in **Fig. 6.12**.

The arch in **Fig. 6.13** seems similar to that in **Fig. 6.12**, but there is a fundamental difference because **Fig. 6.13** is based on an oblique cylinder, that is, a cylinder in which the axis is at an angle (other than 90°) to the circle of the cylinder (see Chapter 16 on oblique solids). The given elevation is a semi-circle and the view from **S** is semi-elliptical.

Method

- Set up the semi-circle in elevation and index as in the previous example, projecting the divisions **a1** to **81** to the face of the wall in plan.
- Line **a-b** will be the hinged line. Project points **1** to **8** at 90° to **a-b**, as shown.
- Take dimension **d** from the elevation and scribe an arc from **a** to projection line **1**, from **1** to projection line **2** and continue to point **8**.
- The rear line will be parallel to **a-8** developed with dimension **a-b** being constant on all ordinates.

Using this method ensures that the developed curved line **a-8** is equal to the semi-circumference in the elevation. Note: In examples, **Figs. 6.12** and **6.13**, the developed curved line must be equal in length to the curve on the face of the wall.

Fig. 6.14 shows a gothic arch in elevation and the skew of the arch in the plan view. The development of one side of the soffit is drawn, using the method explained in **Fig. 6.13**.

Exercise

1. In **Fig. 6.15** the line **a-b** represents the major axis of an ellipse, while point **p** is a point on the ellipse. Reproduce **Fig. 6.15** and complete the exercise.
   - Determine the length of the minor axis and draw the ellipse.
2. **Fig. 6.16** shows a segmental skew arch. Reproduce **Fig. 6.16** and complete the exercise.
   - Develop the shape of the soffit.

6 GEOMETRIC CONSTRUCTION OF ARCHES

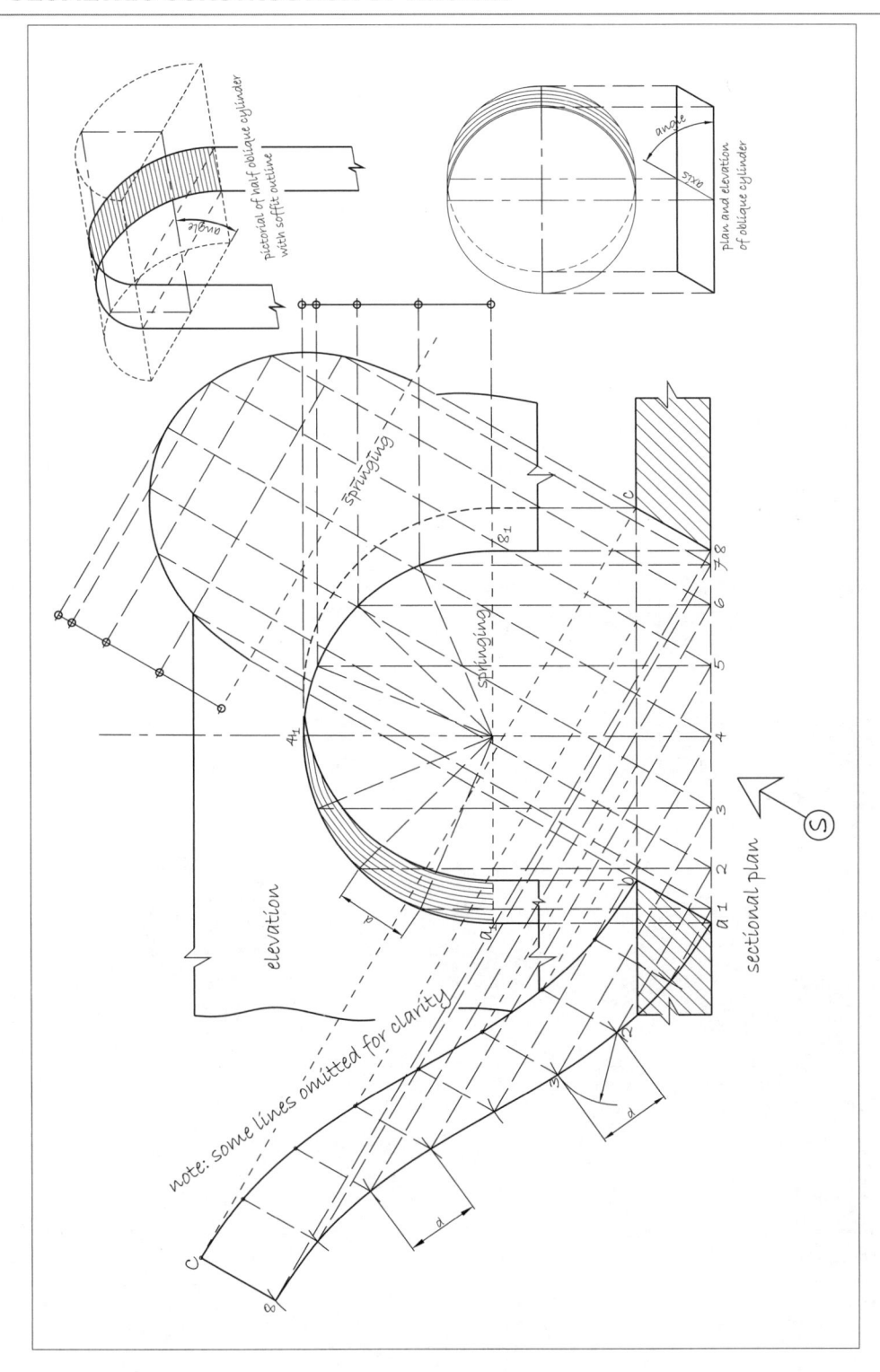

Fig. 6.13

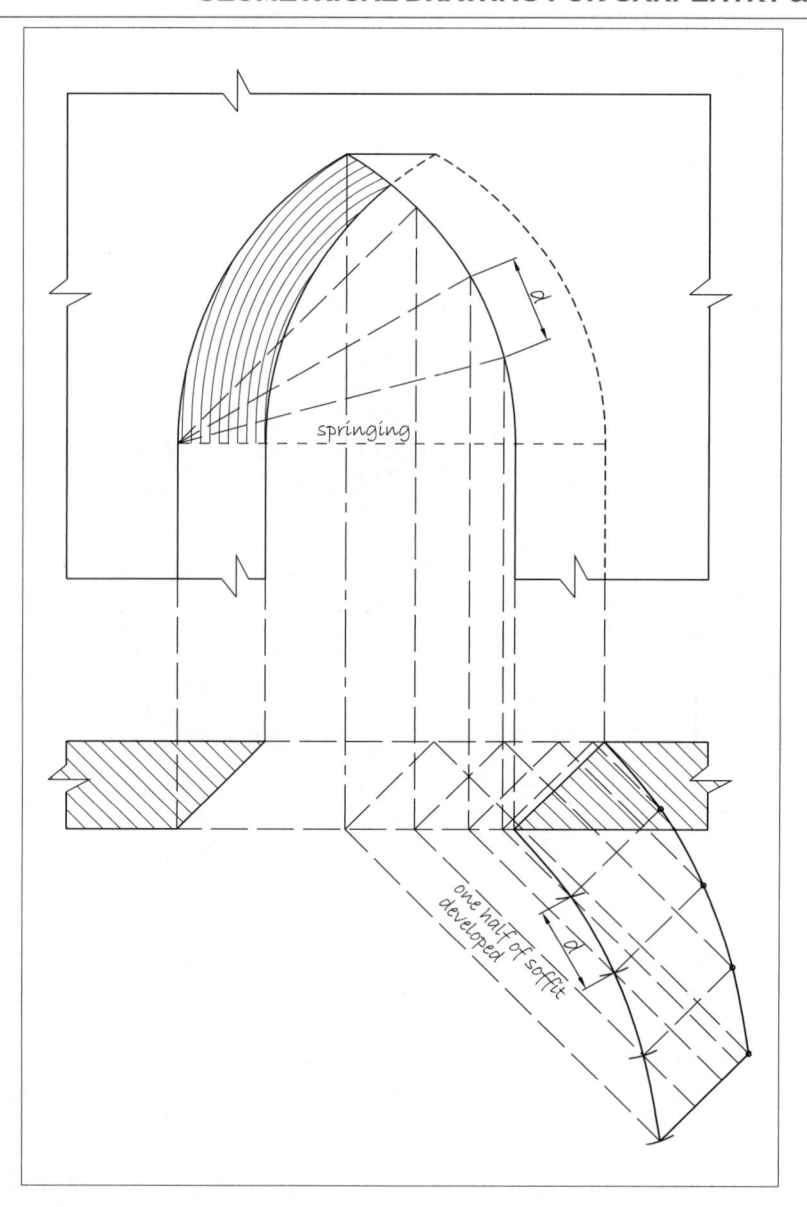

Fig. 6.14

6 GEOMETRIC CONSTRUCTION OF ARCHES

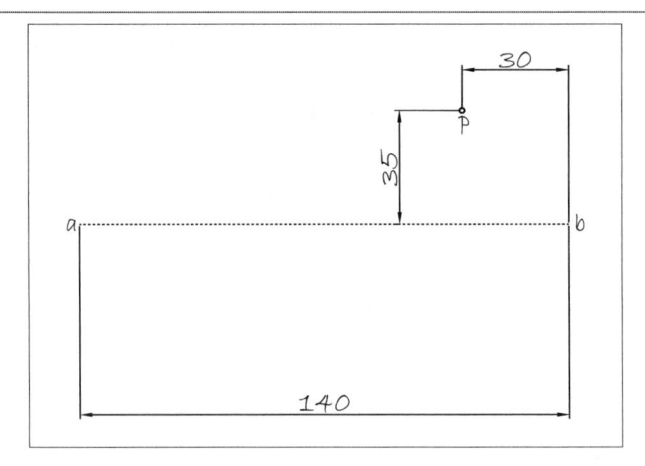

Fig. 6.15

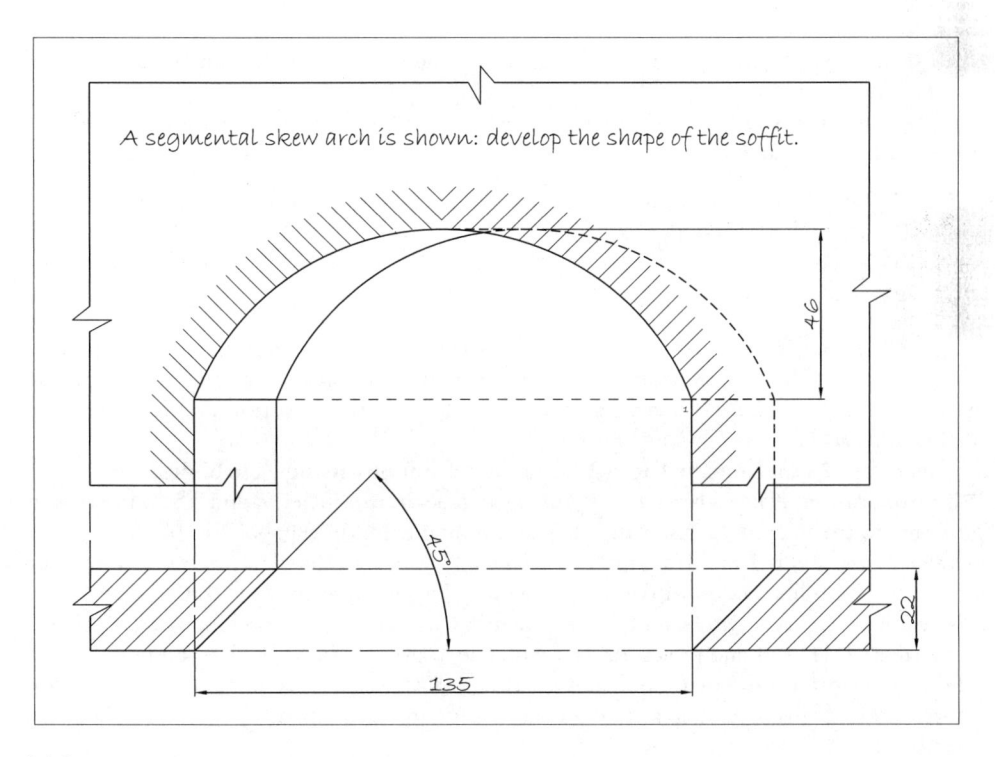

Fig. 6.16

7 CONIC SECTIONS AND CONIC DEVELOPMENT

Sections of cones and developments of conical surfaces are applied to roofing where conical roofs intersect with flat-plane pitched roofs. The conic sections are:

- *elliptical*
- *parabolic*
- *hyperbolic.*

The surface development of cones also has various practical applications.

In **Fig. 7.1** the elevation of a cone is drawn (top right), with three lines representing cutting planes.

- Cutting plane **A-B** will give an *elliptical* section on the cone. Note: If the plane **A-B** was horizontal, the section would be circular.
- Cutting plane **C-D** will give a *parabolic* section of the cone. Note: The line **C-D** is parallel with the generator.
- Cutting plane **E-F** will give a *hyperbolic* section, as will any other angular aspect of **E-F** (to the base) other than a parabola.

Figs. 7.2, **7.3** and **7.4** below show methods of developing the conic sections.

To develop the parabola in **Fig. 7.2**:

Method

- Line **a-b** represents the cutting plane which in this case is parallel with the generator (**01-41**) of the cone. The base of the cone is divided into an equal number of parts (12) and indexed, as shown, with the points then projected to the base of the elevation and joined to the apex of the cone, **o1**.
- The plans of 12 generators, **o-1** to **o-12** and their elevations are now established.
- Where the cutting plane **a-b** intersects with each generator in the elevation, these intersections are projected to their counterparts in the plan to give points on the parabola (in plan).
- Generator **o-1** and **o-7** are treated in a special way because in the plan no intersection is possible as the projection line lies on top of the generators. This is overcome as follows:
    - Where **a-b** intersects with **o1-11** in the elevation, project a line horizontally to meet the outer generator **o1-41** at **e1** and project to the centre line in plan, as shown, giving point **e**.
    - Take radius **o-e** (in the plan) and scribe a circle. Where this circle intersects with **o-1** and **o-7**, it gives points on the parabola. Join all the points now established to complete the plan of the parabola.

Note: The circle, radius **o-e**, in the plan represents a horizontal section of the cone at level **e1** in the elevation. This circle cuts all generators at the same level, but only intersections at **o-1** and **o-7** are required in this case.

7 CONIC SECTIONS AND CONIC DEVELOPMENT

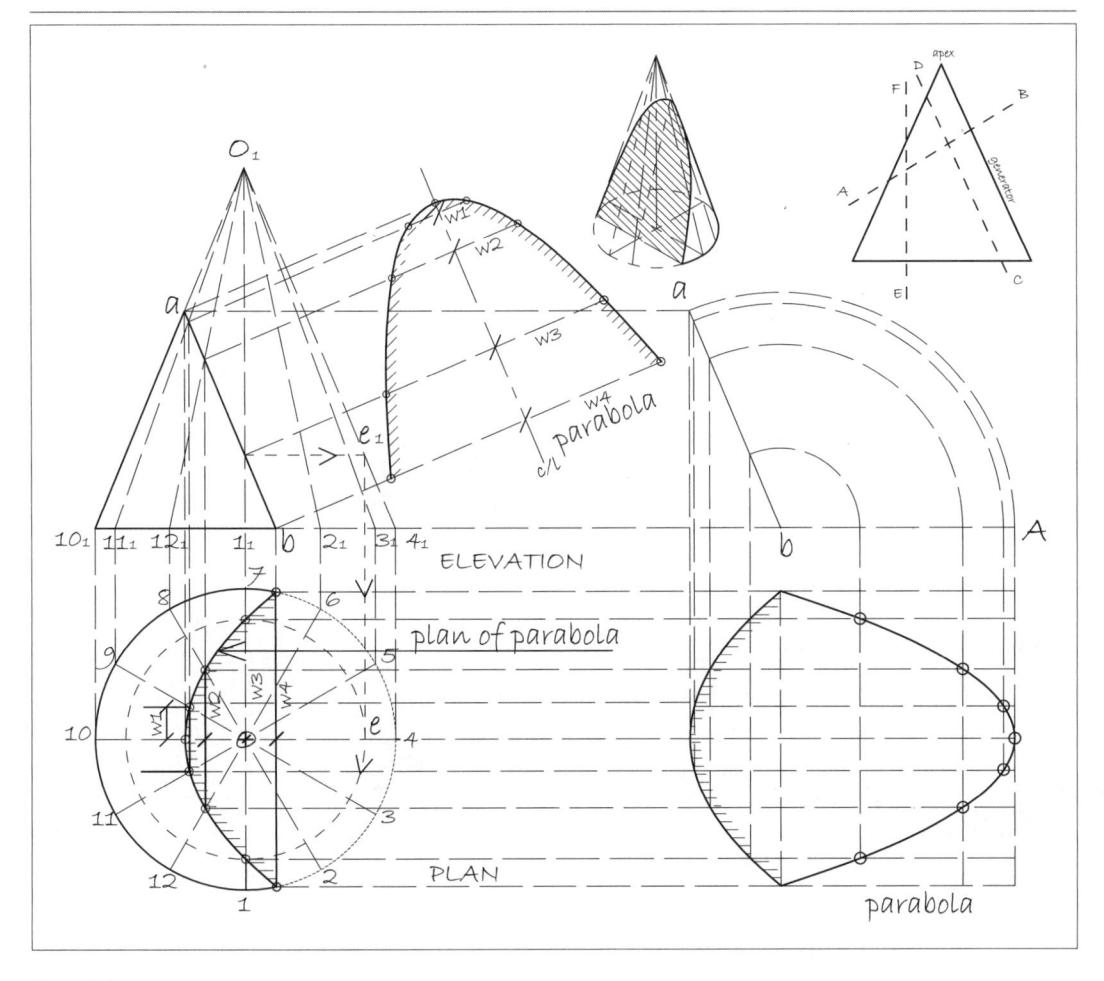

Fig. 7.1
Fig. 7.2

To develop the true shape of the parabola (see **Fig. 7.2**):

Method 1

- In the elevation, construct centre line **(c/l)** parallel with **a-b** and project all points from **a-b** at right-angles to it.
- Take the widths **w1**, **w2**, **w3** and **w4** from the plan and apply them as shown to the left and right of the **c/l**.
- Join all points to give true shape of parabola.

Method 2

On the right of **Fig. 7.2**, the plan and elevation of the parabola are reconstructed for clarity.

- In the elevation, the line **a-b** and all points on it are knocked flat and projected to the plan. This method is used in Chapter 4 on geometric solids.
- The widths are projected from the plan and the grid into which the development fits is now drawn.
- Draw a curve through all relevant points to complete the parabola.

Fig. 7.3 shows the ellipse, as a conic section, developed using the same principles.

Again, level **e1** is determined in the elevation and as **e** in the plan, and in the development the widths **w1** to **w5** are taken from the plan and transferred to it (the development). Use methods 1 and 2 above, as for the parabola.

The third conic section, the hyperbola, is shown in **Fig. 7.4**. In this case the plan and elevation of the hyperbola is a straight line. Note: The developed hyperbola (in this case, a vertical cut) may also be achieved by drawing an end view of the pyramid.

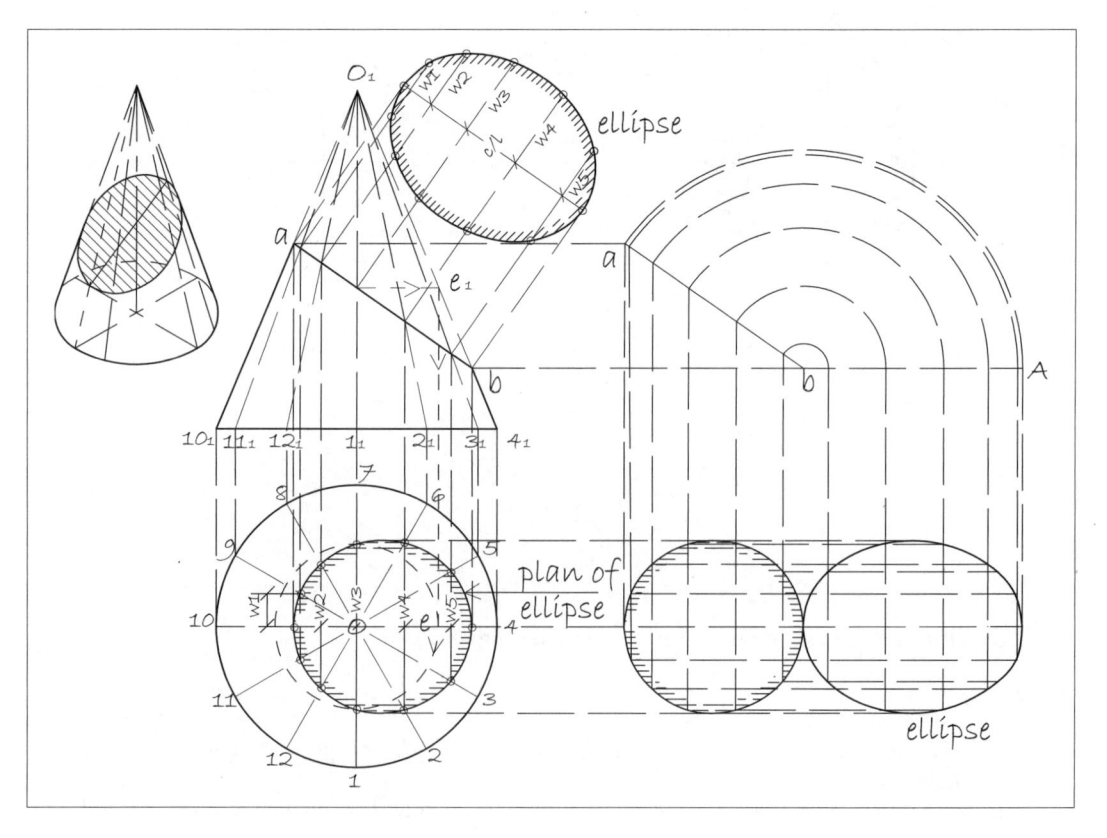

Fig. 7.3

7 CONIC SECTIONS AND CONIC DEVELOPMENT

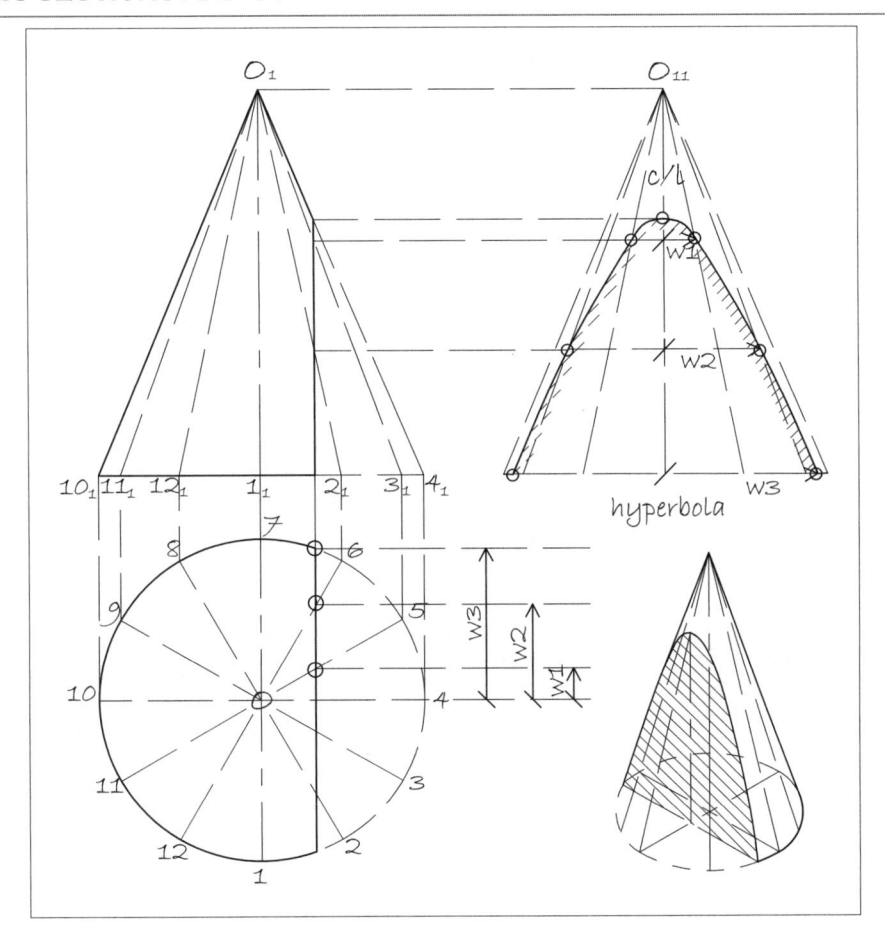

Fig. 7.4

The examples given in **Fig. 7.5** are practical applications of the conic sections: parabola and ellipse. Two cones are shown intersecting with a pitched main roof (in both plan and elevation) that is hipped at both ends. On the right-hand side the cone meets the main roof at line **a-b**, which is parallel with the cones generator **o1-g1**, the true shape of this intersection is a parabola.

The plan and the development of the parabola are drawn in the same way as in **Fig. 7.2**. The development shows the shape of the curb (curved lay-board in this case) which is fitted to the main roof to take the seat cut of the shortened conical rafters; the curb in true shape is a parabola.

The cone on the left intersects with the hipped end of the main roof at the angle given. If the angle of the roof and the generator of the cone are continued to **h**, it will be seen that the resulting conic section is an ellipse, therefore, the roof intersection will be part of a true ellipse. The procedure for obtaining the true elliptical shape is the same as in **Fig. 7.3**.

Fig. 7.5

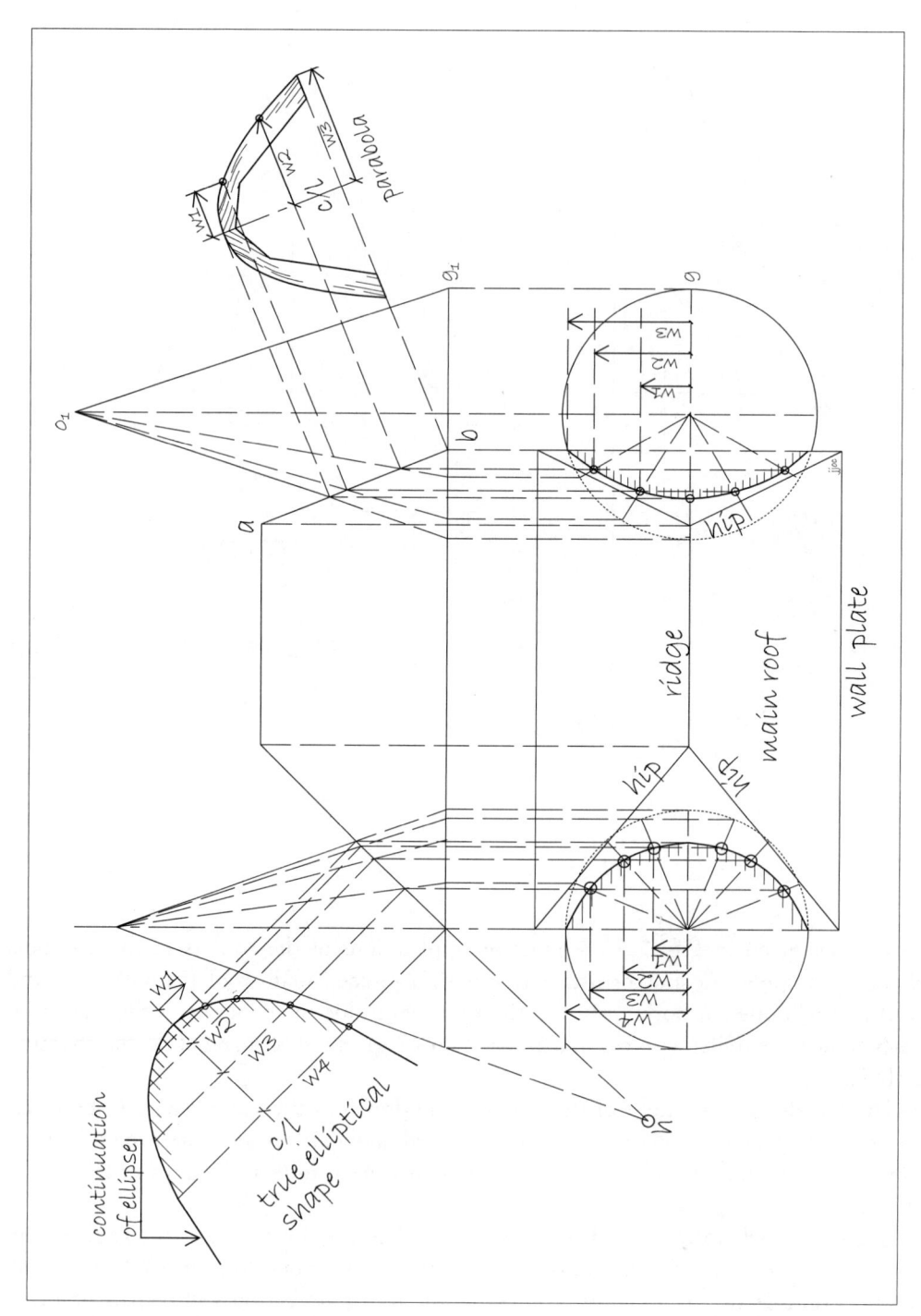

7 CONIC SECTIONS AND CONIC DEVELOPMENT

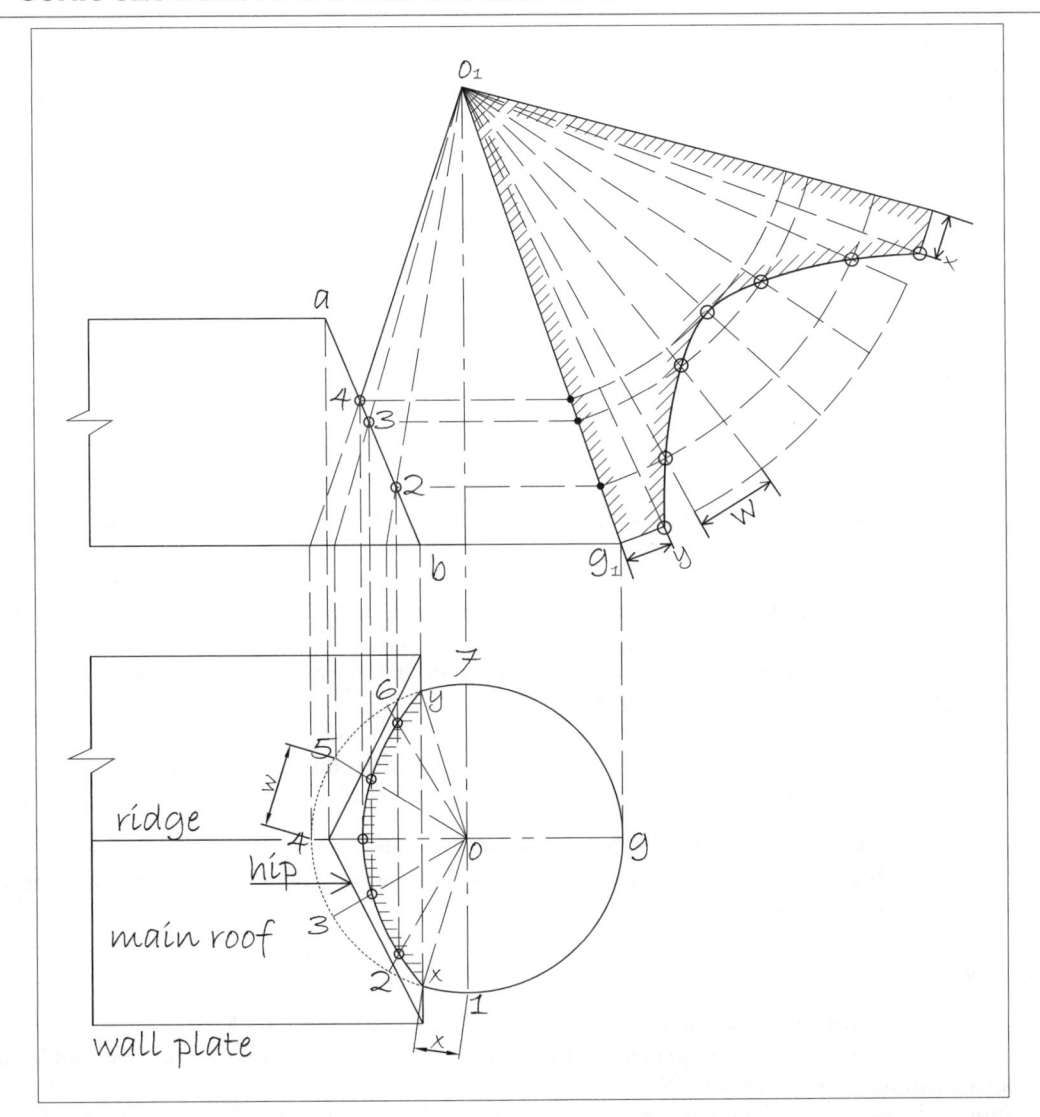

Fig. 7.6

Fig. 7.6 shows part of the same roof arrangement as in **Fig. 7.5**, but this time the curved conical surface is developed.

Method

- In the elevation, swing an arc of **o1-g1** radius and mark off on it distance **w** six times and join back to **o1**.
- Project the intersections **4**, **3** and **2** from the elevation to the outside generator **o1-g1** (these intersections are where the generators cross line **a-b** in the elevation). Note: The outside generator shows a

true length and by transferring the inner generators to it, parallel with the cones base, horizontally in this case, their true lengths are also revealed (on the outside generator).
- Swing all points on the generator **o1-g1**, from **o1**, to complete the grid into which the development fits.
- Establish points **x** and **y**. Join all the relevant points to complete the development of the half-cone that intersects with the main roof.

In **Fig. 7.7** the plan, elevation and end view of a roof to a segmental extension to a building are given. The intersection between the roof and the vertical wall of the main building will be a hyperbola, and, consequently, the shape of the vertical wall piece to which the rafters are attached.

To find the shape of the hyperbola:

Method

- From the plan which is given, project and determine the end view of points **1**, **2**, **3**, **4** and **5**.
- From point **511**, project the pitch of rafter at the angle given (50°) to intersect with **C** projected from the plan to find the apex of the cone.
- Join remaining rafters **2**, **3** and **4** to the apex. This completes the end view of the rafters.
- Project where the rafters (being treated as single lines) intersect with the wall line in the plan, and in the end view, to the elevation to give points on the hyperbola. Join all the points to complete.
- By joining rafters **21**, **31**, etc., to the apex in elevation, it will be seen that the rafter lines pass through these points already established.
- By adding the thickness (**T**) of the vertical wall piece in plan and end view, another hyperbola is realised in the elevation; this time by projecting where the rafters intersect with the thickness of the wall piece (see inset **A**) in the end view on to the rafters in the elevation.

This second hyperbola represents the bevel-line on the wall piece.

In **Fig. 7.8** thickness is added to the rafters, and some rafter lengths and bevels are found. There are two methods shown.

Method 1

- Make **C-a** horizontal in the plan and project **a** to the **X-Y** line to become **A**.
- Join **A** to apex in the elevation to give the true length of the complete rafter for the cone, and add a width to the rafter, as shown.
- Project the heights on the plumb cut (from each side) of each rafter in the elevation to the developed full rafter to give the true lengths of each of these rafters. The edge cuts on the rafters are now visible.
- To determine the true edge bevels, add the thickness (**T**) to the rafter and project each edge cut (at 90°) to this line. Join points **f-g** as shown in the larger detail to give the true edge bevel.

Method 2

This method uses an auxiliary elevation. Rafter **b** is used in the example:

- Project all points on the rafter at 90° to the plan of the rafter.
- Set up **X1-Y1** and take heights **H1** and **H2** from the elevation and apply as shown.

To determine the edge bevel, apply the same procedure as in method 1.

7 CONIC SECTIONS AND CONIC DEVELOPMENT

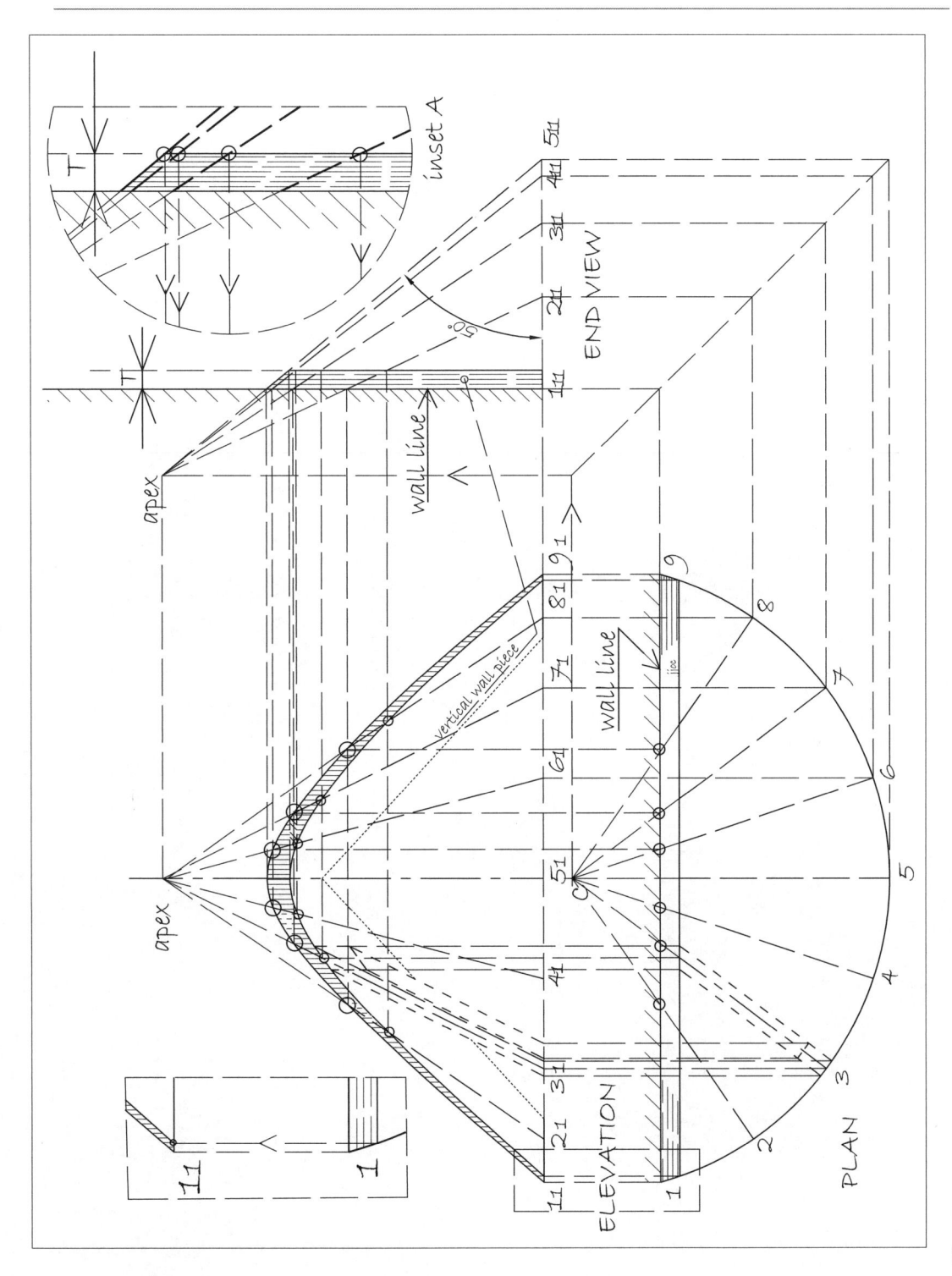

Fig. 7.7

74 GEOMETRICAL DRAWING FOR CARPENTRY & JOINERY

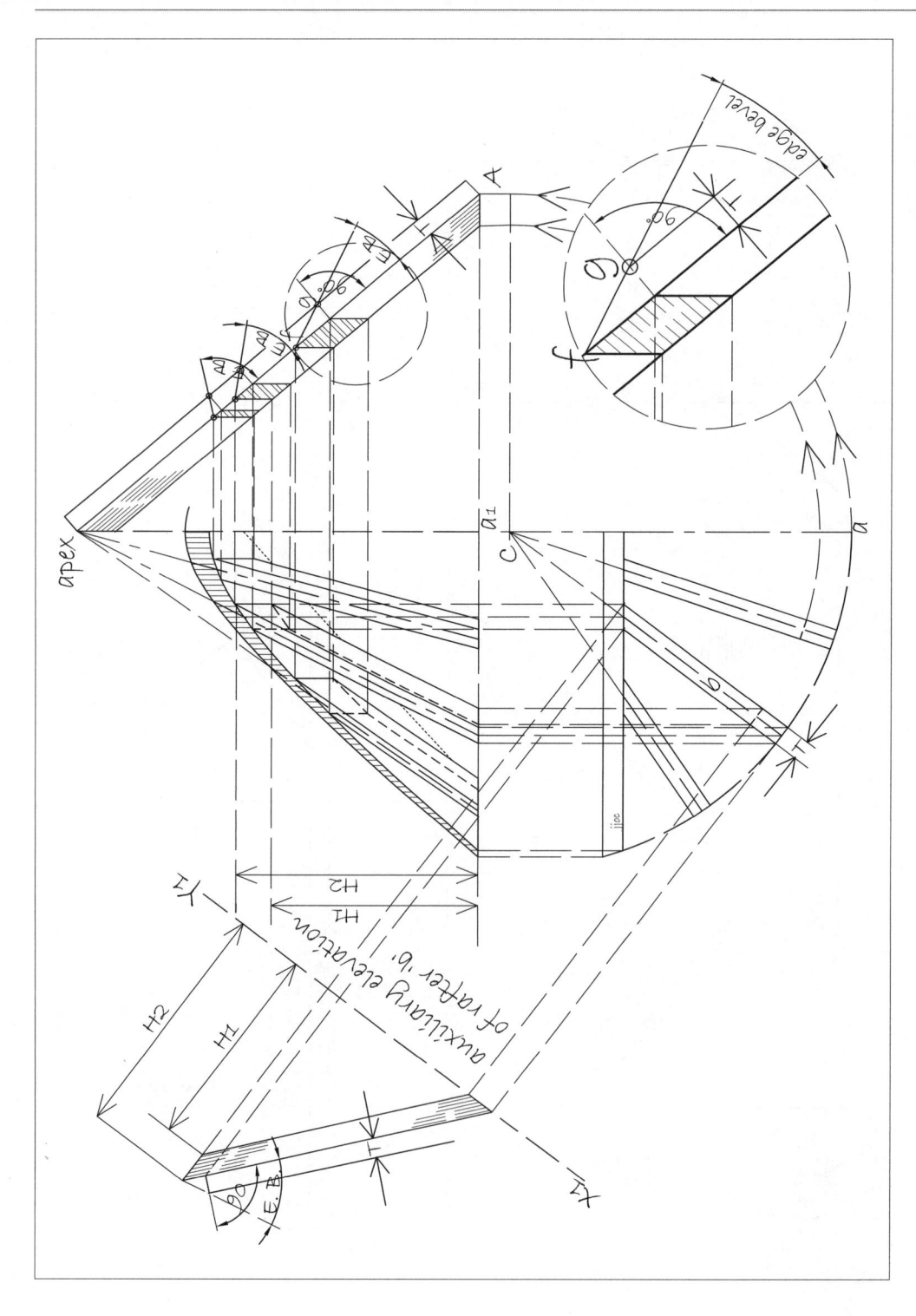

Fig. 7.8

7 CONIC SECTIONS AND CONIC DEVELOPMENT

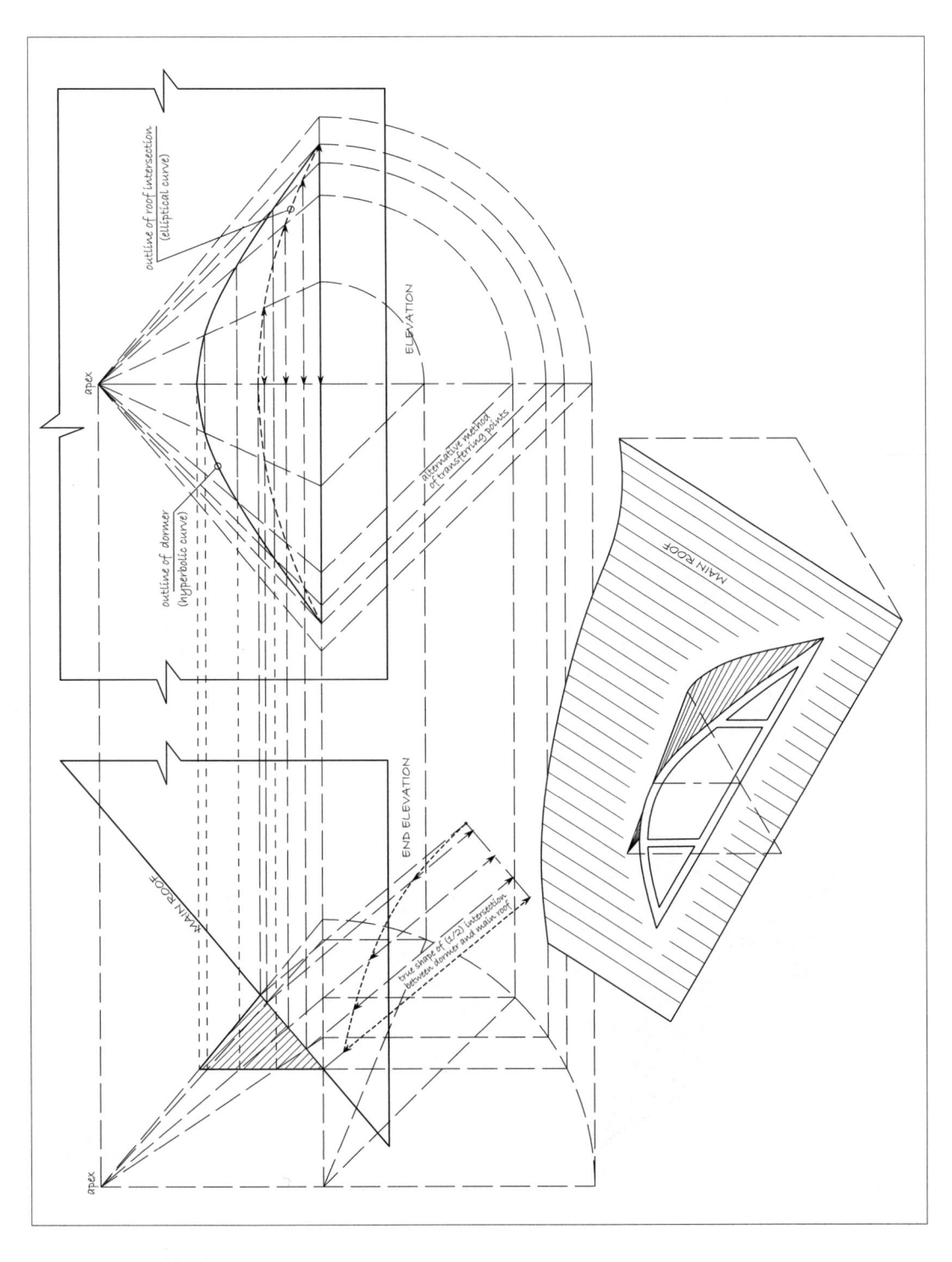

Fig. 7.9

GEOMETRICAL DRAWING FOR CARPENTRY & JOINERY

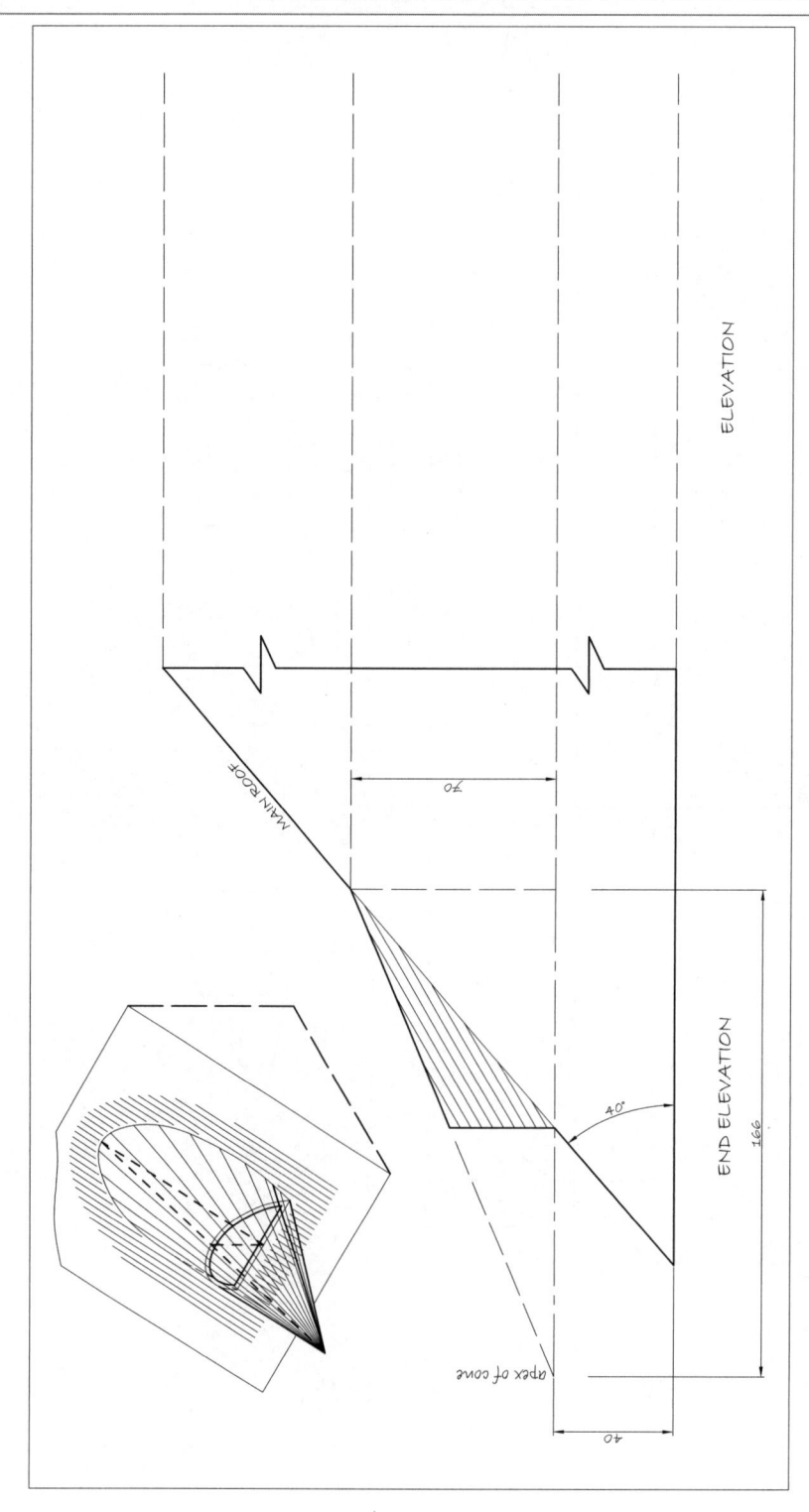

Fig. 7.10

7 CONIC SECTIONS AND CONIC DEVELOPMENT

Fig. 7.9 represents an applied example where both the parabola and the hyperbola are used in a roofing context. The pictorial view shows a dormer roof on a main pitched roof. The front window head of the dormer is hyperbolic and the intersection between the main roof and the dormer roof is part of an ellipse.

The end elevation shows the cone on which the construction is based, and the elevation reveals the true shape of the hyperbola as well as the intersection line between the dormer roof and the main roof.

The true shape of half the intersection between the two roofs is drawn as a projection from the main roof in the end elevation. This shape represents the shape of the lay board on which the dormer rests.

Exercise

A dormer roof, which is part of a cone, resting on a pitched roof is shown in the end elevation of **Fig. 7.10**. This example is based on a sloping plane cutting a cone across its axis. The resulting section will be part of an ellipse. Reproduce **Fig. 7.10** and complete the exercise.

- Draw an elevation on the right-hand side.
- Determine the true shape of the intersection between main roof and dormer roof.
- Develop the dormer roof surface.

8 ROOF GEOMETRY

SECTION I: CONVENTIONAL PITCHED ROOFS

In the setting-out of the common rafter shown in **Fig. 8.1**, the *measuring triangle* is outlined. An understanding of this right-angled triangle is important to roof geometry. It could be said that all pitched roof geometry is based on the right-angled triangle. Some roofing terminology is also given.

Fig. 8.1 shows the plan and elevation of a roof with a gable end and a hipped end, and the setting-out of a common rafter. Note: The pitch of a roof may also be expressed as a fraction, e.g., 1/2, 1/3, meaning the rise over the span. Take the example of a roof of 1/3 pitch with a span of, say, 7.5 metres. The rise would be: 1/3 × 7.5 = 2.5.

Fig. 8.2 shows the plan and elevation of a hipped roof with a splayed wall plate at one end. All roof surfaces have the same pitch. In **Fig. 8.2** the true pitch is shown to the left of the plan at **D** as well as the plumb and seat cuts for the common rafter. When all roof surfaces

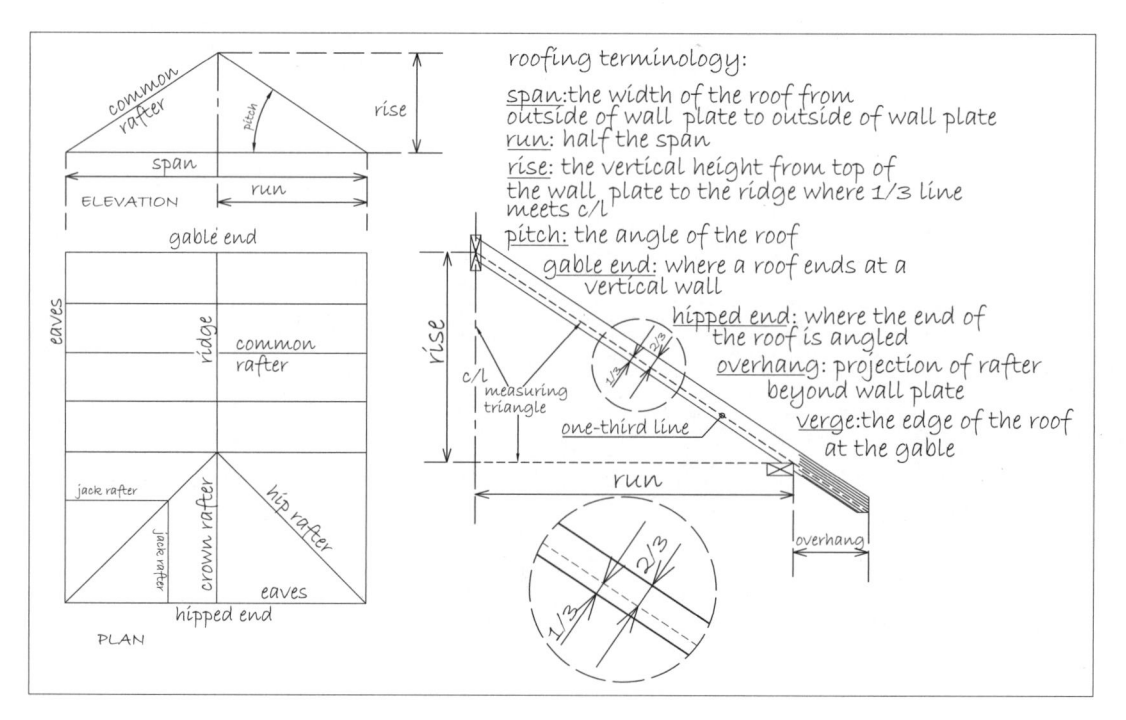

Fig. 8.1

8 ROOF GEOMETRY

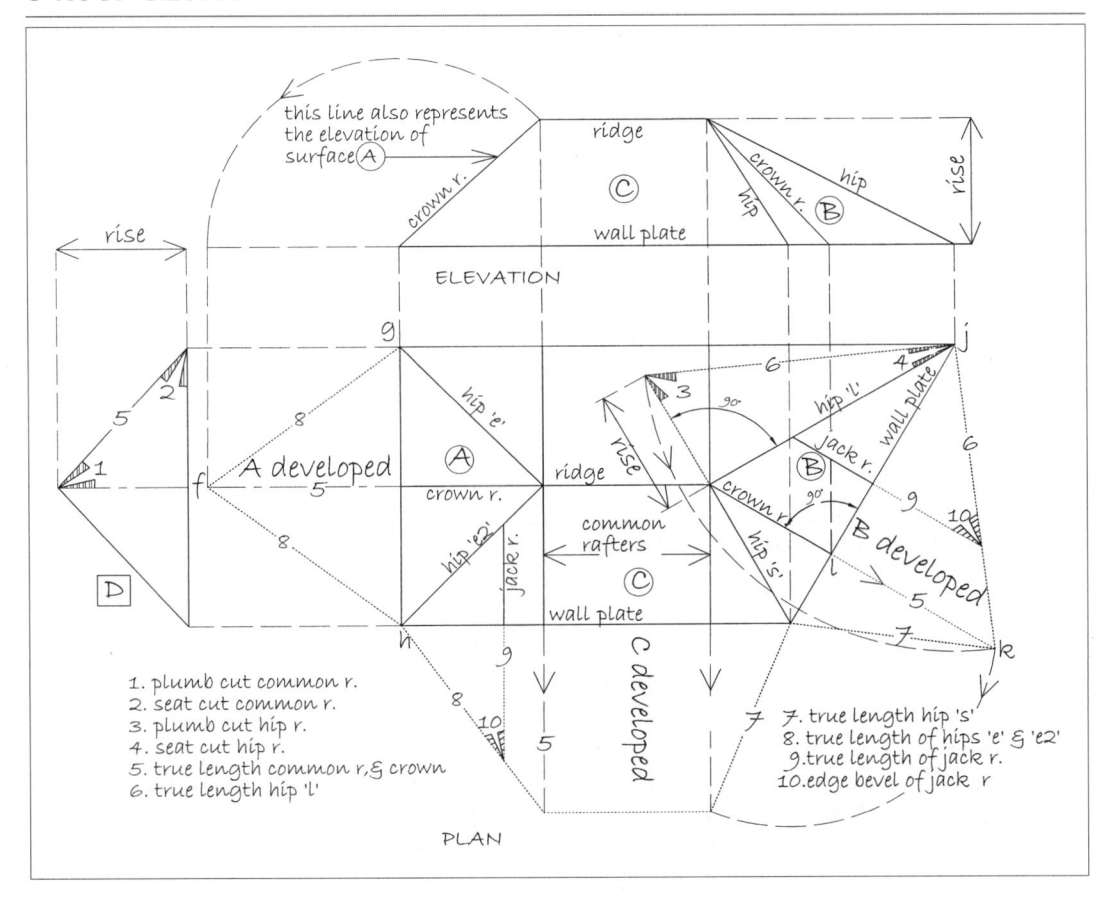

Fig. 8.2

have the same pitch, the hips in the plan will always bisect the corner angles. This is evident in the plan presented at **Fig. 8.2**.

To find the true length, plumb and seat cuts for the long hip (hip 'l'):

Method

- Set off the rise at 90° to the plan of the hip (from the ridge) and join to the wall-plate corner, giving **6** as the true length, and cuts **3** and **4**.

To develop roof surface **A**:

Method

- Pivot the surface to a horizontal position in the elevation and project to the plan at **f**.
- Join **f** to **g** and **h** to complete the development, in which the true length of hip **e** is realised at **8**.

To develop roof surface **B**:

Method

- Extend from the crown rafter at 90° to the wall plate and pivot the true length **6** (from **j**) to meet the extension line at **k,** resulting in true lengths **6** and **7** when the development is completed.
- Line **l-k** is the true length of the crown.

The development of surface **C** is derived as shown.

Note: All surface developments are 'hinged' from their respective wall plates.

Fig. 8.3a shows the plan of a roof which has a hip and valley rafter. The common rafters are developed to the right, and an auxiliary elevation showing the development of the hip and valley rafters is also shown.

Method

- In the auxiliary elevation a baseline representing the top of the wall plate is set up.
- The rise is transferred as shown, and lines drawn from the rise to the corners of the wall plate give the counterpart of the 1/3 line on the hip and valley rafters.

To find the top line of the hip:

Method

- Transfer the upstand **u** from the common rafter and apply as shown in inset **B (Fig. 8.3b)**.
- Then draw a line parallel with the 1/3 line to give the top of the hip. (Note: To avoid confusion, on the hip the 1/3 line is best called the measuring line.)
- Apply the width of the hip square-off this line, see inset **B (Fig. 8.3b)**.

To compensate for the corner of the wall plate being cut off to house the hip, the upstand **u** is moved inwards, as shown at inset **B**. This distance is equal to half the thickness of the rafter, and a line parallel with the hip at the top of the upstand at this point gives the *backing (dihedral) angle* for the hip (see inset **D** and also see **Fig. 8.5** below).

Inset **C** shows the arrangement for the valley rafter, the essential difference being that, in this case, the upstand is moved outwards from the corner of the wall plate. The dihedral angle for the valley is the obverse of that for the hip (see inset **D**), however, in practice the top of the valley rafter is left square. (Exceptions to this are in situations where the underneath construction of the roof is visible and adds to the aesthetics of the structure, such as in churches and other public buildings.) When doing this, the top line of the valley rafter is at the top of the inner upstand (see inset **C**).

Note: When transferring the rise to the auxiliary elevation, it must be marked off above the top of the wall plate along the projected centre line.

Fig. 8.4 shows the plans of four roofs, which all have one thing in common: the roof surfaces have the same pitch. This is so because all hips and valleys bisect the wall-plate corner angles, which in this case are 90°.

Fig. 8.5 shows the method of finding the dihedral (backing) angle for a hip rafter.

8 ROOF GEOMETRY

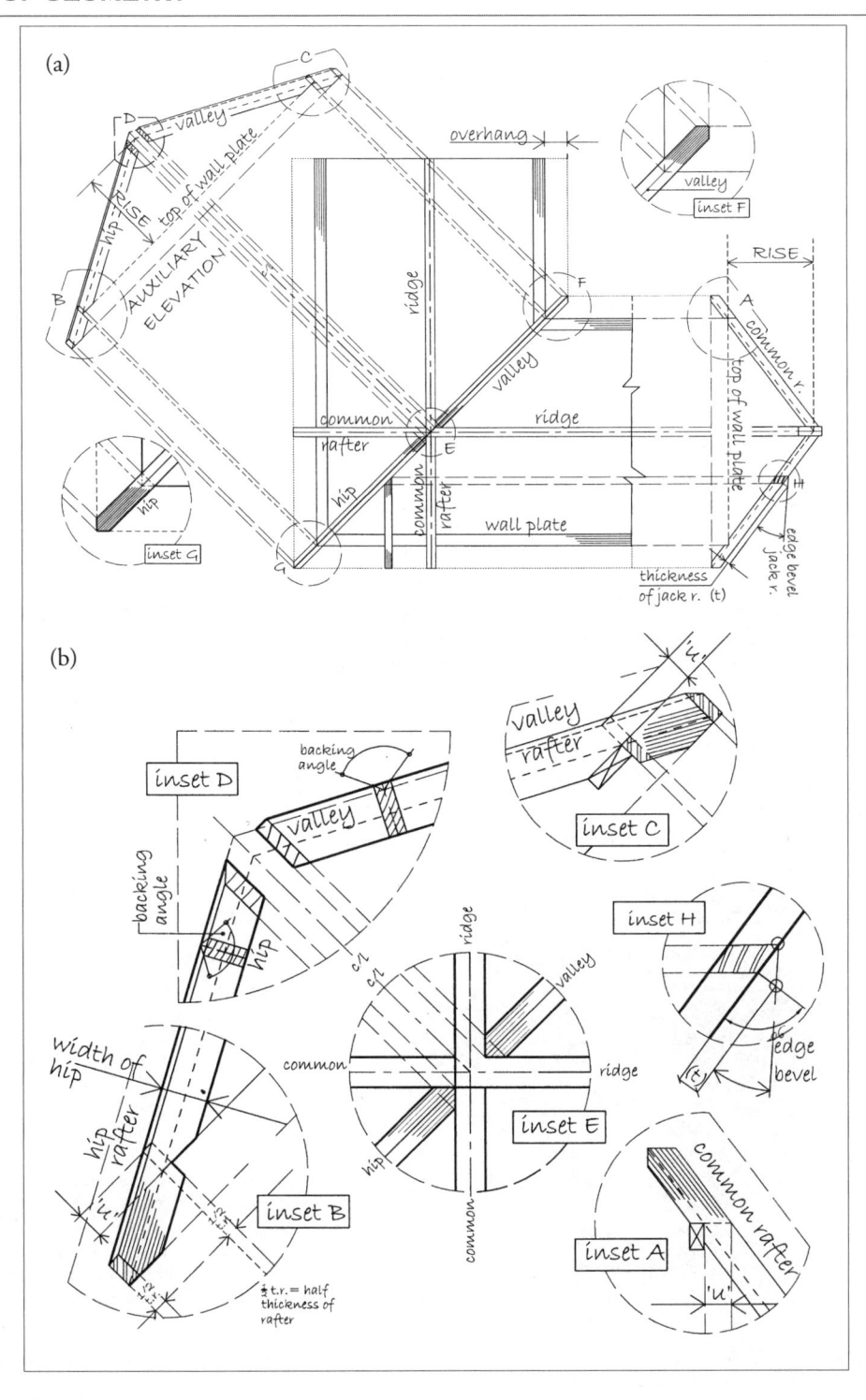

Fig. 8.3a
Fig. 8.3b

82   GEOMETRICAL DRAWING FOR CARPENTRY & JOINERY

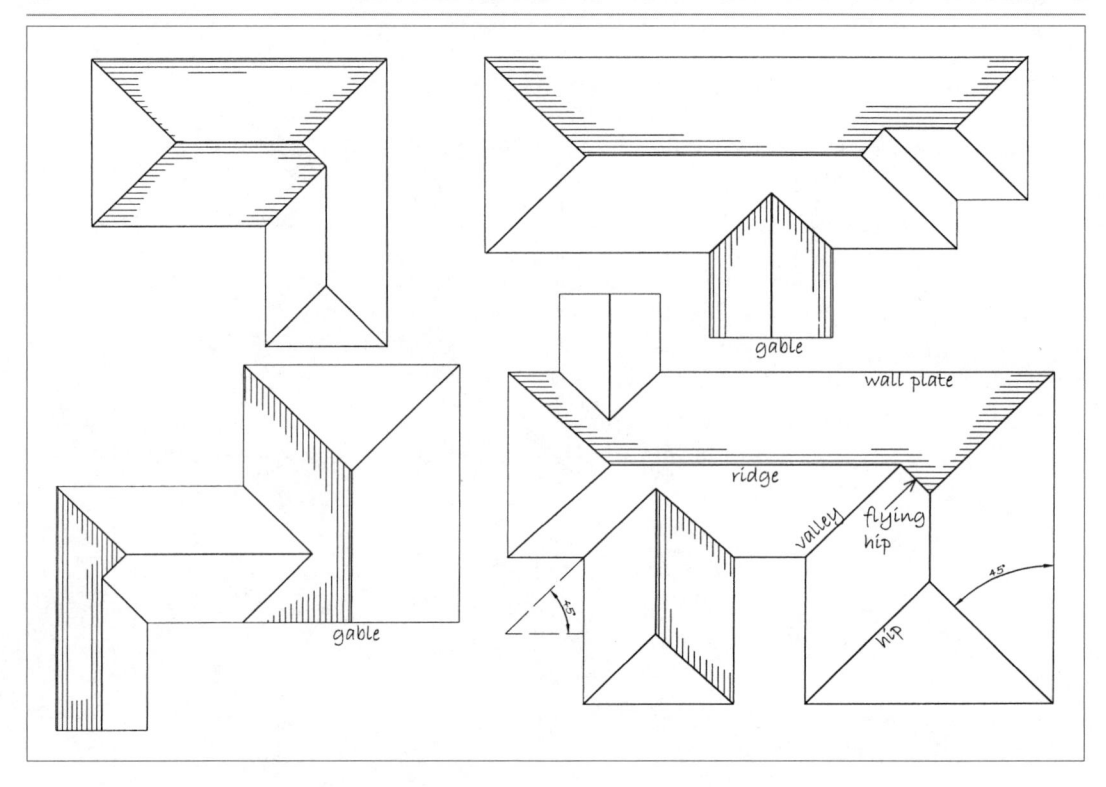

Fig. 8.4

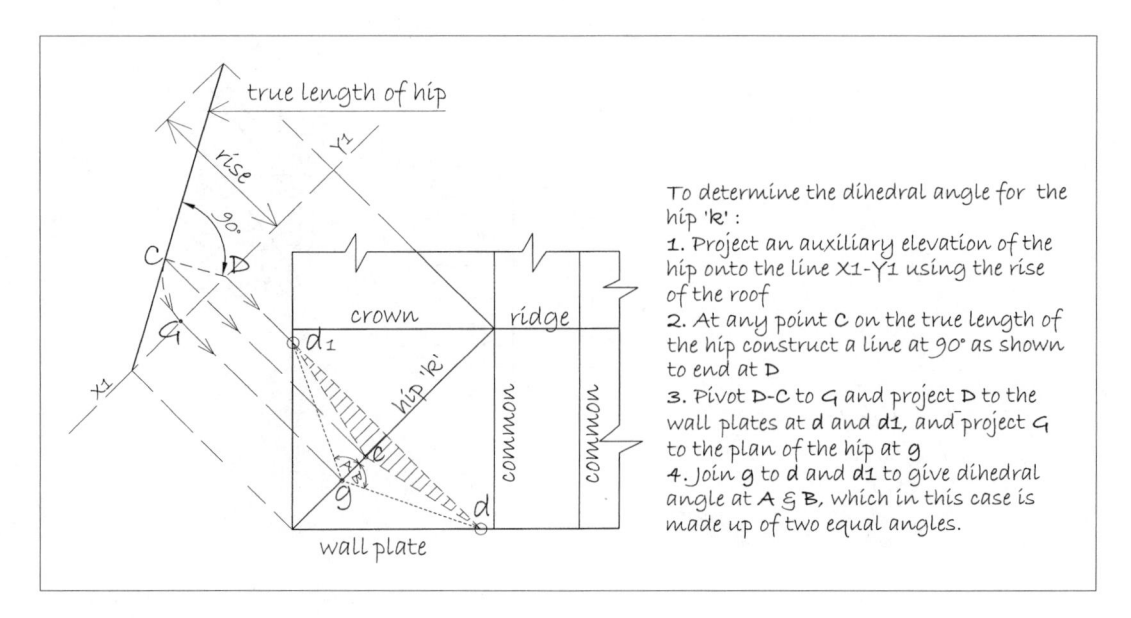

To determine the dihedral angle for the hip 'k':
1. Project an auxiliary elevation of the hip onto the line X1-Y1 using the rise of the roof
2. At any point C on the true length of the hip construct a line at 90° as shown to end at D
3. Pivot D-C to G and project D to the wall plates at d and d1, and project G to the plan of the hip at g
4. Join g to d and d1 to give dihedral angle at A & B, which in this case is made up of two equal angles.

Fig. 8.5

8 ROOF GEOMETRY

Method

- Using the rise of the roof, project an auxiliary elevation of the hip.
- At any point **C** on the true length of the hip, construct a line at 90° to end at **D**.
- Pivot **D-C** to **G** and project **G** to the plan of the hip at **g**.
- Join **g** to **d** and **d1** to give the dihedral angle at **A** and **B**, which in this case is made up of two equal angles.

Fig. 8.6 is a line drawing of a roof with unequal pitches. The main roof is pitched at 40° and the hipped ends are pitched at an angle of 60°. In this case, the hips do not bisect the corner angles in the plan. On the right-hand side where there is a splayed wall plate, the intersection where the hips meet the ridge in the plan is found as follows:

Method

- Set up **X1- Y1** at 90° to the splayed wall plate.
- Draw an auxiliary elevation at 90° to the splayed wall plate.
- Set up the ridge line using the rise and draw the roof pitch at 60° to the **X1- Y1**, which represents the top of the wall plate.
- Where the 60° line intersects with the ridge line, project to the ridge in the plan and join to the wall-plate corners, giving the long and short hips for this end of the roof.
- The crown rafter may now be drawn in the plan at 90° to the splayed wall plate.

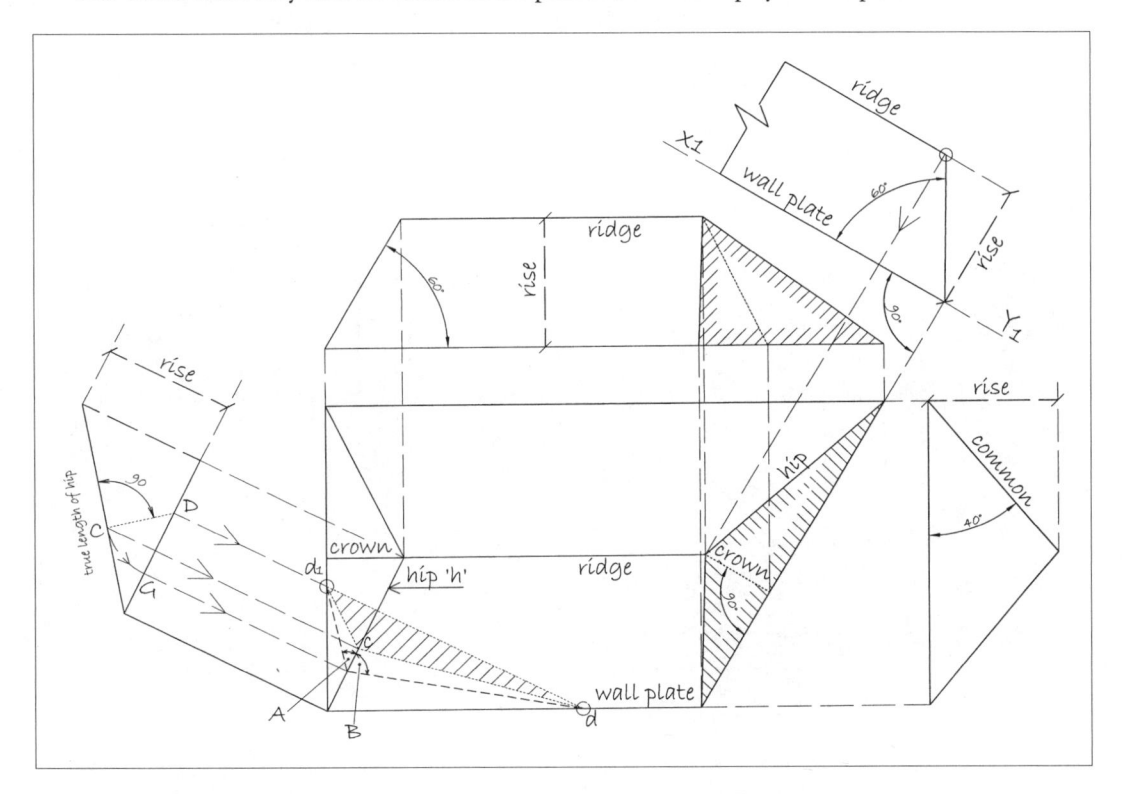

Fig. 8.6

The procedure for finding the dihedral angle for hip **h** is the same as for hip **k** in **Fig. 8.5**, but this time angles **A** and **B** are not equal because the roofs have different pitches.

Note: The shaded triangle **d-d1-c** in the plan (which is **D-C** in the auxiliary elevation) represents a triangle square-off the hip, under the two planes of the roof.

In **Figs. 8.7**, **8.8** and **8.9** bevels relating to purlins are explored. A purlin has three associated bevels: *face*, *edge* and *lip bevels*. To ensure correct fitting against the hip, these must be known. The *face* and *edge bevels* are relatively easy to comprehend, but the *lip bevel* requires a considerable understanding of applied geometry and geometric planes (see Chapter 17).

Fig. 8.7 deals with the face and edge bevels which are found as follows:

Method

- Construct a horizontal line at **a1** and pivot the face and edge upwards as shown, giving points **b1** and **c1**.
- Project **b1** and **c1** to the plan and project **b** and **c** horizontally to become **B** and **C**.
- Join **a-B** and join **a-C** to give the face and edge bevels.

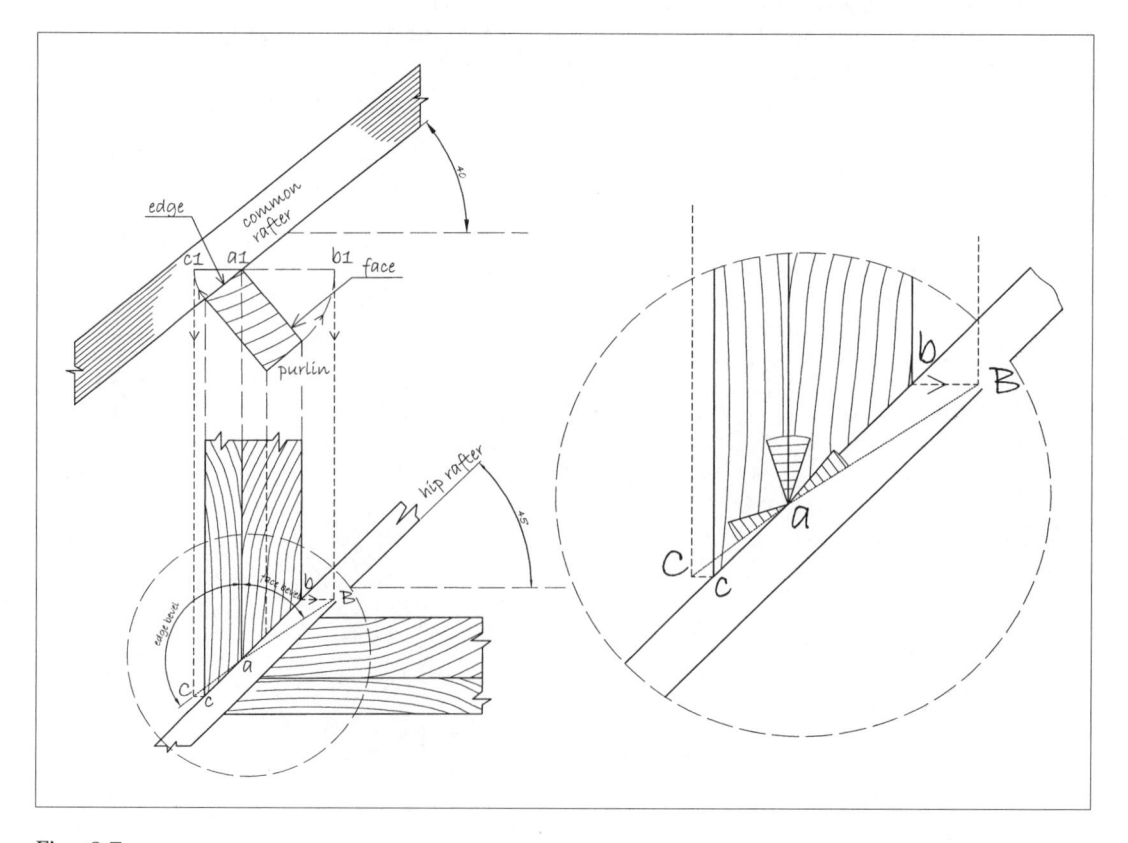

Fig. 8.7

8 ROOF GEOMETRY

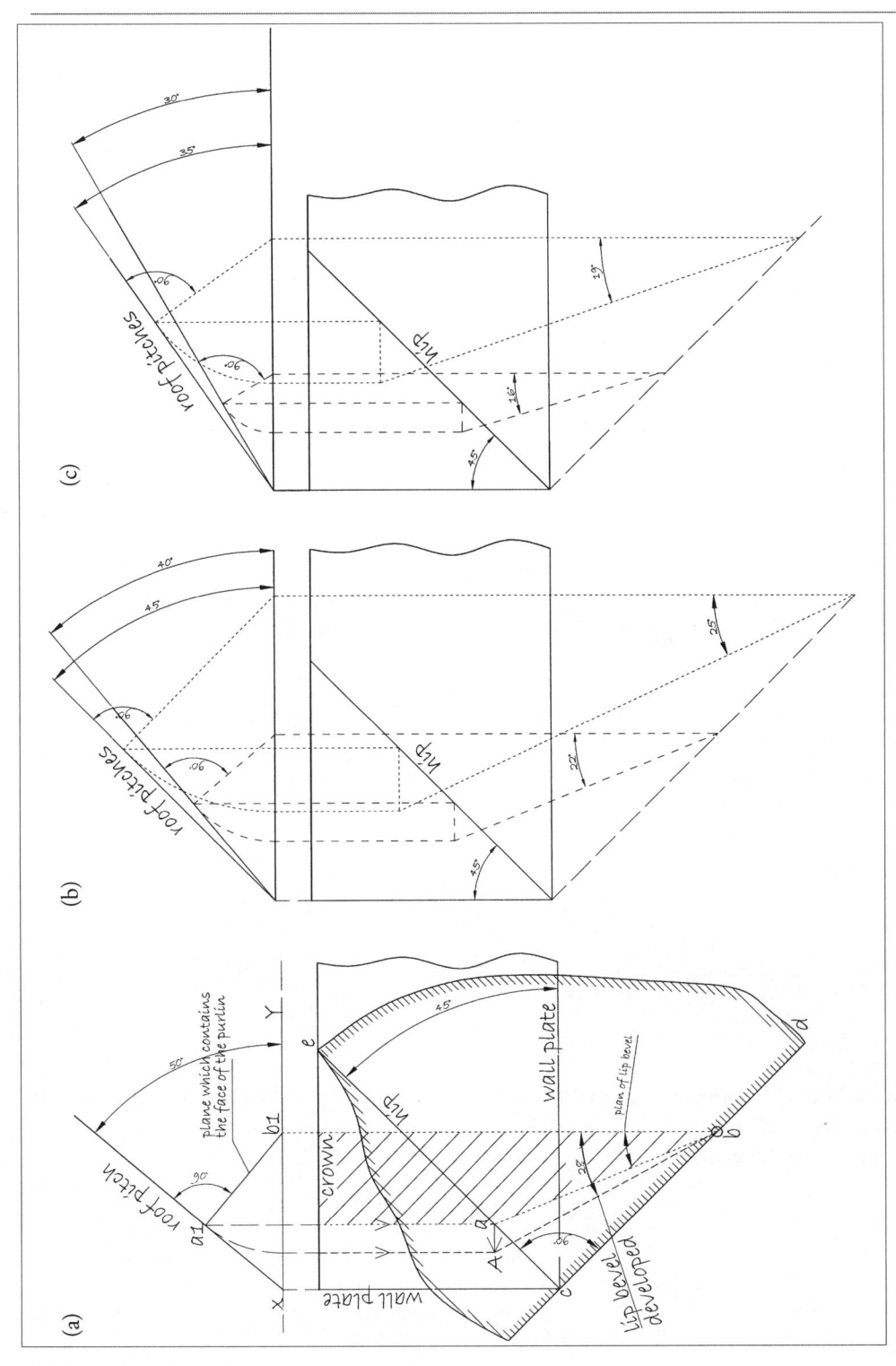

Fig. 8.8a
Fig. 8.8b
Fig. 8.8c

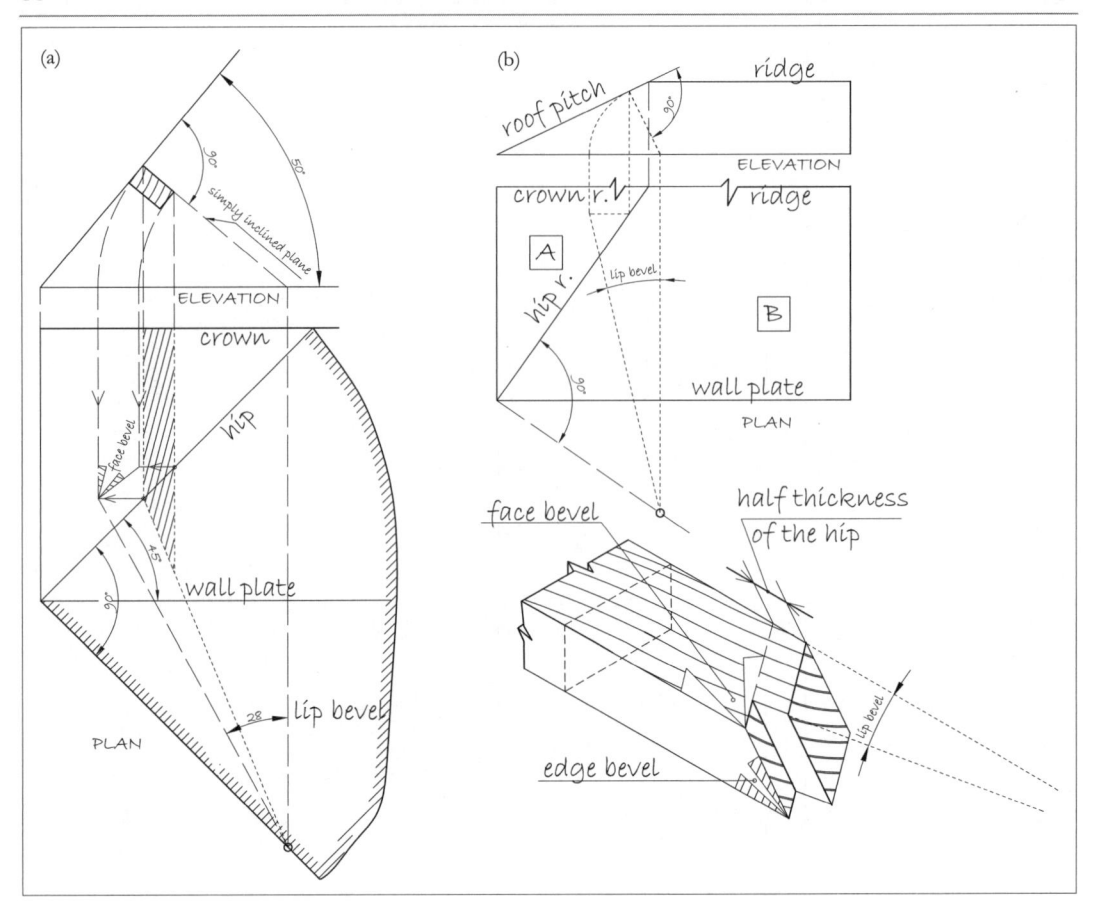

Fig. 8.9a
Fig. 8.9b

Fig. 8.8a outlines the associated geometry for the purlin lip bevel. The outline part plan and elevation of a pitched roof is shown. In this case, the roof is pitched at 50° and the plan of the hip is at 45° to the wall plates.

Method

- In the elevation, a line **a1-b1** is drawn at right angles to the pitch of the crown rafter to touch the **X-Y** line. This line represents the sloping plane on which the purlin lies; specifically, this is a *simply inclined plane* (see Chapter 17) because it is at an angle to the horizontal plane only, which is represented in the elevation by the **X-Y** line.

8 ROOF GEOMETRY

- In the plan, a line is drawn at 90° to the plan of the hip, giving **c-d**; **c-d-e**, which, it is now established, represents the plane on which the underneath edge of the hip lies. This is an *oblique plane* (see Chapter 17) because it is inclined to both the horizontal and vertical planes. The intersection between the oblique plane **c-d-e** and the simply inclined plane **a1-b1** gives the lip bevel for the purlin.
- In the elevation, pivot **b1-a1** to the **X-Y** line and project to the plan where **a** moves to **A**. Join **A** to **b**.
- The plan of the lip bevel is shaded and the developed lip bevel is between line **b-b1** and line **b-A**.

Figs. 8.8b and **8.8c** show the principle applied to roofs of varying pitches where the hip remains at 45° in the plan. The resulting table gives the lip bevels for the pitches in question. It can be seen from the table below that with every increase of five degrees in the roof pitch, there is an increase of three degrees in the lip bevel. It is thus possible to extrapolate that a roof having a pitch of 55°, with a hip at 45° in the plan, would have a lip bevel of 31° on the purlin.

Roof with a 45° hip in plan (square wall plate)

Roof pitch in degrees	*Lip bevel in degrees*
30	16
35	19
40	22
45	25
50	28

It is important to remember here that hips are not always at 45° in the plan of a roof drawing. We shall deal with that scenario later.

Fig. 8.9a has the same roof as in **Fig. 8.8a**. On the simply inclined plane, the face of the purlin is lying. This is the portion of the plane which concerns us here, and the method of developing it to give the lip bevel as well as an alternative method of developing the face bevel.

At **Fig. 8.9b** the lip bevel for a roof with a hip at a random angle in the plan is given. It can be seen that the same principles are applied in obtaining the bevel, but it must be emphasised here that the resulting lip bevel found is for the purlin on roof surface **A** only. The purlin on roof surface **B** will have a different lip bevel.

In the lower part of **Fig. 8.9b** is a sketch showing how the three bevels are applied to the purlin.

Exercise

Fig. 8.10 shows an outline plan of wall plates for a roof with an off-centre ridge. The pitches of the roof are also given to the left of the plan. A key plan, not to scale, is also shown. Note: The hipped ends of the roof are pitched at 45°. Reproduce **Fig. 8.10** and complete the exercise.

- Complete the plan to show: ridge, hips and crown rafters.
- Determine the true length, plumb and seat cuts for hip **A**.
- Develop the true shape of roof surface **B**.

Fig. 8.10 shows an outline plan of wall plates for a roof with an off-centre ridge, the pitches of the roof are also given on the left of the plan. A key plan, not to scale, is also shown.
Note: the hipped ends of the roof are pitched at 45°.
Q.1. complete the plan to show: ridge, hips, and crown rafters; also determine the true length, plumb and seat cuts for the hip A.
Q.2. develop the true shape of roof surface B.

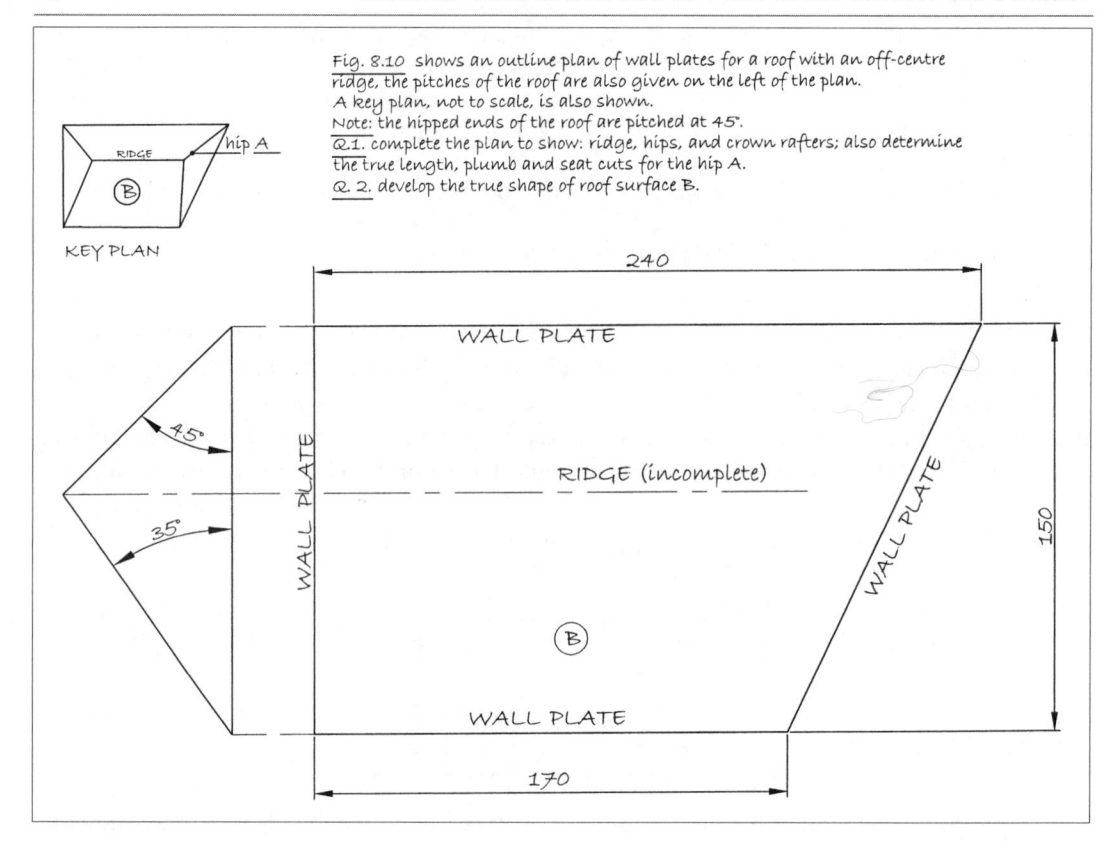

Fig. 8.10

SECTION II: PYRAMID ROOFS

Pyramid roofs are pitched just like conventional roofs and will be treated using the same principles. These types of roofs are generally made up of hips, crown and jack rafters and have an apex instead of the conventional ridge.

Fig. 8.11 represents the plan and elevation of a roof with octagonal wall plates. The roof rises to a square level and from there to an apex. It is a transition roof: from octagon to square to apex. The roof will be glazed on all surfaces. There are basically three roof surfaces, each of which is replicated four times. The surfaces in question, **A**, **B**, and **C**, are shown developed from the plan, using principles already covered. In the developments, the true lengths of all relevant rafters will be found.

The dihedral angle for rafter **a-c** is found and it is evident that only one half of the rafter thickness will have a backing angle. The other half, impacting roof surface **A1**, will be square (90°). Rafter **b**, at the intersection between two pyramid roof surfaces, will have a backing angle on both sides of the rafter, and in this case the angle will be equal on both sides.

8 ROOF GEOMETRY

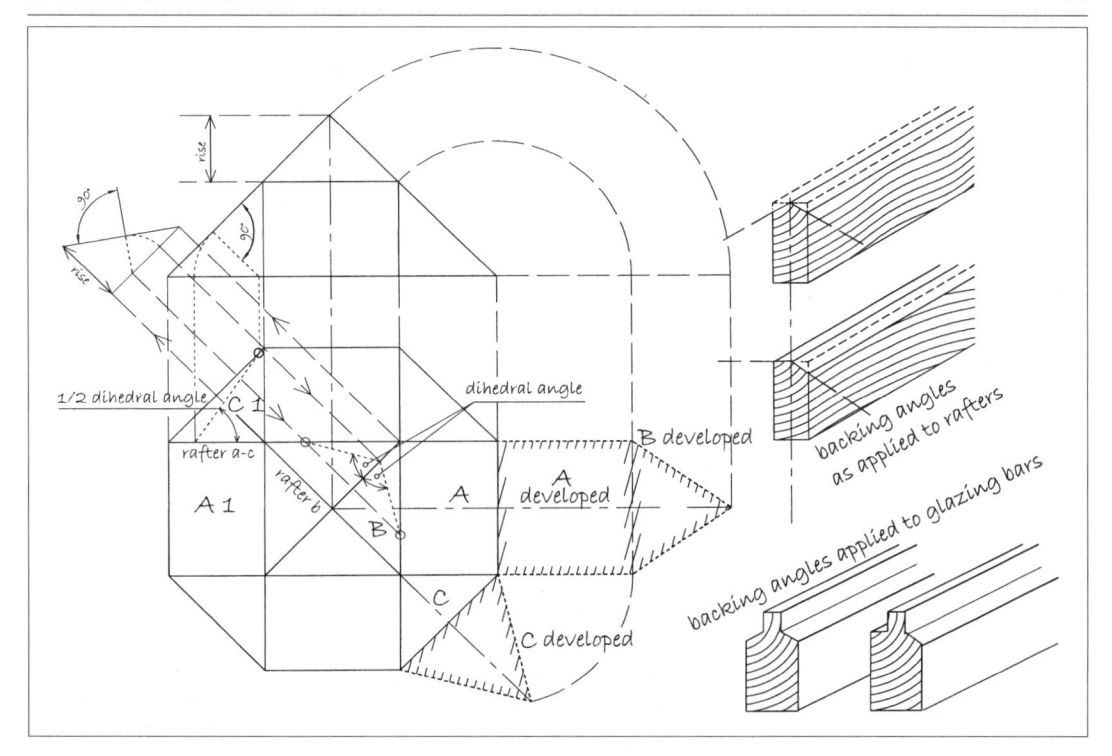

Fig. 8.11

Fig. 8.12 shows a hexagonal pyramid roof in plan and elevation. In the plan, lines **o-a**, **o-b**, **o-c**, etc., represent hips on the pyramid, and lines **o-g** and **o-h** represent crown, or king, rafters. In the elevation, hip **o1-f1** shows a true length. This line is reproduced for clarity on the left, and the plumb and seat cuts are indicated on the figure.

The crown rafter is developed in auxiliary elevation, giving its true length **O-G**; also marked are the true lengths for the respective jack rafters, **G-j** and **G-k**. In the plan, the development of surface **b-c-o** also reveals the true lengths of crown and jack rafters as well as the edge bevels for these. Surface **a-f-o** in the plan has thickness added to the rafters.

In **Figs. 8.13a** and **8.13b** the dihedral (backing) angle for the pyramid is found using three methods: two geometric and one practical.

Method 1

In the **Fig. 8.13a** in the elevation, **o1-f1** is the true length of the hip and therefore can be used in determining the dihedral angle. Follow the procedure as for **Fig. 8.5** above, but note that when **w1** is projected to the plan, it will be seen that the projection line does not intersect with the wall plates. When this occurs, the wall plates are extended, as shown at **f-e-w** and **f-a-w**, to intersect with the projection line.

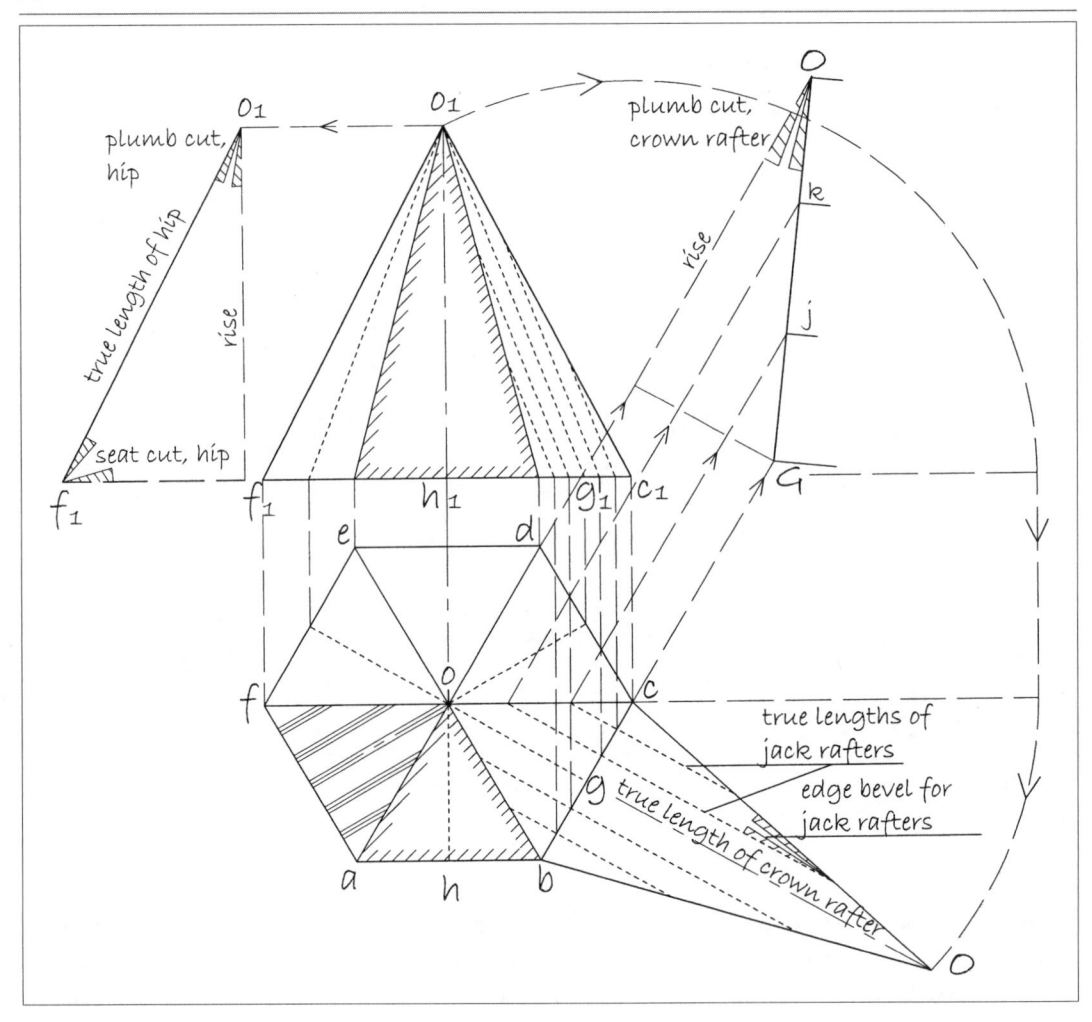

Fig. 8.12

Method 2

The second method uses an auxiliary elevation of hip **o-b**. **Fig. 8.13b** has thickness added to the members, and a practical method of acquiring the backing angle is given. Inset **h** shows the rafter upstand moved inwards to where the wall-plate corner is cut and a line parallel with the top of the rafter is drawn. This line represents the backing angle and the hip is bevelled to this line. To highlight this, a superimposed section of the hip is drawn at inset **j**. **Fig. 8.13c** shows a plan view of hip and crown rafters coming together at the apex of the pyramid where a finial is used.

8 ROOF GEOMETRY

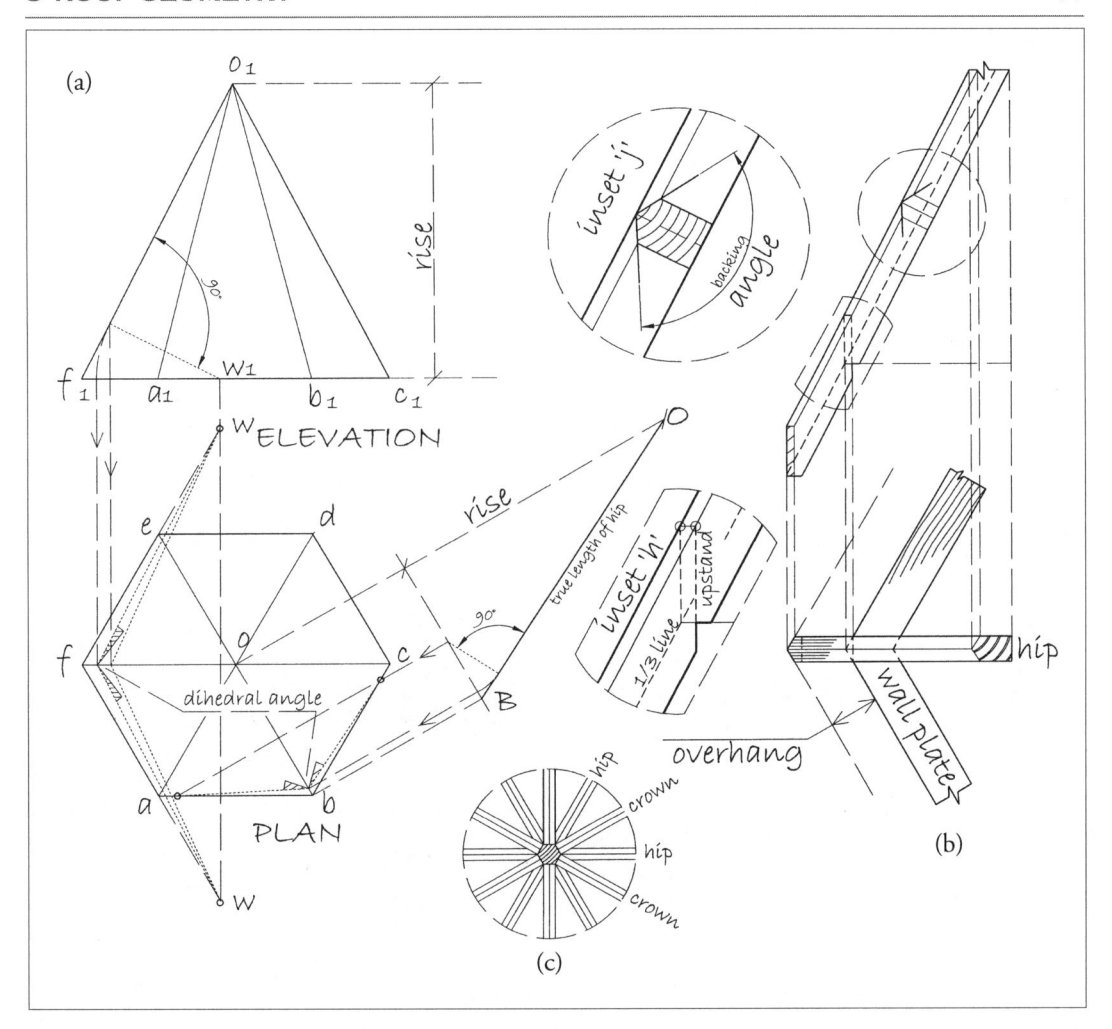

Fig. 8.13a
Fig. 8.13b
Fig. 8.13c

A roof based on a regular *tetrahedron* is shown in **Fig. 8.14a**. The tetrahedron is one of the platonic solids, the first one in the series. It has four sides made up of equilateral triangles. The next solid in the series is the cube, with sides made up of squares, six in all. The ochtahedron is next with eight sides, and the list goes on.

The tetrahedron is the only solid in the above series which is a pyramid and, consequently, may be applicable to roof construction. The plan in **Fig. 8.14a** represents the equilateral triangle on which this tetrahedron is based. Triangles **abc**, **oab**, **obc** and **oca** are, therefore, all identical in true shape. In the plan, lines **o-a**, **o-b** and **o-c** represent the hips, while lines **o-d**, **o-e** and **o-f** represent the crown rafters. These rafters are also drawn in the elevation.

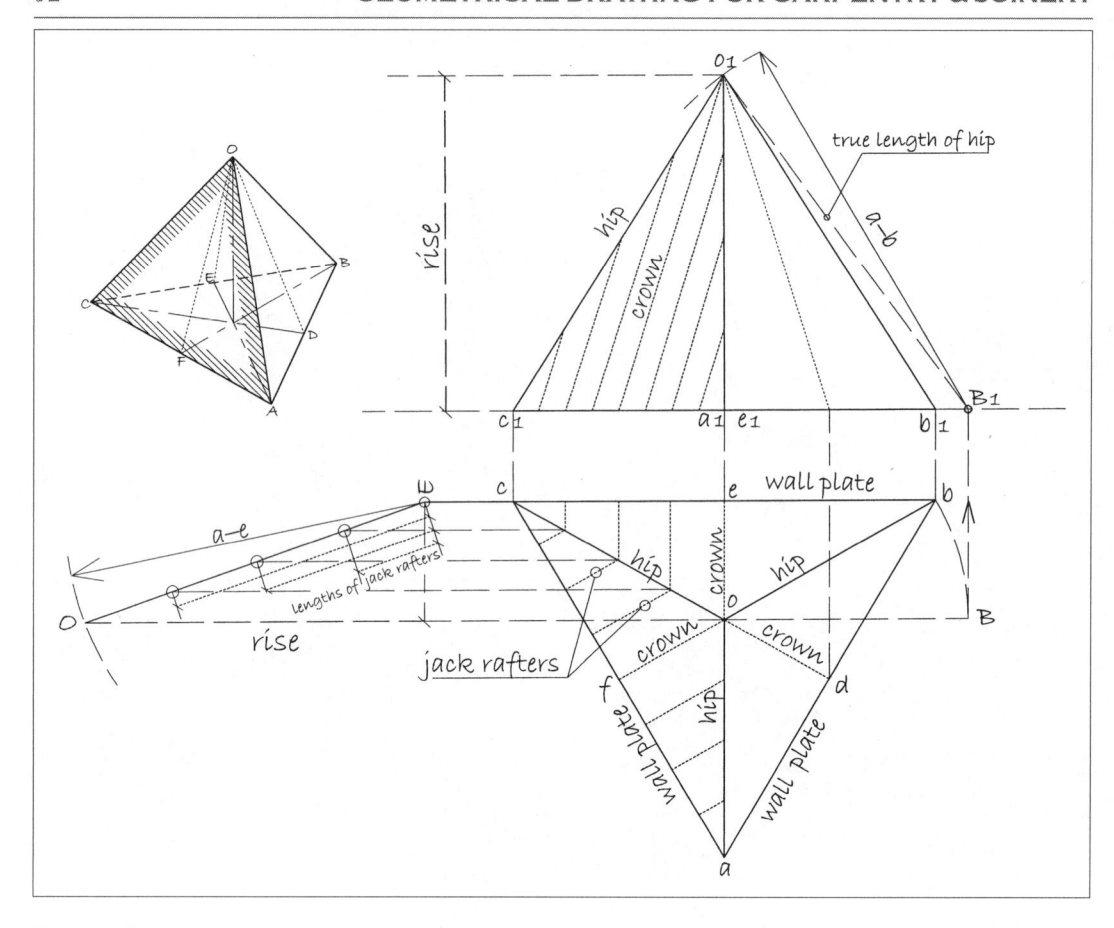

Fig. 8.14a

It should be stressed here that the elevation as given is not an equilateral triangle. To draw the elevation, there are two methods shown:

Method 1

- Pivot **o-b** in the plan to become **o-B** and project to **B1** in the elevation.
- Take length **a-b** (the length of side of the equilateral triangle) and strike from **B1** to intersect at **o1**, projected vertically from the plan. This gives the height of the pyramid in the elevation.
- Join **o1** to **c1** and **b1** to complete the elevation.

Method 2

- Draw an auxiliary elevation of rafter **o-e** by striking **a-e** (the true length of the crown rafter) from **E** to intersect with **O** projected from the plan, resulting in **E-O** = **a-e**.
- This gives the rise which is transferred to the elevation.

8 ROOF GEOMETRY

[Figure 8.14b: Plan view showing triangle with vertices c (top), a (bottom left), b (right), and point f on the left. Interior point o with points e and d. Labels include "crown rafter" and "hip rafter".]

Handwritten note in figure: FIG. 8.14b shows the plan of a roof in the form of a tetrahedron: draw the elevation and determine the dihedral angle for one of the hips.
Note: choose a convenient length of side for the equilateral triangle.

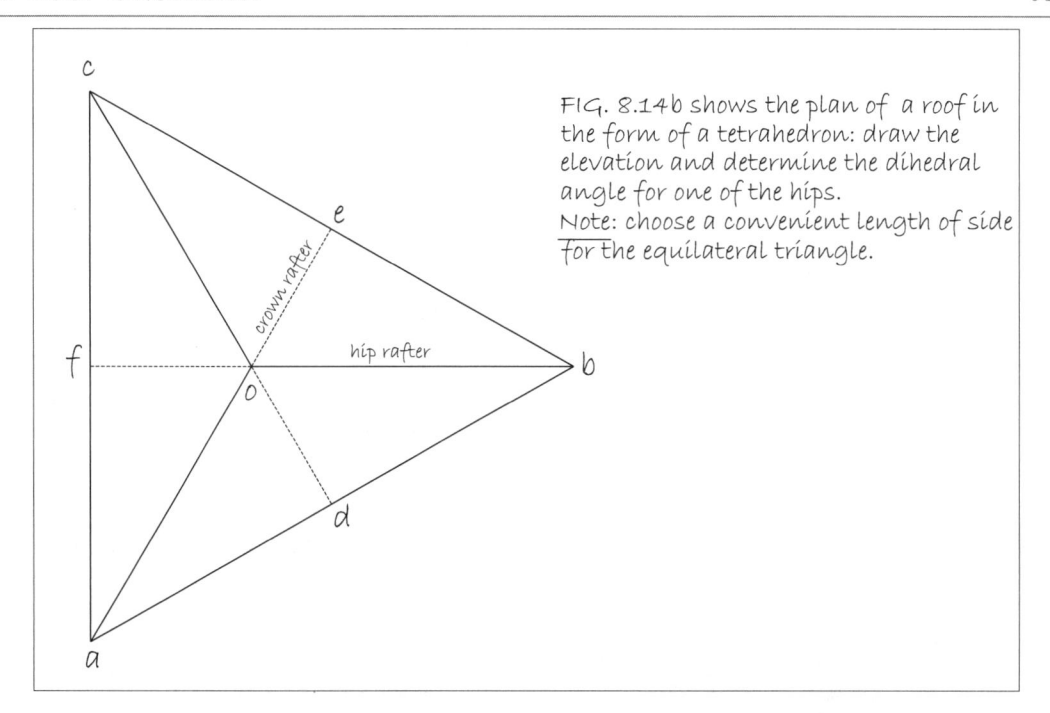

Fig. 8.14b

Exercise

Fig. 8.14b shows the plan of a roof in the form of a tetrahedron. Reproduce **Fig. 8.14b** and complete the exercise.

- Draw the elevation.
- Determine the dihedral angle for one of the hips.

Note: Choose a convenient length of side for the equilateral triangle.

In **Fig. 8.15** a square-based pyramid roof resting on a pitched roof is shown in plan, elevation and auxiliary elevation. This example is included in order to emphasise the value of using auxiliary views to find true lengths and true bevels.

In the elevation, the lengths and bevels for the crown rafters are clearly visible, but to find the same information for the hip rafter it is necessary to look elsewhere.

Method

- Index the plan as shown.
- In the plan, a line **j-k** is drawn as an extension of the plan of hips **o-b** and **o-d**, this line (**j-k**) is on the pitched roof surface.
- Project point **k** to the elevation at **k1**.
- Construct line **X1-Y1**, which is parallel with **j-k**, at a convenient distance.
- Transfer the heights **h1**, **h2** and **h3**, as shown, from the elevation.

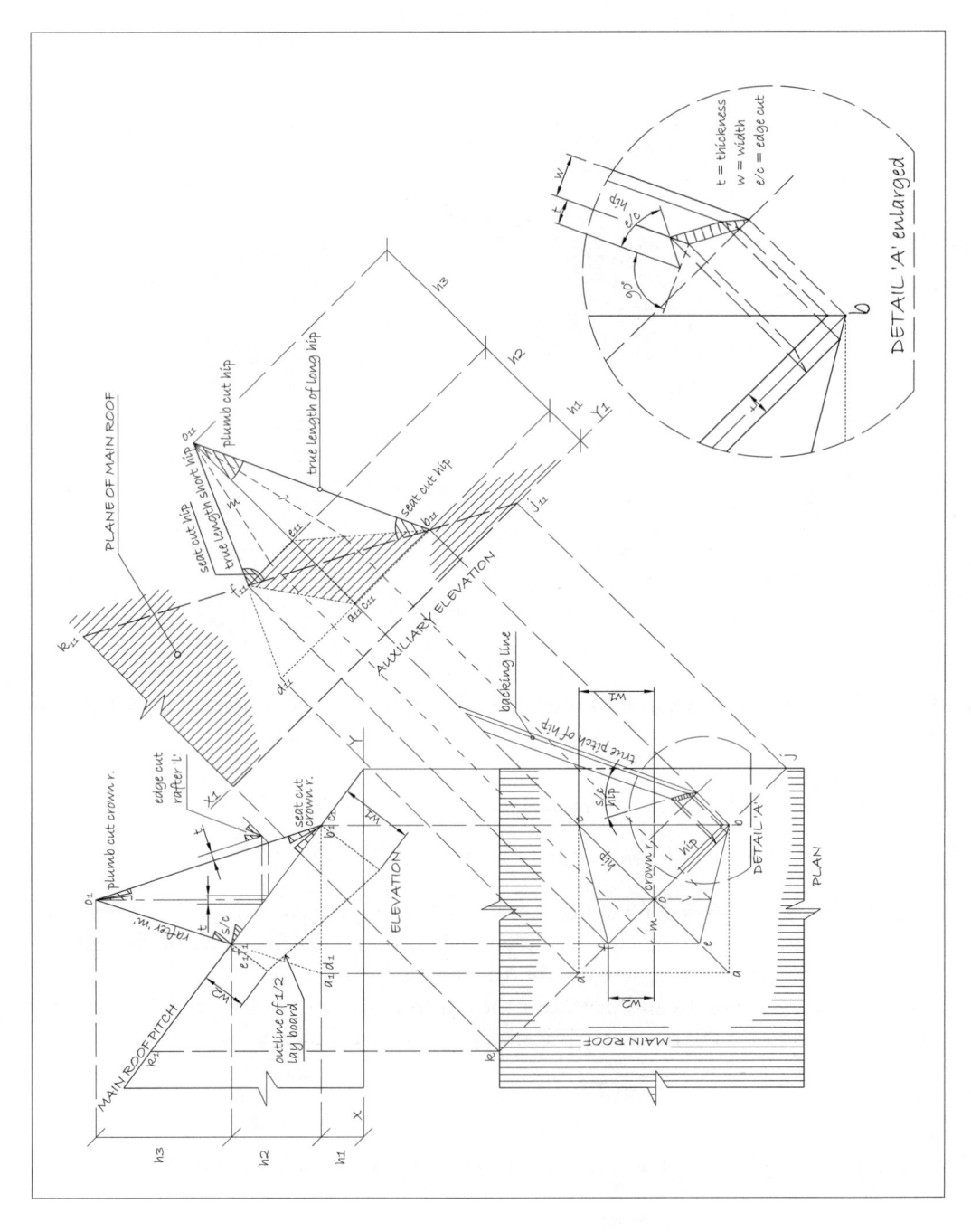

Fig. 8.15

8 ROOF GEOMETRY

- Join the projection points from the plan, with their corresponding heights, to give the outline of the pyramid.
- Transfer the height of **k1** from the elevation to **k11** (in auxiliary) and join to **j11**.

In the auxiliary elevation, the line **k11-j11** and the lines of the hip rafters are on the same plane (a plane that is square-off the viewer's line of vision), thus revealing lengths and bevels in their true form. Additional information is given at detail **A** enlarged. For a more comprehensive understanding of auxiliary elevations, see Chapter 3.

SECTION III: ROOFS TO BAY WINDOWS

There are four basic types of bay window: square, angled, triangular and segmental. The first three will be covered in this section, and the segmental bay window is dealt with in Chapter 7 on conic sections because the roof is part conical. See **Fig. 8.16**.

Fig. 8.17 shows the plan and elevation of a roof to a square, or right-angled, bay window with its associated geometry. It must be emphasised here that when bays have hips, the line of the hip in plan bisects the corner angle, which in this case is an angle of 90°.

In the example in **Fig. 8.18**, it can be seen at corner **h** that the angles on each side of the hip are equal. In both cases, the shape of the vertical wall piece (the member to which the rafters are secured at the wall) is given.

In **Fig. 8.19** the geometry for the triangular bay window is presented. From the plan, an auxiliary view of the crown rafter is projected, and the rise of the roof is determined by measuring the roof pitch from **X1-Y1**, the rise is then transferred to the elevation as shown.

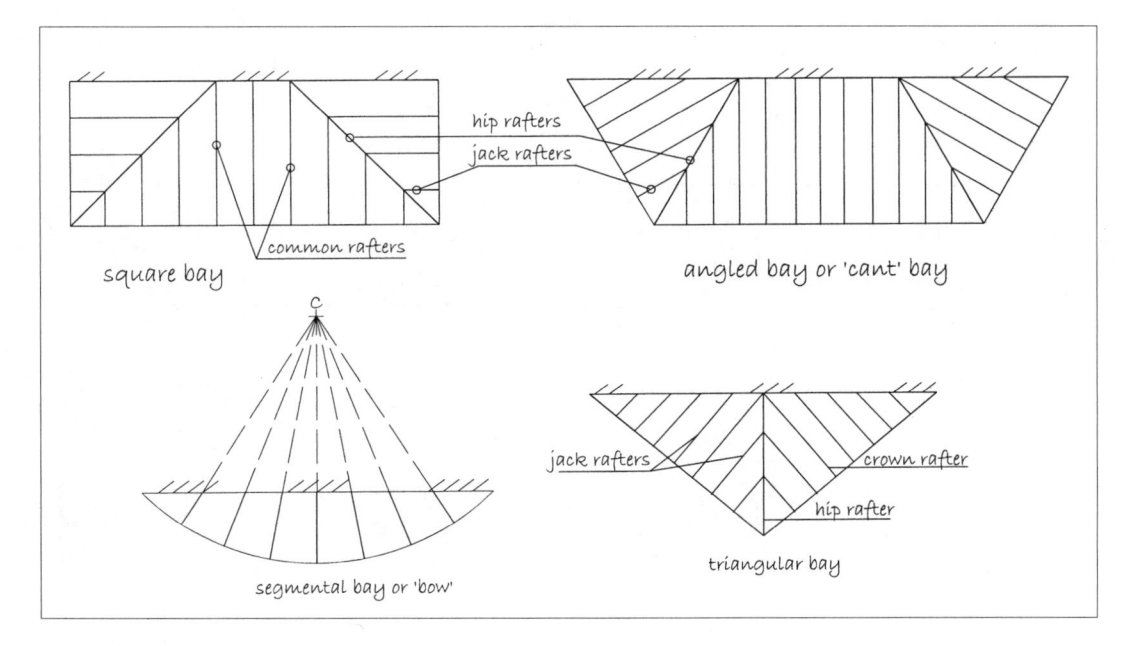

Fig. 8.16

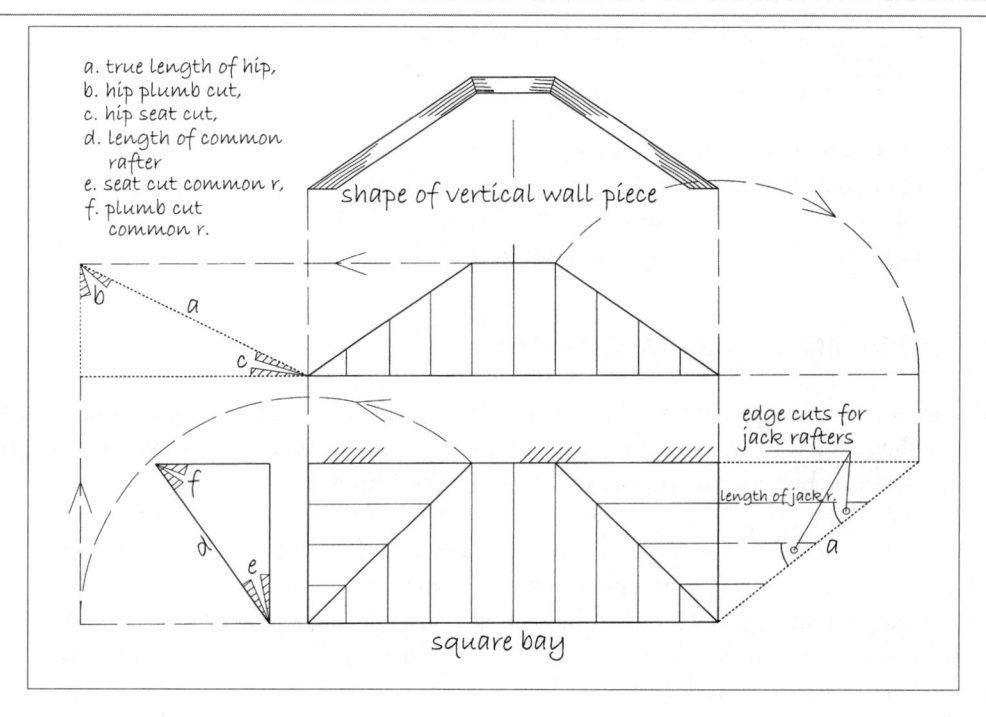

Fig. 8.17

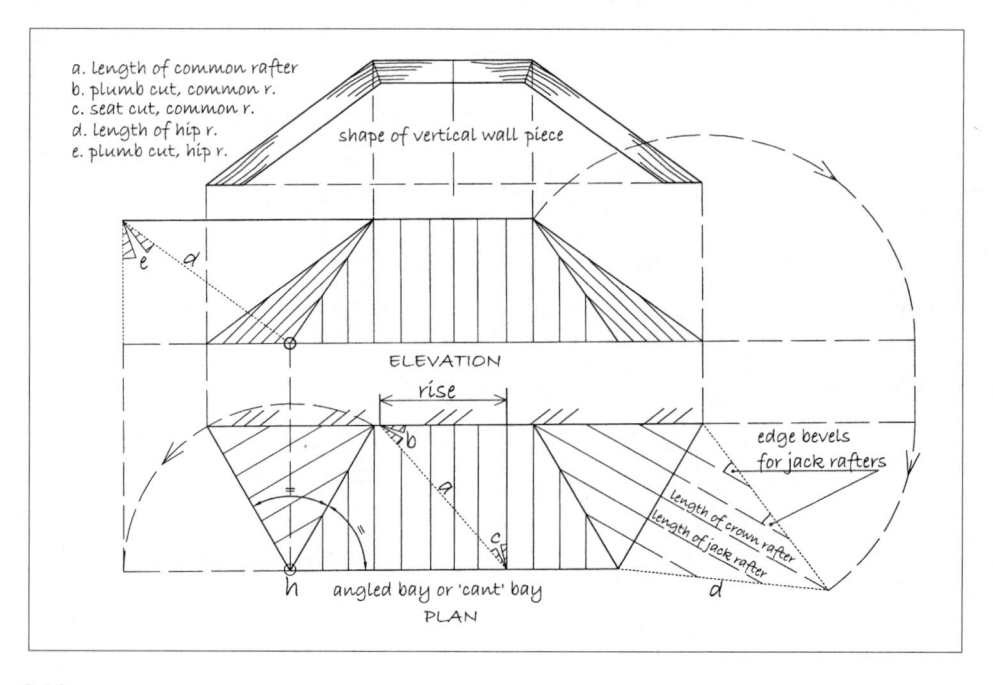

Fig. 8.18

8 ROOF GEOMETRY

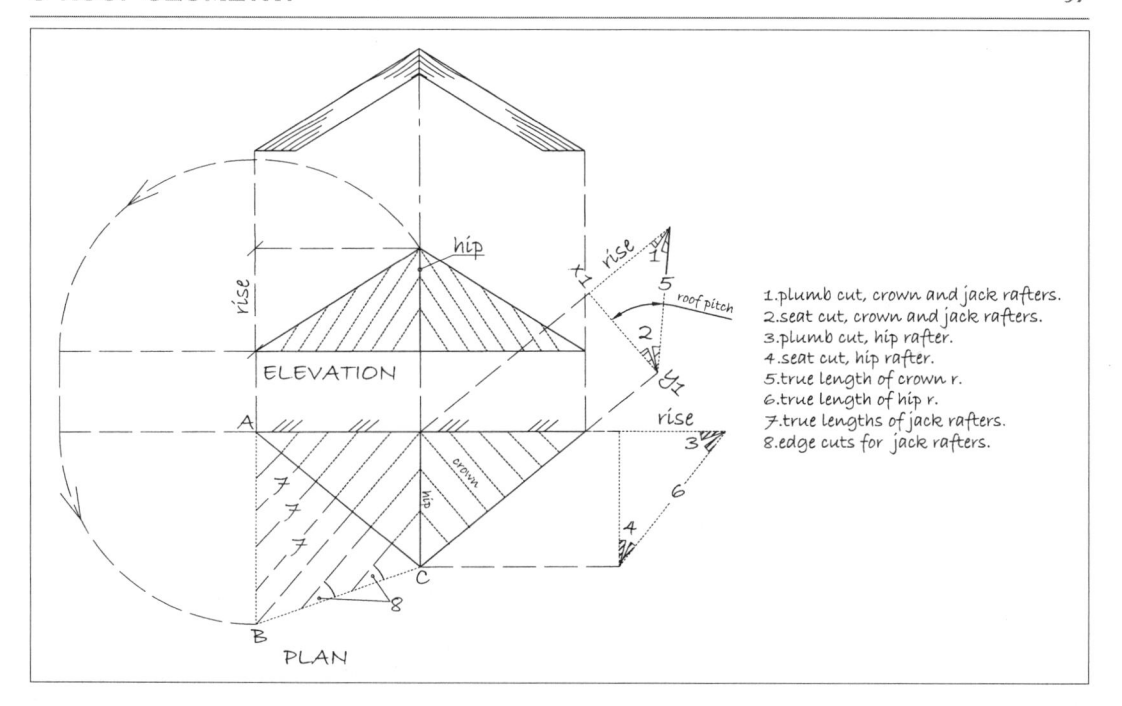

Fig. 8.19

1. plumb cut, crown and jack rafters.
2. seat cut, crown and jack rafters.
3. plumb cut, hip rafter.
4. seat cut, hip rafter.
5. true length of crown r.
6. true length of hip r.
7. true lengths of jack rafters.
8. edge cuts for jack rafters.

It is important to emphasise here that the roof pitch in these examples is always measured off the run of the common, or crown, rafters and that jack rafters, as is the case with conventional roofs, are always fitted square-off the wall plate. Plumb cuts and seat cuts are found by combining the run, rise and true length of any rafter. Edge cuts are determined by developing the true shape of a roof surface, as in **ABC** in **Fig. 8.19**.

Exercise

Fig. 8.20 shows the outline plan of a roof to an angled bay window. The roof will have four surfaces, all are pitched at an angel of 45°. Reproduce **Fig. 8.20** and complete the exercise. There will be hips at **e**, **f** and **g**.

- Complete the plan and elevation.
- Develop roof surfaces **A** and **B**.

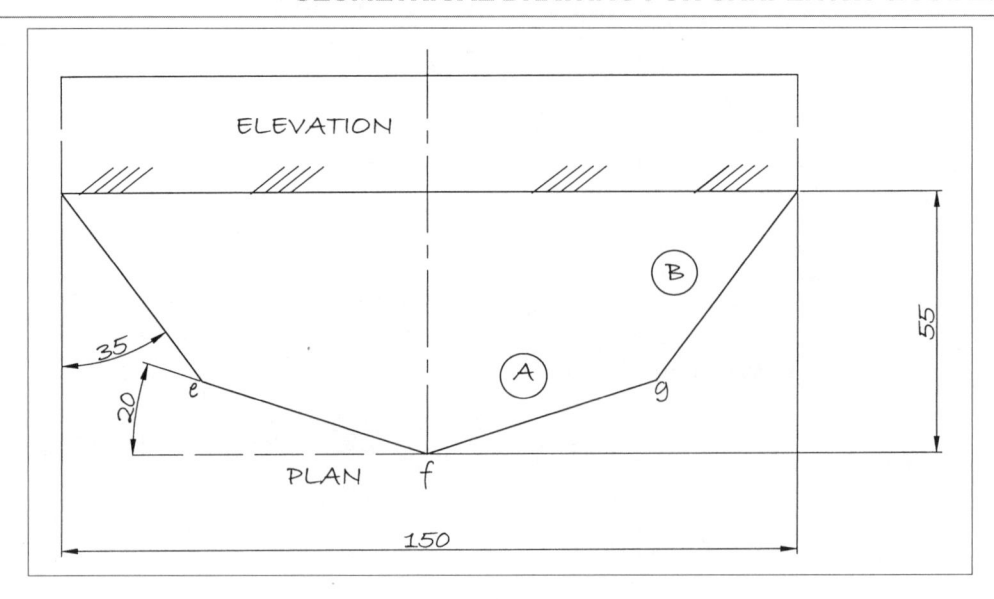

Fig. 8.20

SECTION IV: TURRETS AND DOMES

This section deals with the geometry of curved rafters (also known as ribs) and the development of curved surfaces for domed and turret roofs. The example at **Fig. 8.21** is a hexagonal dome. This means the plan is hexagonal and the elevation is domical (meaning hemispherical and also semi-circular). The lines going through corners **a**, **b**, **c**, **d**, **e** and **f** represent hips. The lines at **g**, **h** and **j** represent crown or common ribs.

In the elevation, the dome is divided into four equal parts from **0** to **4**. At these parts, we may consider that they are levels or heights, levels **1**, **2**, **3** and **4**. The plans of these levels are found as shown and may be likened to contour lines on a map. If the dome is sliced horizontally at any height of the elevation, the resulting plan of that slice will be a hexagon. Dividing up the dome in this way is helpful for solving the geometrical problems associated with the dome.

One quadrant of the elevation represents the true shape of the crown rib for the dome. The true shape of the hip rib is found by projecting an auxiliary elevation of the hip in question.

8 ROOF GEOMETRY

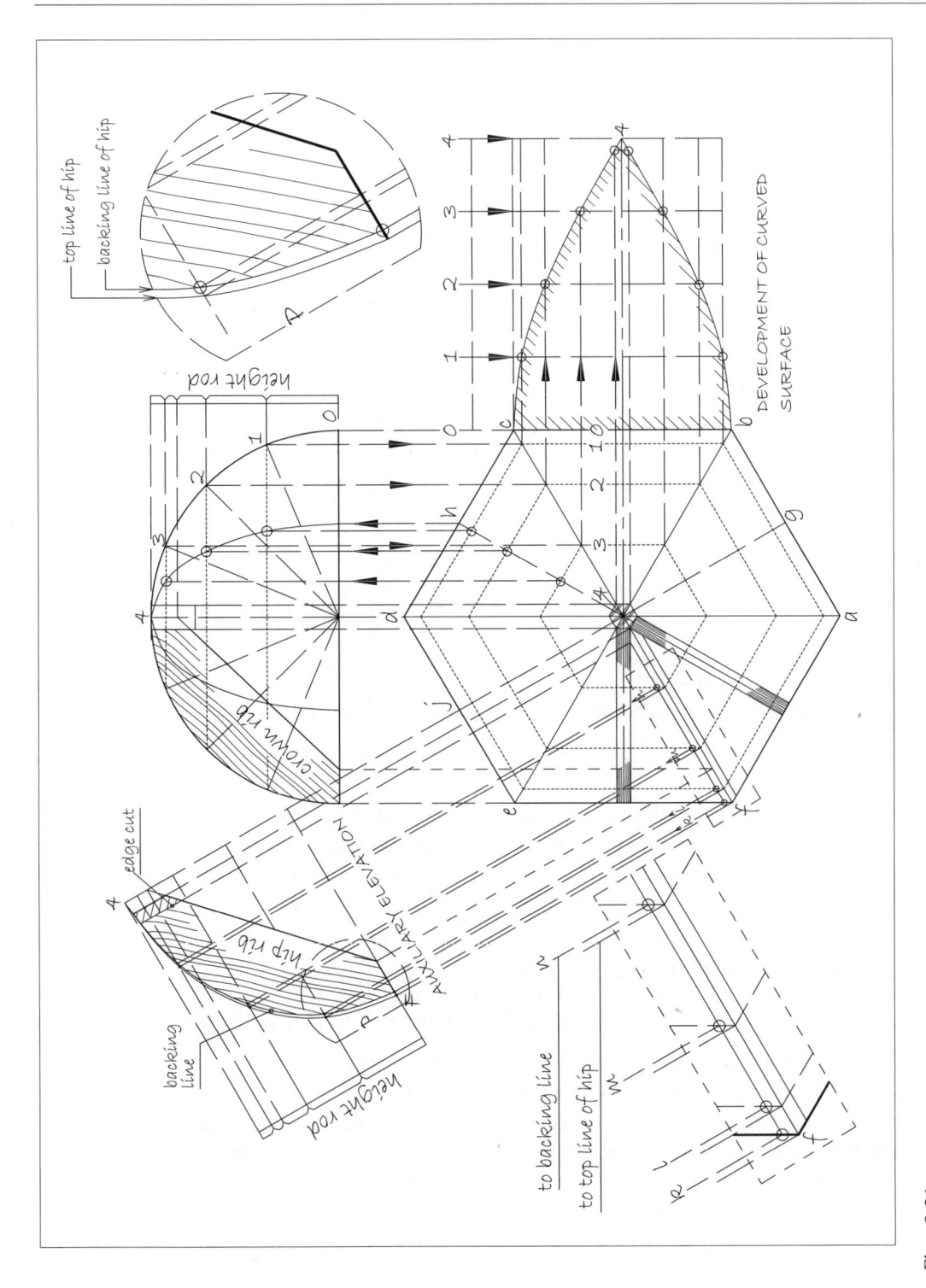

Fig. 8.21

Method

- A height rod is set up in the elevation and transferred to the auxiliary elevation as shown.
- After finding the outline of the hip, it is necessary to apply the backing line, representing the backing angle to the hip. This is done as follows:
  - Project points **k**, **l**, **m** and **n** from the plan to their relevant height lines in the auxiliary view and join the points, to terminate at **4**.

The curved surface **b4c** is developed on the right-hand side. This is done by stepping out the distances **0** to **4**, taken from the elevation curve as shown and projecting the relevant hip intersections to make up the grid into which the development fits. The resulting true shape is that of the sheet of copper or lead which would cover that roof surface.

It is important to emphasise here that when determining the true shape of the hip, the heights are combined with the plan run, resulting in an auxiliary elevation. When finding the shape of the curved surface, the length of the curve in the elevation (in this case, one quadrant of the semi-circle) **0** to **4** is used 'flattened-out'.

Fig. 8.22a shows the plan and elevation of a hemispherical dome, which is divided into sections similar to the example above in **Fig. 8.21**. In this case, the sections are called *zones* and *gores*. For the purpose of surface development, the zones are assumed to be portions of cones.

Method 1

- In the elevation, **b-c** and **c-d** are extended to **a** and **a1**, respectively; **a** and **a1** are the apexes of the cones in question.
- For zone **b-c**, arcs **a-b** and **a-c** are drawn from **a**, and the distance **1-2** (taken from the plan) is stepped along the arc from **b** three times to give one quarter of that zone.
- For zone **c-d**, arcs **a1-c** and **a1-d** are drawn, with distance **3-4** stepped along arc **c** in the elevation.

To complete the development in each case, a radiating line is drawn back to **a** and **a1**, respectively.

A second method of developing the curved surface of the dome is given in **Fig. 8.22b**. This may be likened to 'peeling off' the surface in sections (gores) from the top. A gore is highlighted in plan and elevation.

Method 2

- The centre line **c/l** of the gore is extended in the plan to **J**.
- Arcs are drawn from **J** through **H**, **G**, **F** and **E**, as shown.
- The intersections at the projections, from levels **e**, **f**, **g** and **h** in the plan, complete the development.

In **Fig. 8.23** the part plan and elevation of a roof to an aquatic centre is shown. The main roof is conical in shape and topped by a glass dome, which is made up of 160 glass panels on four levels. The structure of the roof is laminated curved rafters and common conical rafters which meet at a circular laminated curb (or ring), as shown to the right of the elevation. Two of the glass panels are developed by knocking them flat in the elevation and projecting to the plan. Inset **A** is enlarged to illustrate this. The plan view shows four of the glass panels shaded. When these shapes are developed, they represent templates for all the glass panels for the dome; 40 of each will be needed.

8 ROOF GEOMETRY

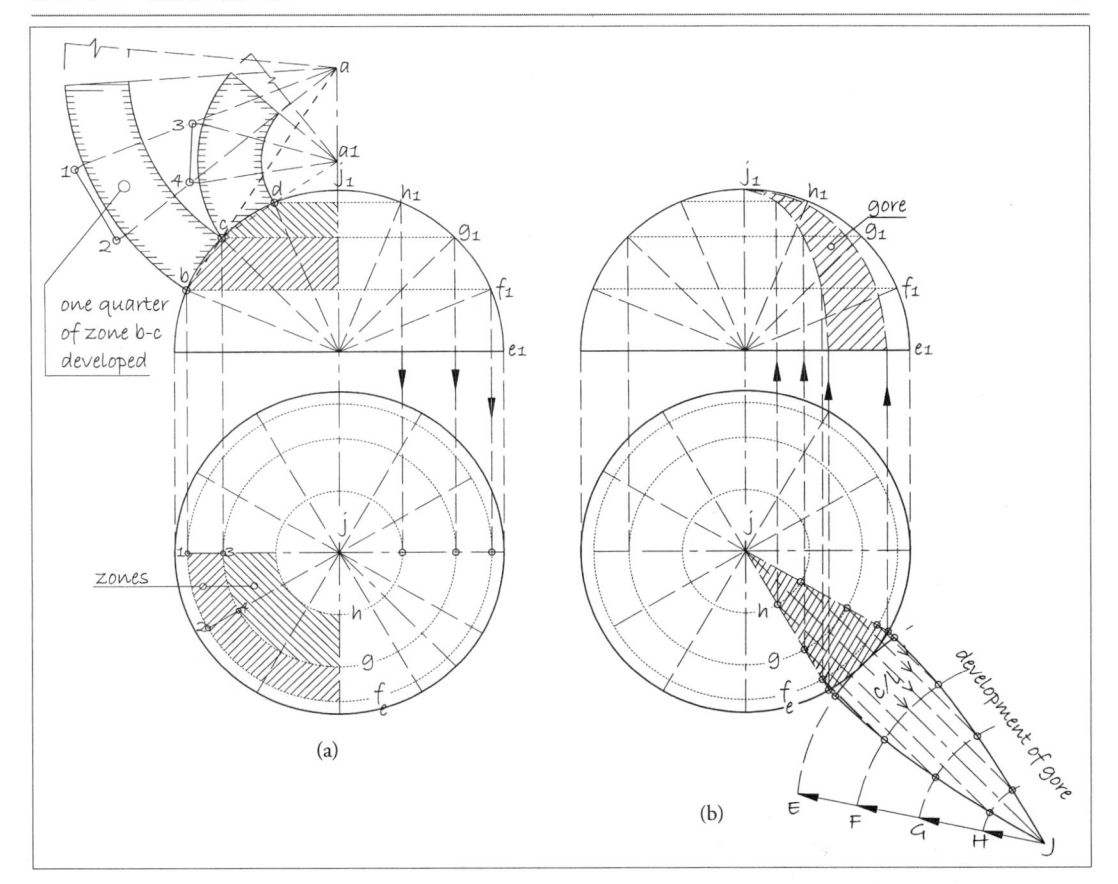

Fig. 8.22a
Fig. 8.22b

Fig. 8.24 represents the plan and elevation of a square-based turret, which is ogee-shaped in the elevation. The development of the hip rib is given in auxiliary elevation, and one of the curved surfaces is also developed. As with the two examples in **Figs. 8.21** and **8.22a**, the turret is divided into a series of levels in the elevation which are then transferred to the plan.

Imagine **level 1** in the plan being a line painted on the turret and that you are walking around the turret along that line. As you do so, you will be remaining at that height, neither going up or down.

In **Fig. 8.25**, an alternative method of developing the hip rib is given. This entails revolving the hip and all points on it to a horizontal position in the plan and projecting to the elevation, where the levels are projected to the right to intersect with their counterparts from the plan. In this example, width is added to the hip to help visualise its make-up.

A turret based on a hexagonal pyramid, on a pitched roof is given in **Fig. 8.26**. The upper portion is glazed. The surfaces of the pyramid are developed from the elevation.

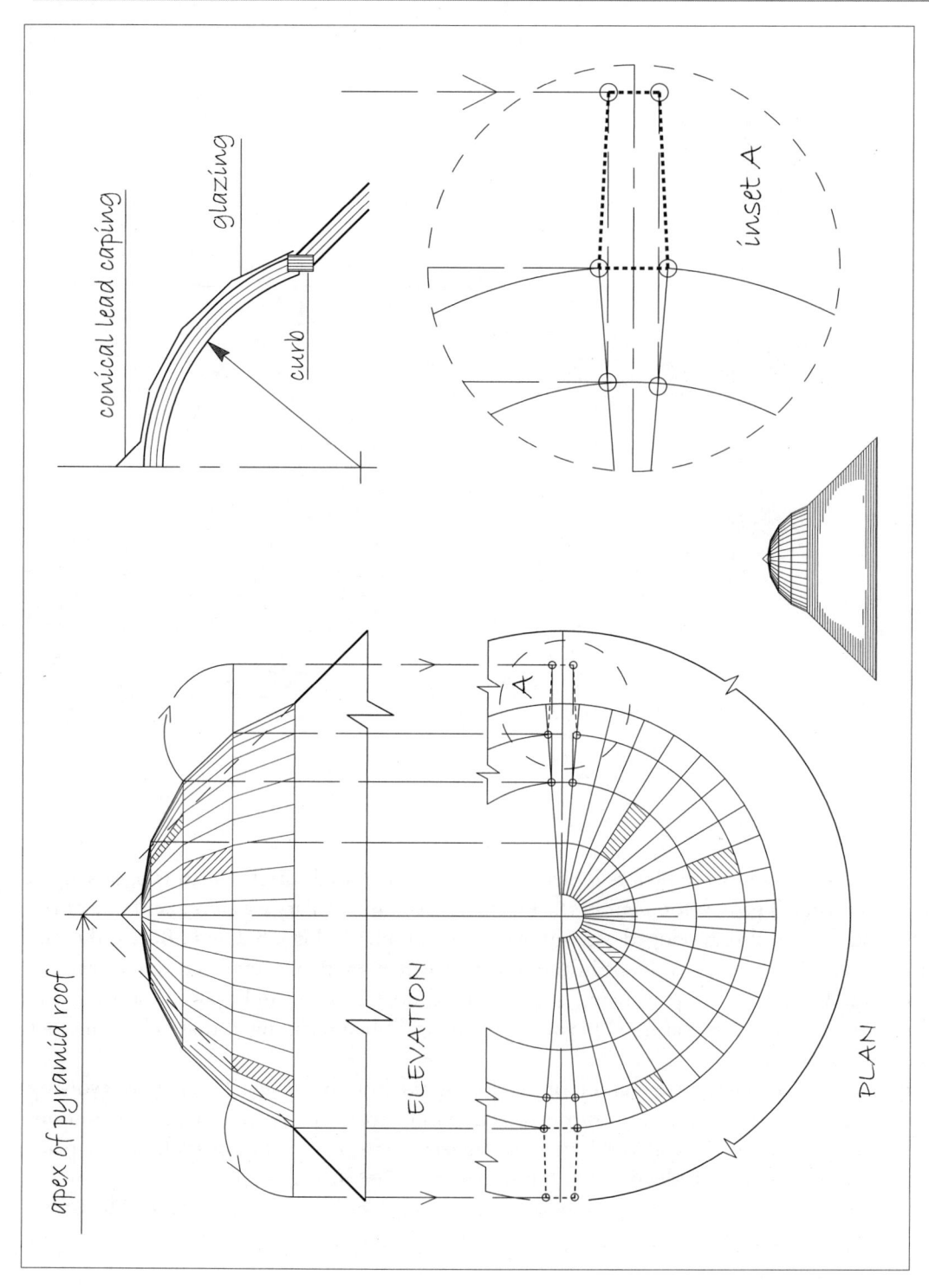

Fig. 8.23

8 ROOF GEOMETRY

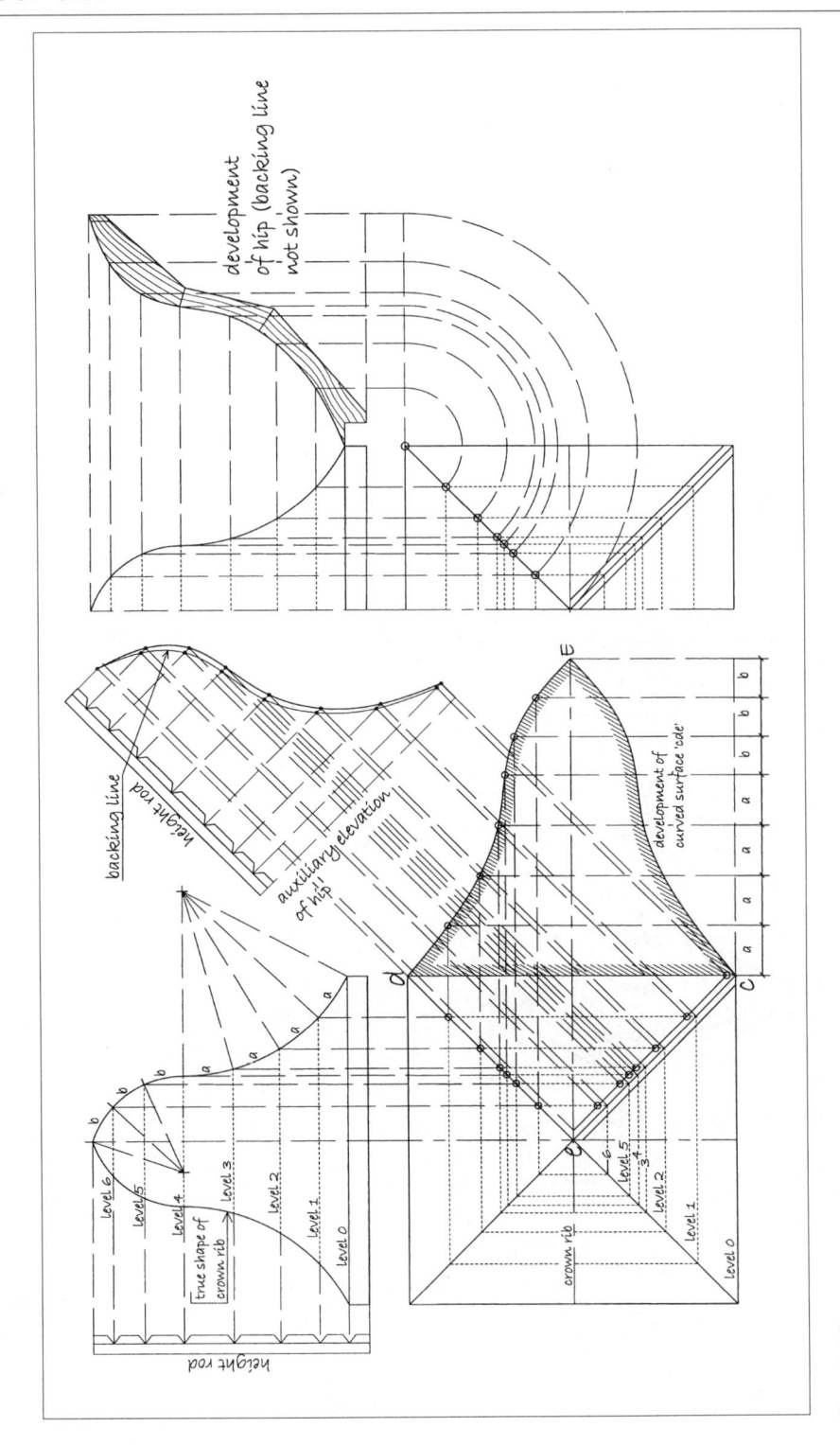

Fig. 8.24
Fig. 8.25

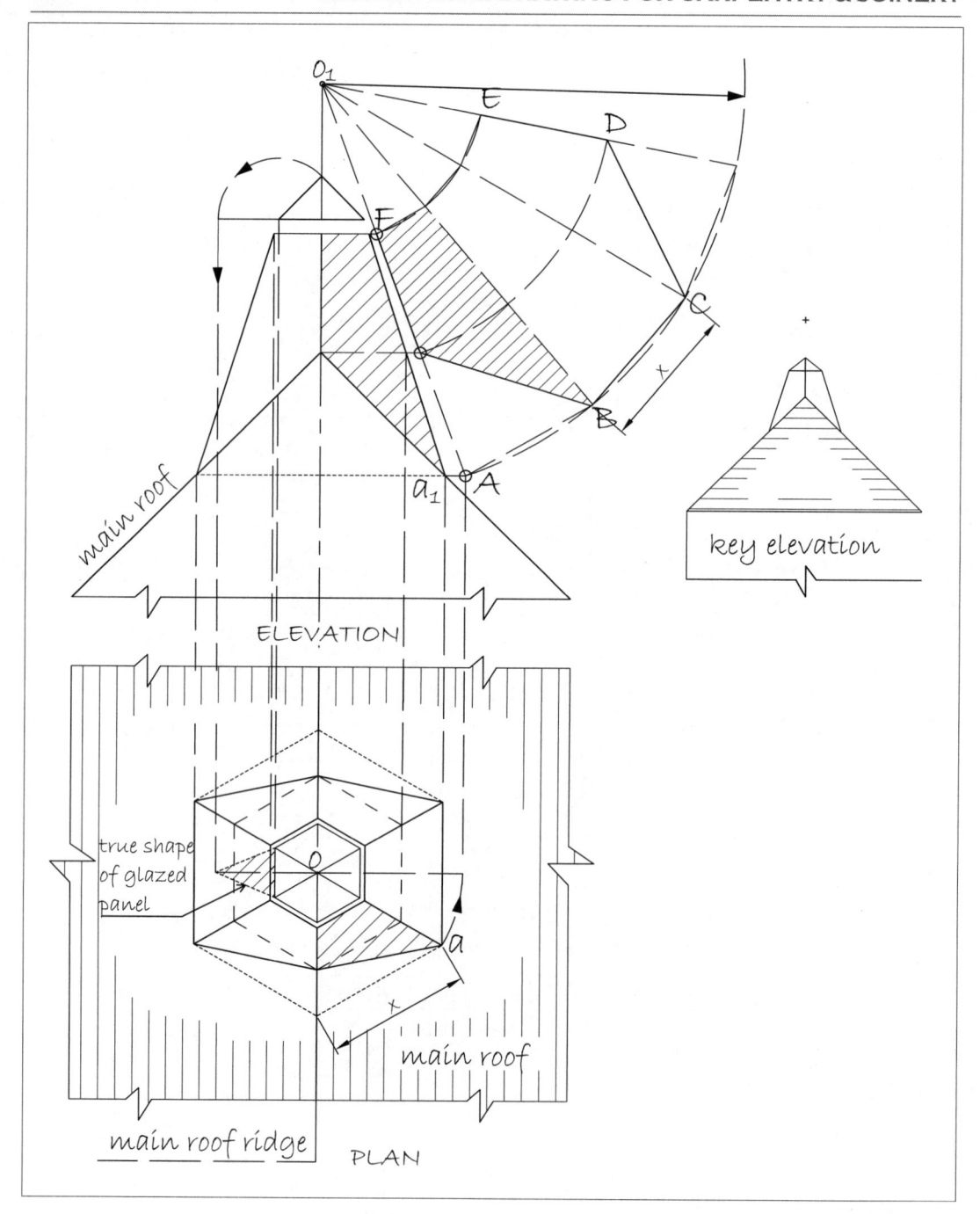

Fig. 8.26

8 ROOF GEOMETRY

Method

- **o-a** is pivoted to the horizontal in plan and projected to become **o1-A** in the elevation, which is the true length of the hip for the pyramid.
- An arc of radius **o1-A** is scribed from **o1** and along it the length of side of the hexagon base (**X**) is marked off three times, giving the developed sides of the pyramid.

One of the glazed panels of the top of the pyramid is developed on the left-hand side. In the elevation, the panel is knocked flat and projected to the plan, where point **o** is drawn horizontally to intersect with the projection. This is an alternative to development from the apex, which was done on the main pyramid.

In **Fig. 8.27** the same drawing as in **Fig. 8.26** is reproduced with the addition of an end view, where the true length of the hip **o-a** can be seen as **o11-a11**. In this case, the pyramid surfaces may be developed from the edge **o11-a11**, with an arc scribed from the apex **o11**.

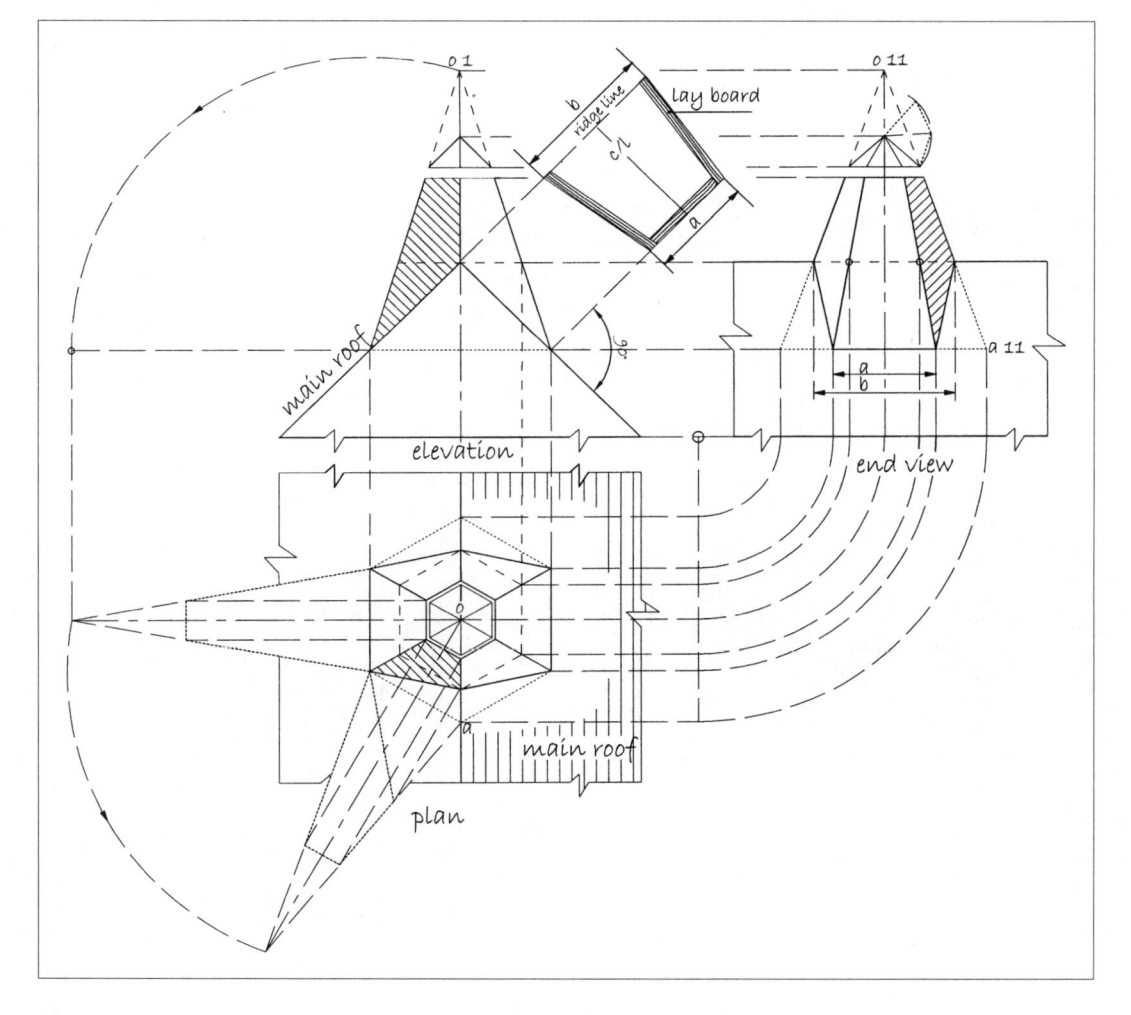

Fig. 8.27

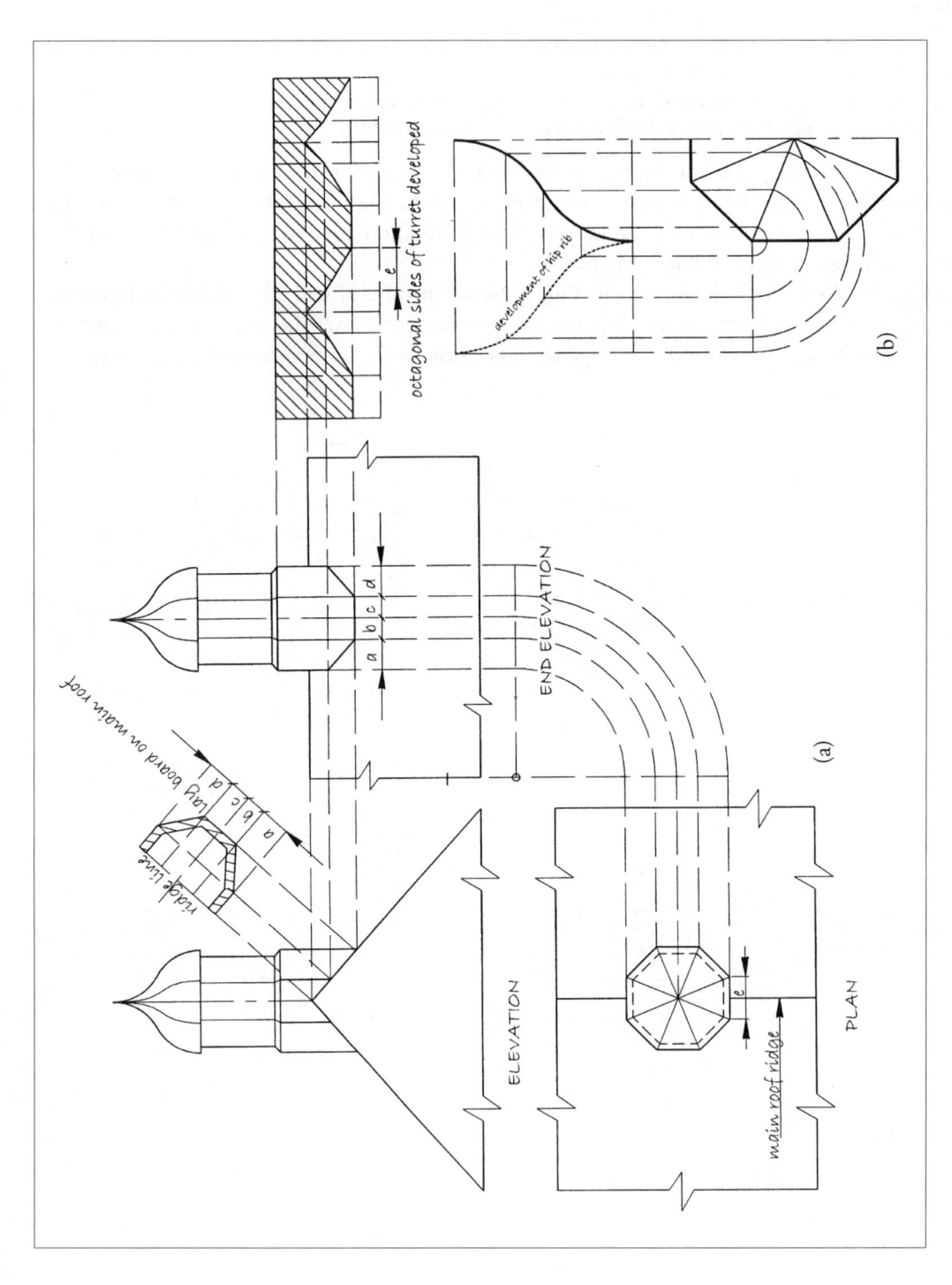

Fig. 8.28a
Fig. 8.28b

8 ROOF GEOMETRY

From the apex of the glass pyramid two surfaces are developed, as shown. The true shape of the lay board, on which the pyramid rafters rest, is shown projected from the elevation. The dimensions **a** and **b** are transferred from the end view to achieve the shape. This outline is the portion of the main roof on which the pyramid rests. The plan view shows two of the pyramid roof surfaces developed; hinged from the pyramid base.

The orthographic projection of an octagonal turret is given in **Fig. 8.28a**. To ventilate the main roof, the prismatic portion of turrets such as this may contain louvers. As with the example in **Fig. 8.27**, the sides of the turret (walls) are shown developed, and also the true shape of the lay board, which receives the turret on the main roof. At a larger scale, **Fig. 8.28b** shows the hip rib from the turret roof developed.

In **Fig. 8.29**, the plans and sectional elevations of various turrets are shown. In this case, they all have square bases and their elevations are gothic, ogee and cavetto, respectively. The turret bases (or plans) may take various polygonal shapes, as in some of the examples already covered in this Chapter.

Exercise

Fig. 8.30 shows the plan and elevation of an ogee roof turret. Reproduce **Fig. 8.30** and complete the exercise.

- Determine the true shape of hip rib **A**.
- Develop the shaded area of the roof.

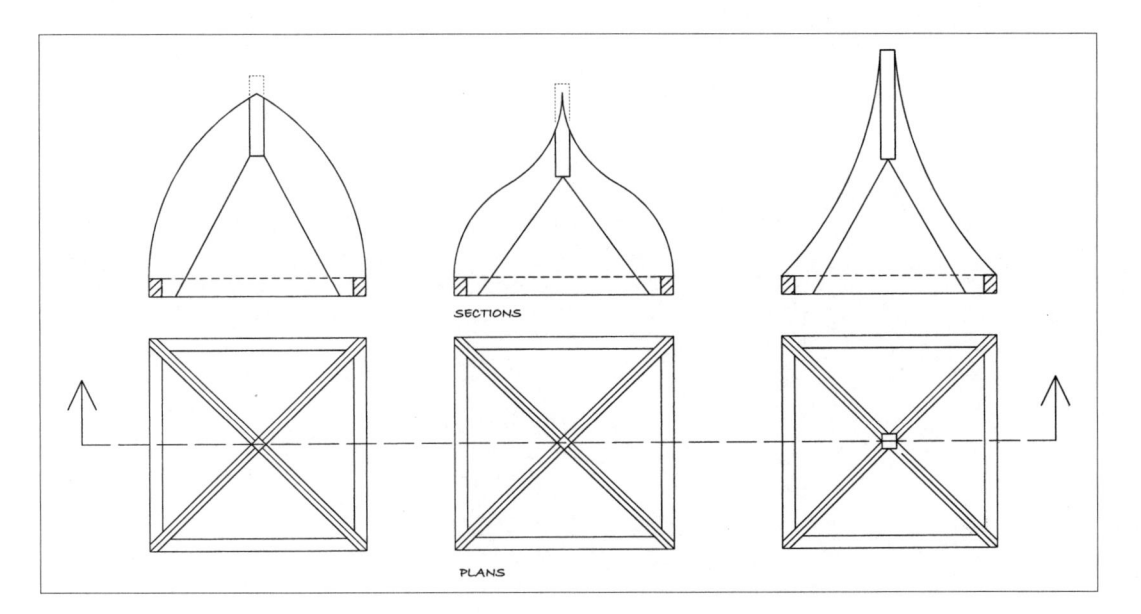

Fig. 8.29

Fig. 8.30 shows the plan and elevation of an ogee roof turret. Determine the true shape of hip rib A and develop the shaded area of the roof.

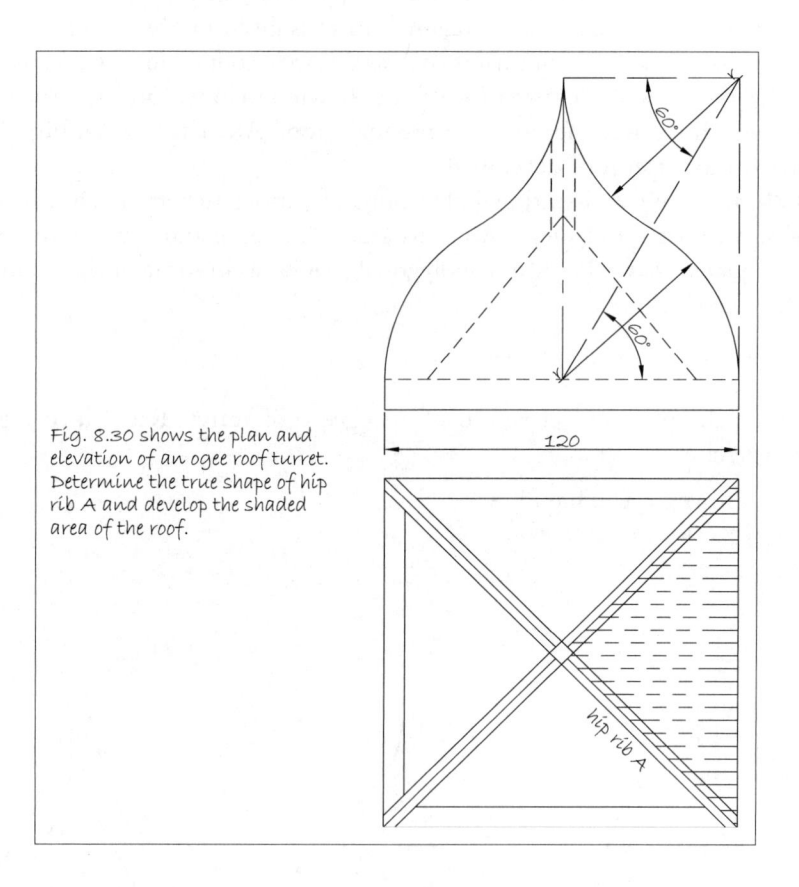

Fig. 8.30

8 ROOF GEOMETRY

SECTION V: DORMER ROOFS

Dormer roofs have some form of aperture to allow light or ventilation into the interior. These types of constructions are commonly called dormer windows. The word 'dormer' derives from 'dormitory', which means 'a sleeping area'.

Fig. 8.31 shows a pitched main roof with a triangular dormer window. On the drawing, the shape of the lay board is developed using principles already covered in this Chapter. The development of one of the sides of the dormer at **A** is also shown.

Fig. 8.32 (top) shows a dormer in the shape of a circle segment. The roof of this dormer is, of course, part of a cylinder. The plan, elevation and end view show clearly how the intersection between main roof and dormer is achieved. The line of intersection is shown in the plan. This line represents the curved valley between the two roof surfaces.

Fig. 8.33 (bottom) presents the same roof as **Fig. 8.32**. The development of the dormer roof is shown along with the development of the lay board for the dormer; which may also be called the true shape of the main roof cut-out.

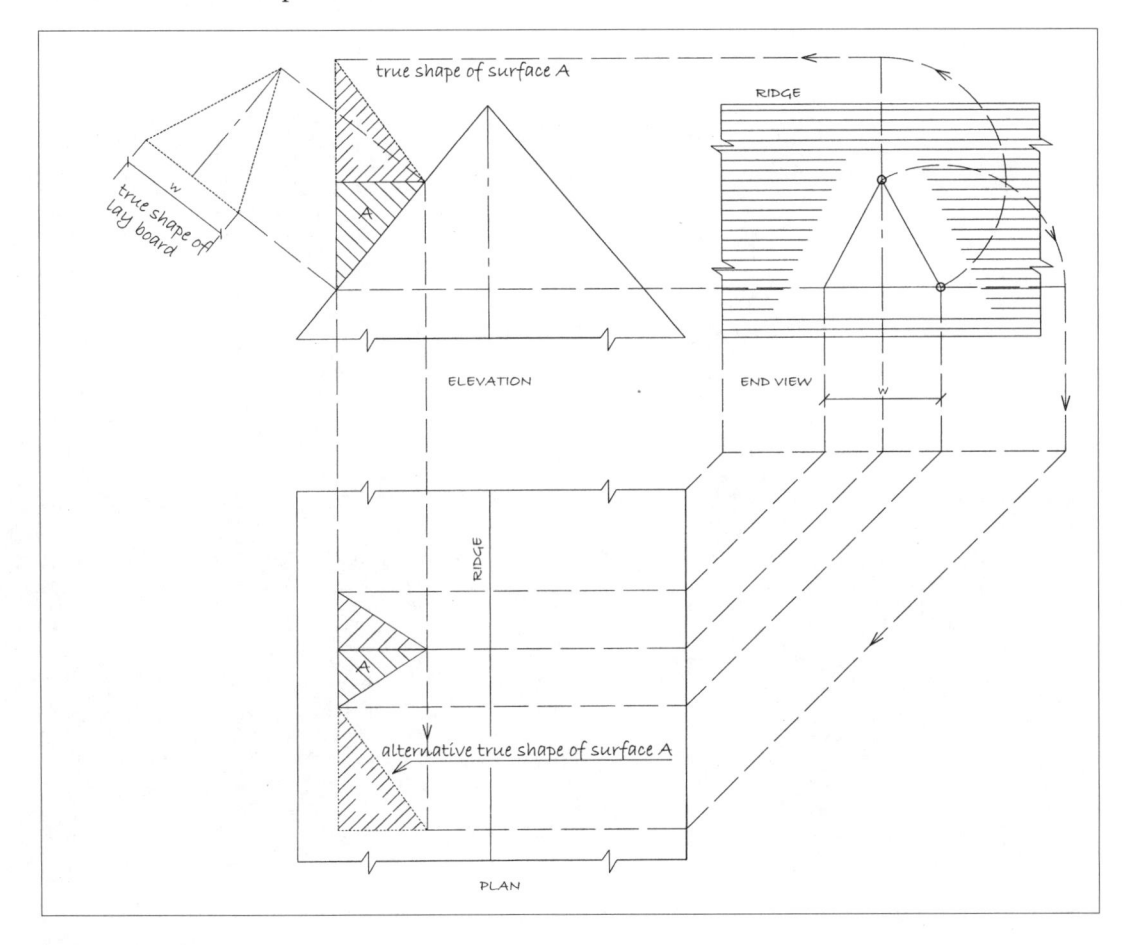

Fig. 8.31

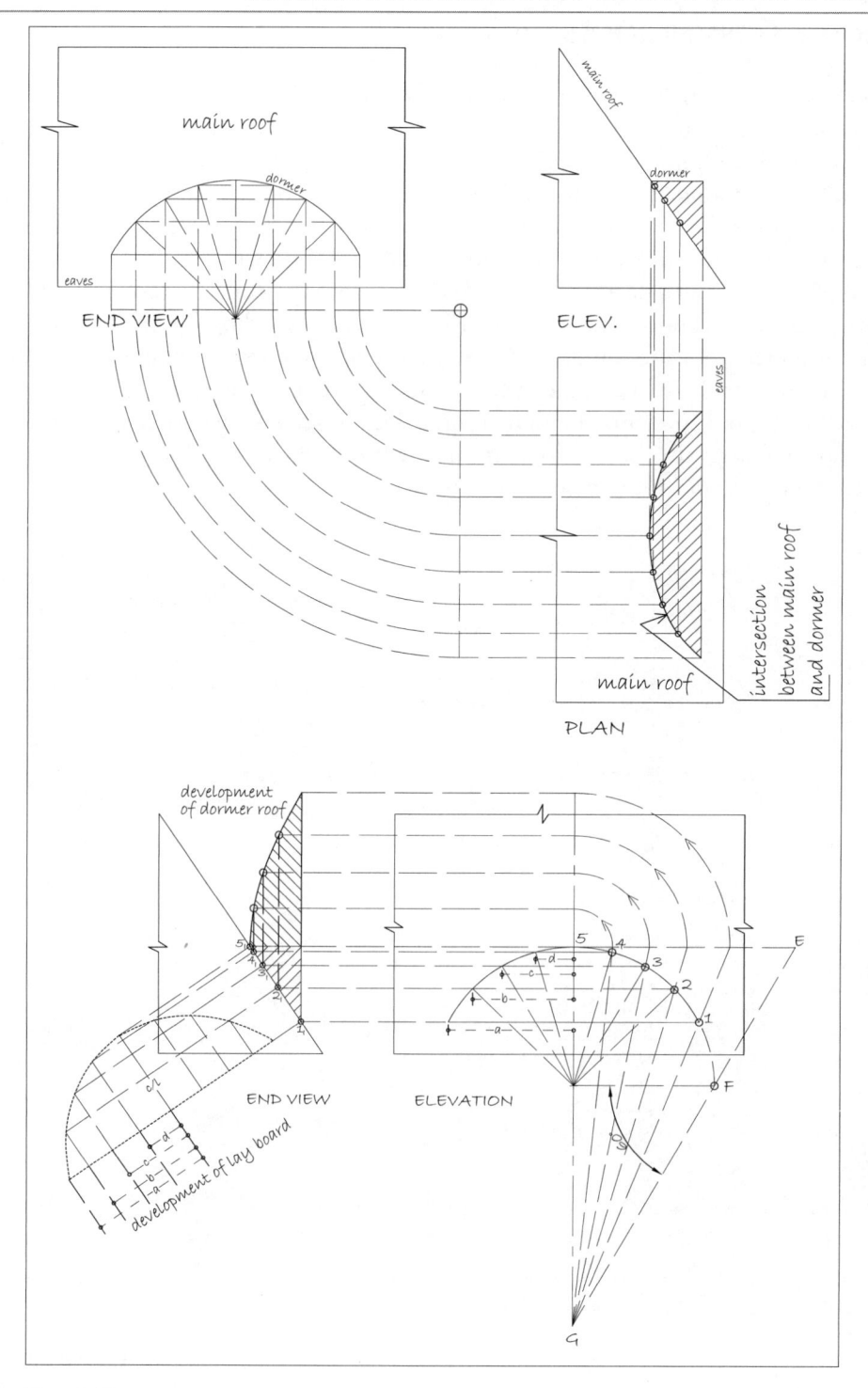

Fig. 8.32
Fig. 8.33

8 ROOF GEOMETRY

Method

- In the elevation in **Fig. 8.33**, extend the circle segment to **F** to become a semi-circle (see Chapter 17 on constructions), and construct an angle of 60° from the horizontal at **F** to meet the centre line (**c/l**) at **G**.
- Draw line **5-E** and project from **G**, lines through **1** to **5**. This gives on **5-E** the correct lengths of the curved segments.
- From **5**, pivot these true lengths to the vertical and project these to the end view to intersect with the vertical projections from **1¹** to **5¹**.
- Draw a fair curve through the intersections to give the true shape of half the dormer roof.

To determine the true shape of the lay board:

Method

- Project lines from **1¹** to **5¹** at 90° to the main roof and step off from the centre line (**c/l**) lengths **a**, **b**, **c** and **d** taken from the elevation.
- Join the intersections of these lines to give the true shape of the lay board.

Fig. 8.34a shows the orthographic projection of a conical roof with a triangular dormer. The geometric principles necessary to complete this type of drawing will be covered in greater detail in Chapter 14 on interpenetrations. For the example in **Fig. 8.34a**, the procedure is as follows:

Method

- Horizontal sections of the cone and dormer are taken at the four levels, **1**, **2**, **3** and **4**.
- In the plan, the sections of the cone are represented by circles **1**, **2**, **3** and **4**, and the sections of the dormer are represented by the projection lines **1**, **2**, **3** and **4**.
- Where line **1** intersects with circle **1**, a point of interpenetration is found; in other words, this is where the dormer and cone meet at this level.
- It is likewise at levels **2**, **3** and **4**.

The end view is completed by projecting the interpenetration points from the plan to intersect with the corresponding levels as shown.

Fig. 8.34b has one quadrant reproduced, from which the quarter cone is developed. The dormer roof development is also given in **Fig. 8.34a**.

Fig. 8.35 shows a lean-to roof, which is based on a half-cone. The conical roof has a dormer roof over a segmental window. The roof of the dormer has a roof, which is in turn based on another cone. The dormer roof surface is developed off the end elevation as shown.

Exercise

A semi-conical roof with a segmental dormer is shown in **Fig. 8.36**. In this case, the dormer roof is part of a cylinder. Reproduce **Fig. 8.36** and complete the exercise.

- Complete the plan and end elevation.
- Develop the shape of the dormer roof.

Note: In Chapter 7 on conic sections, some dormer roof examples are also given. These have geometrical applications particular to that section.

GEOMETRICAL DRAWING FOR CARPENTRY & JOINERY

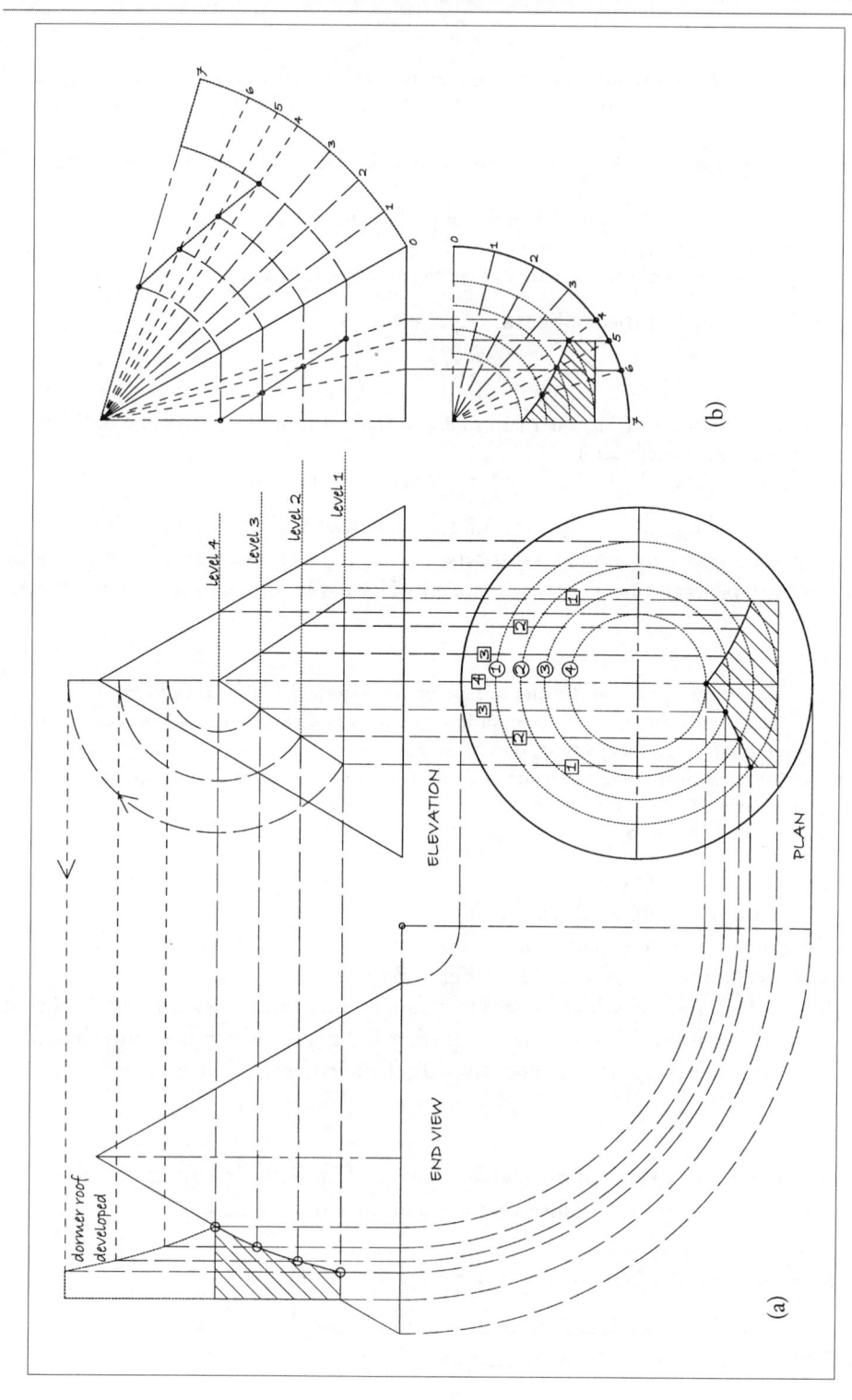

Fig. 8.34a
Fig. 8.34b

8 ROOF GEOMETRY

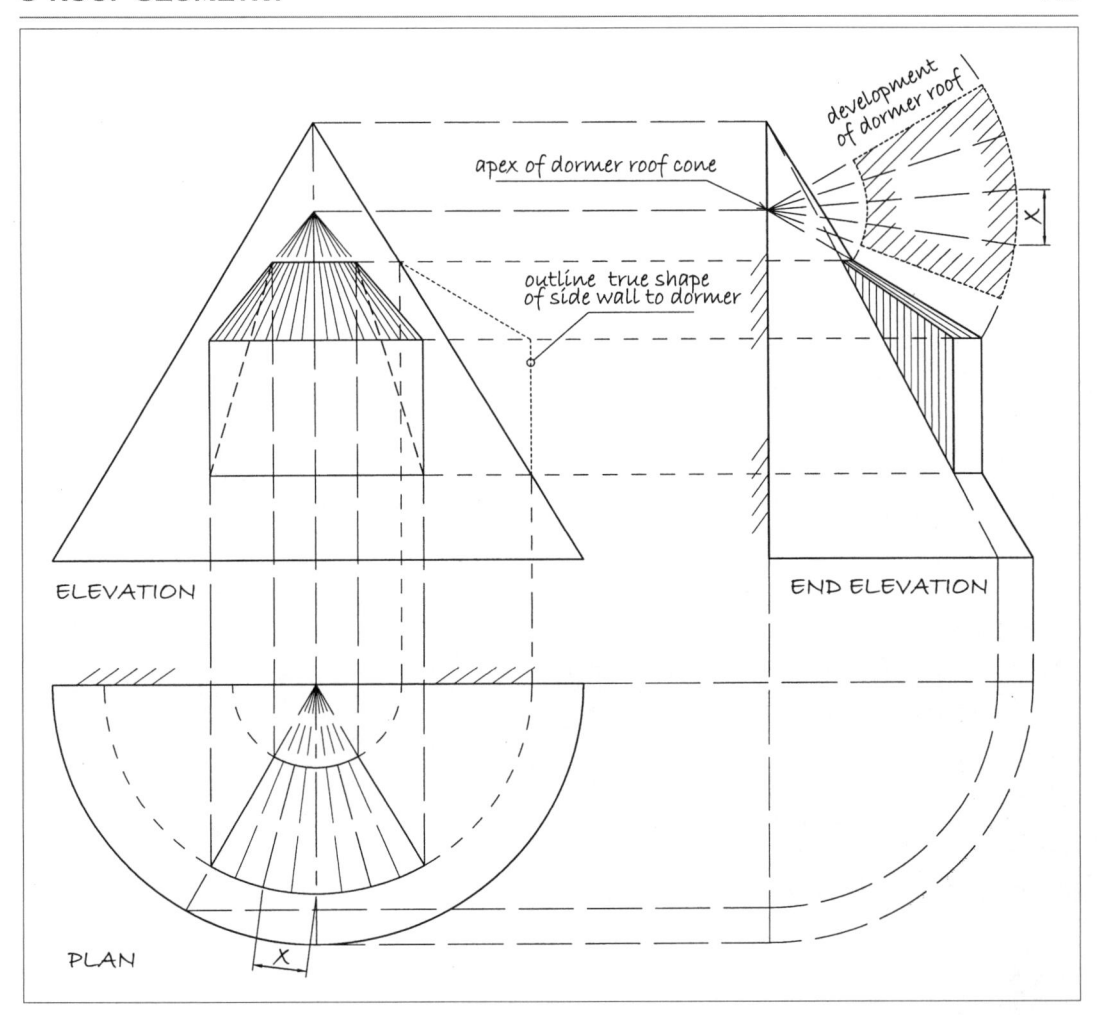

Fig. 8.35

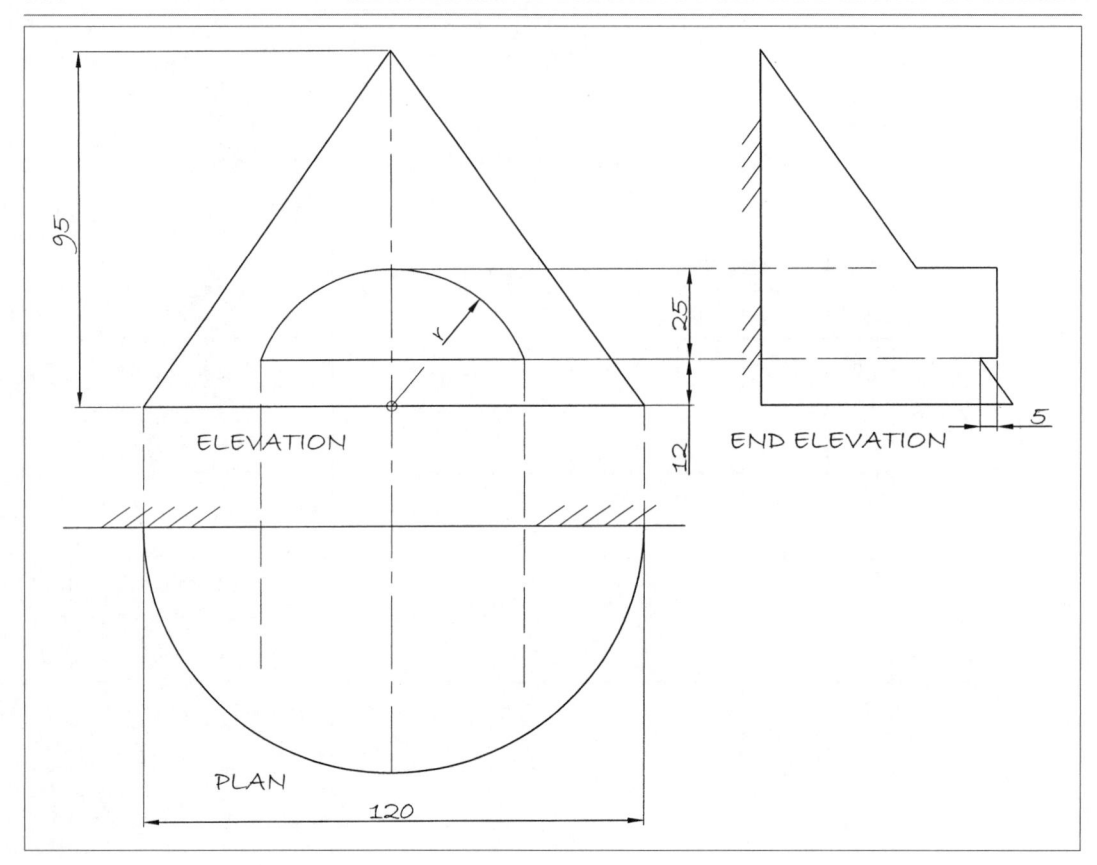

Fig. 8.36

9 HANDRAILING

SECTION I: RAMPS AND KNEES

At **Fig. 9.1**, drawing **A**, the pitched handrail connects to the newel at level **1** via a vertical turn. This may be done as shown in drawing **A**, but it involves a sharp change of direction in the handrail, which would not be free-flowing and comfortable to the hand when going up the stairs. To overcome this, a device called a ramp, a curved portion of handrail, is used. The setting-out of the ramp is given in drawing **B** (upper).

Method

- Having decided on the joint line on the vertical turn, a line is set off at 90° (to the vertical turn).
- An arc **c-j** is swung from **c** to meet the pitched handrail at **j1**.
- A line is set off at 90° to the pitched handrail from **j1** to intersect at **C1**. This gives the centre for the ramp curves, with the springing lines at **j** and **j1**.

In **Fig. 9.2** two options for the ramp are shown. As can be seen, the sweep of the ramp varies, and this variation depends on where the initial joint line on the vertical turn is determined.

Fig. 9.3 shows a combination of a ramp and knee. This shape is called a swan neck, and two variations on the arrangement are given at **A** and **B**. The setting out for **A** is shown, unencumbered by thickness of material, at **C**.

Method

- A vertical is dropped from the joint line at **e**, of the top-level rail.
- The top line of the pitched rail is extended to intersect at **d**.
- From **d**, a horizontal is drawn; quadrant **def** is constructed and a vertical is drawn from **f** to **h**.
- Then **h-f** and **h-g** are made equal. From **g**, a perpendicular is drawn to intersect **d-f** extended at **j**.

The arrangement at **B** shows the knee reduced and the ramp enlarged. This results in the ramp coming closer to the vertical joint line.

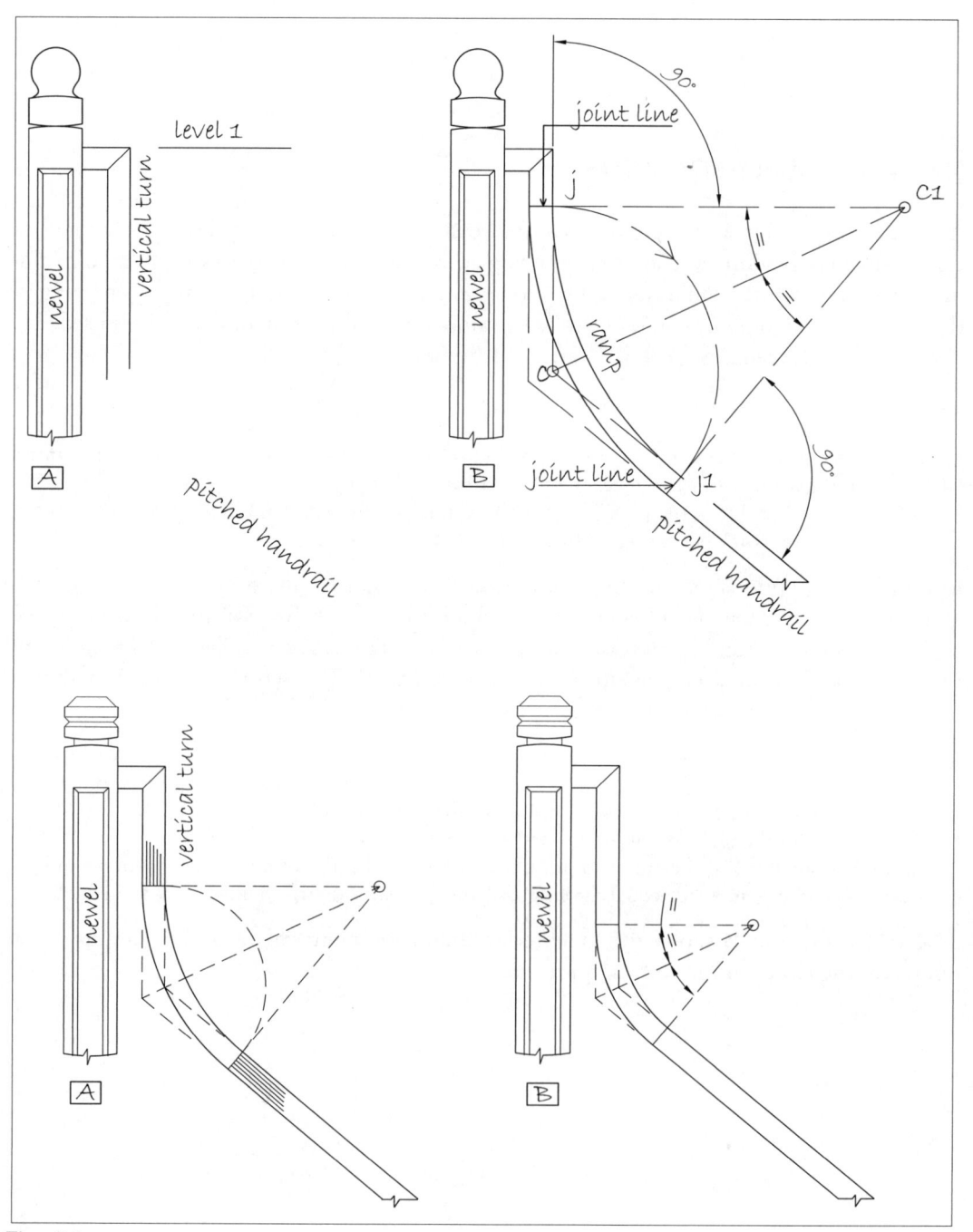

Fig. 9.1
Fig. 9.2

9 HANDRAILING

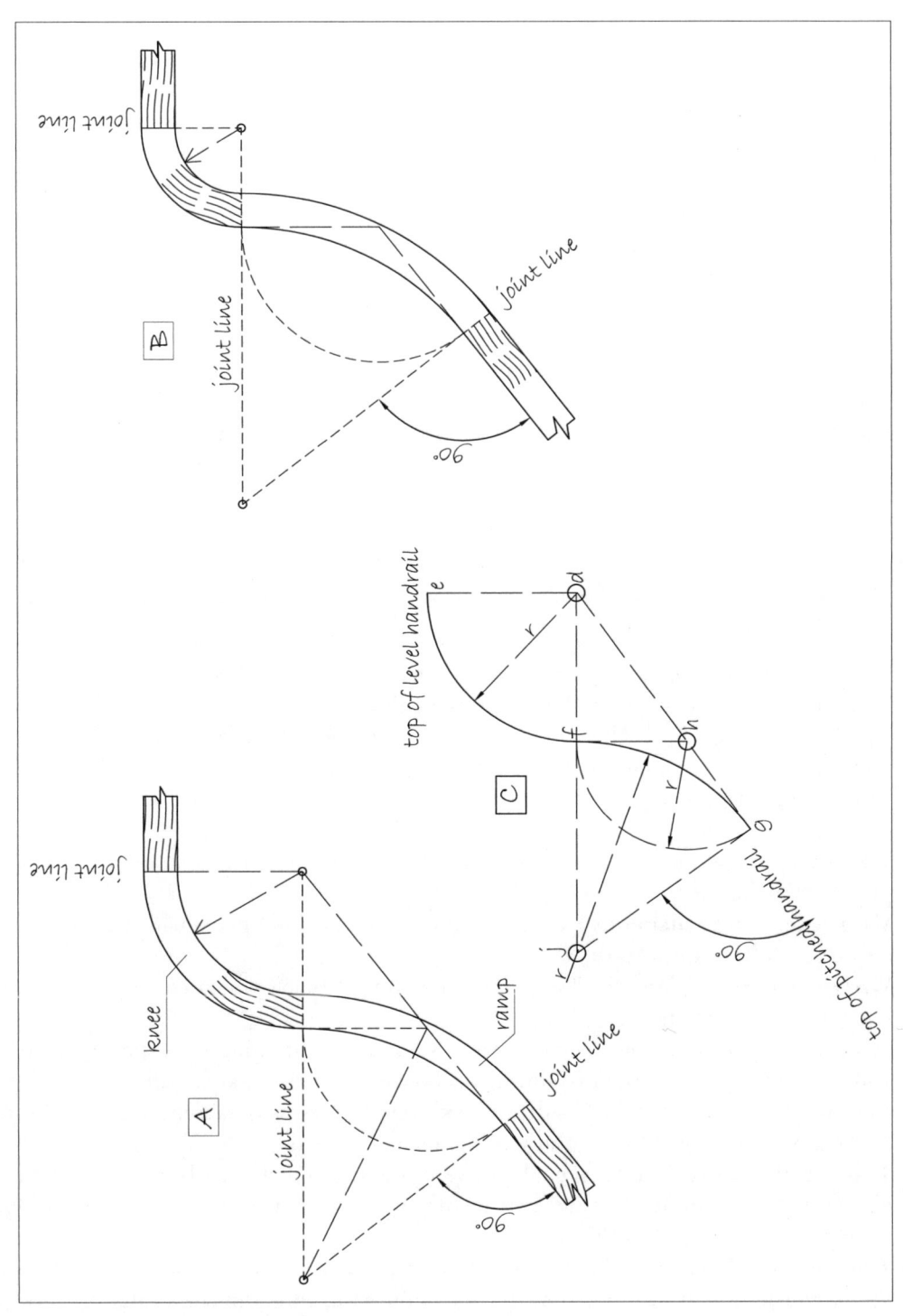

Fig. 9.3

SECTION II: WREATHS

Section I dealt with handrails in single curvature, meaning all parts were on one plane. This section deals with handrails as they traverse different planes, so that some of their parts will have double curvature. It is widely considered within the trade that understanding the geometry associated with handrail wreaths is very taxing. Before any aspiring joiner tackles the subject, study of inclined and oblique planes (see Chapter 17 on constructions) as well as the cylinder (see Chapter 4 on geometric solids) is recommended.

As handrails 'climb' to different levels, they may change direction depending on their plan view. The plan of a handrail can turn most commonly through an angle of 90° as it changes direction onto a higher level and, in some cases, through 180°. These changes in direction are effected by including a wreath, which effectively sees the handrail rounding a cylinder or being part of a cylinder. Handrail wreaths are used especially where there is a wreathed string incorporated in the stair design.

Fig. 9.4 shows the plan and elevation of a handrail wreath (in this case, two 90° wreaths) that turns 180°. The continuity of the handrail is evident from this and the way the wreath emerges from a hollow half-cylinder can be visualised. A wreath that turns through 90° is defined as a quarter-turn, and one that turns through 180° is defined as a half-turn.

Here is some terminology for reference while studying the geometry associated with handrail wreaths. It is used in the drawings for Examples 1 and 2.

Lower springing plane: the vertical plane that defines the lower limit of the wreath minus the shanks (in plan, the lower springing line)

Upper springing plane: the vertical plane that defines the upper limit of the wreath minus the shanks (in plan, the upper springing line)

Lower tangent plane: the vertical plane that contains the centre line (**c/l**) of the lower pitched straight rail

Upper tangent plane: the vertical plane which contains the centre line (**c/l**) of the upper pitched straight rail

Wreath centerline plane: in Example 1, the simply inclined plane which contains the centre line (**c/l**) of the wreath

Wreath centerline plane: in Example 2, the oblique plane which contains the centre line (**c/l**) of the wreath

Shanks: straight extensions of the wreath, beyond the springing lines, normally 50 to 100 mm in length (to facilitate jointing of the wreath to the straight rails)

Face mould: a template which is determined from the geometrical setting out, and used to mark out the shape of the wreath on the plank of wood

Twist bevel: the (true) dihedral angle between the simply inclined plane and the vertical tangent plane (Example 1), and the (true) dihedral angle between the oblique plane and the vertical tangent planes (Example 2)

Thickness of plank: the minimum plank thickness required to contain the wreath

Plank width: not critical, but allow plenty and use the grain direction carefully

9 HANDRAILING

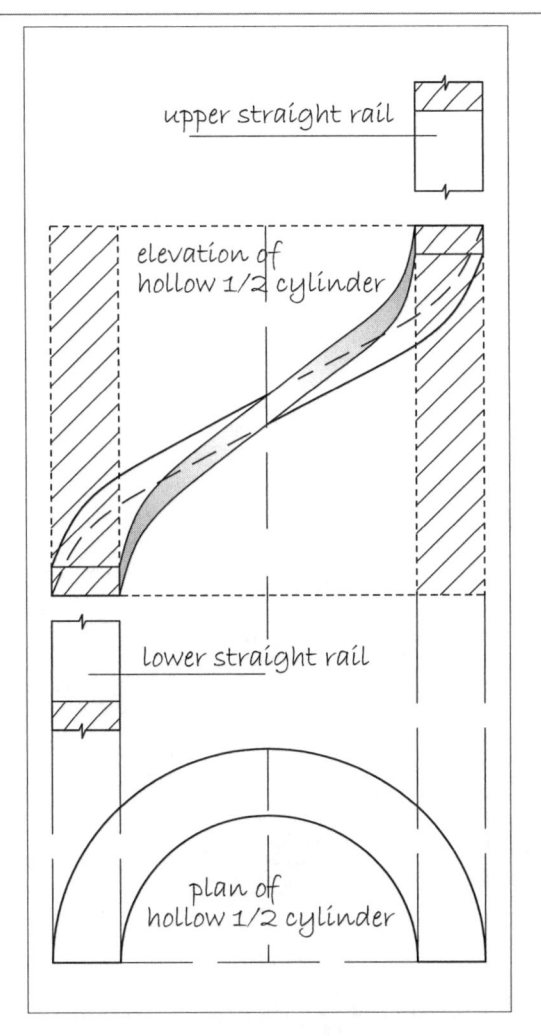

[Figure: diagram showing upper straight rail, elevation of hollow 1/2 cylinder, lower straight rail, and plan of hollow 1/2 cylinder]

Fig. 9.4

Note on twist bevels: In Example 1 where the geometry of the wreath is based on a simply inclined plane, there will be no twist bevel at the lower end because the lower pitched straight rail is on the same plane as the wreath.

In Example 2, the upper and lower twist bevels are the same because the straight pitched rails have equal pitches. If one or the other of the straight rails were pitched differently, the twist bevels would vary. This scenario is not covered in this section.

To construct handrail wreaths it is necessary to have determined:

- The face mould
- The plank thickness
- The associated 'twist' bevels.

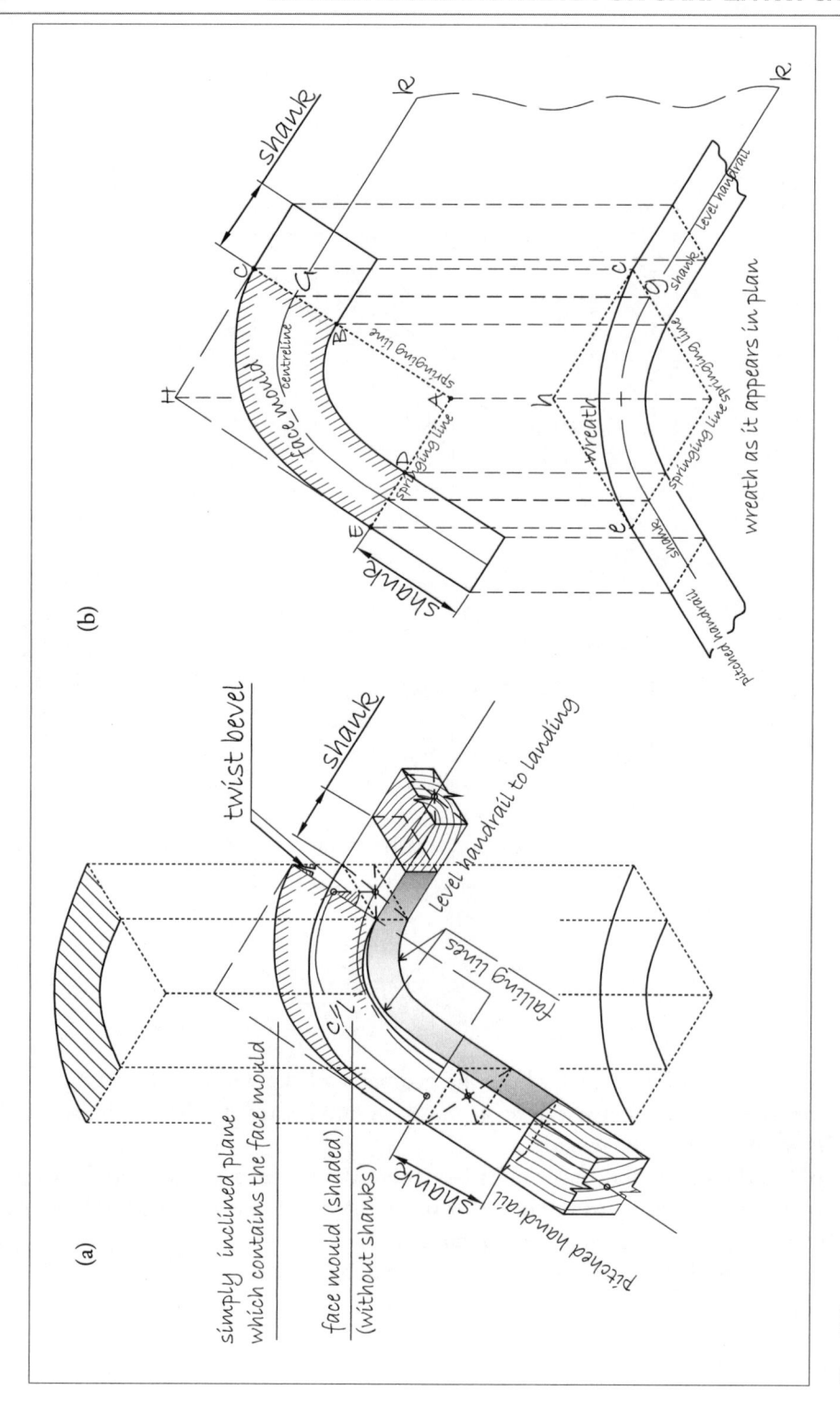

Fig. 9.5a
Fig. 9.5b

9 HANDRAILING

Do not worry about all the terminology at this stage. This is a gradual learning topic where one step leads to next, and the associated drawings to follow will reveal much.

Example 1

In **Fig. 9.5a** a pictorial view of one-quarter of a hollow cylinder is shown, with a wreath joining a pitched handrail and a level quarter-turn handrail to a landing. Also shown is a simply-inclined plane cutting the cylinder, which contains the face mould for the wreath in question. The pitch of this plane is the same as the pitch of the raking (or pitched) handrail.

The wreath is the portion that is contained within the quarter-cylinder. Added to the wreath and continuations of it are shanks, which join up with the straight handrail. Shanks are normally 50 to 100 mm in length and allow for forming the joint between wreath and handrail with relative ease, without impinging on the curved portion.

The twist bevel at the upper end of the wreath is given and it can be seen that the bevel is the angle between the two planes: the plane that contains the centre line (**c/l**) of the wreath and the vertical plane of the level handrail; this is known as the vertical tangent plane, which is **gGkk** in **Fig. 9.5b**.

At the lower end of the wreath, there is no twist bevel, in this case because both the wreath and the pitched handrail are on the same plane. We can, therefore deduce from this: that when a straight pitched (inclined) handrail joins a wreath that returns to a level handrail, no twist bevel will occur at the lower end.

Fig. 9.5b shows a line drawing that illustrates further the geometrical basis for the wreath. The curves on the face mould will be elliptical because whenever a cylinder is cut with an inclined plane, an ellipse results in the cut surface.

- The minor axis of the outer elliptical curve is **A-E**. (This is the outer radius if the wreath in plan.)
- The major axis of the outer elliptical curve is **A-C**.
- The minor axis of the inner elliptical curve is **A-D**. (This is the inner radius if the wreath in plan.)
- The major axis of the inner elliptical curve is **A-B**.

Fig. 9.6 shows a wreath similar to the one in **Fig. 9.5**, but drawn in orthographic projection. This must be done full size in actuality. The plan is first drawn, once the width of the handrail and radius has been determined. The centerline (**c/l**) is also drawn as shown.

Method

- In the elevation, the pitch of the lower (straight) handrail centre line (**c/l**) is drawn, and the **c/l** projected from **g** in the plan intersects with this.
- The section of the handrail is then constructed around the intersection.
- The plank thickness is then drawn from the corners of the rail section parallel with the inclined centre line (**c/l**).
- The twist bevel for the wreath is indicated and it can be seen that, in this case, it is the angle between the inclined plane and the vertical plane.

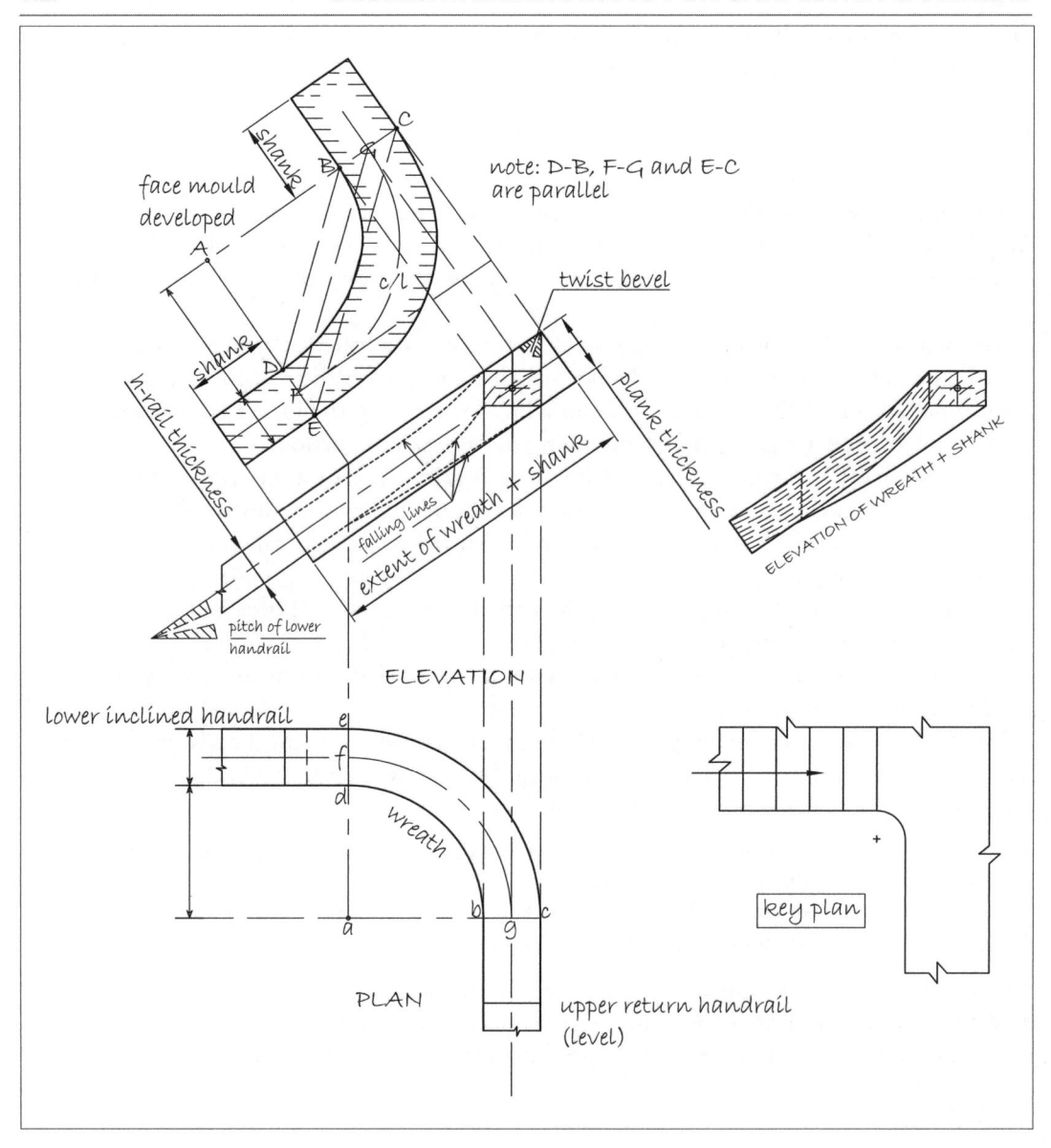

Fig. 9.6

The extent of the wreath and shank is now clearly determined, with the wreath itself shown in dotted lines. These dotted lines represent the *arrises* (the corners where face and edge meet) of the wreath proper and are called falling lines.

The face mould is developed by *rebatment* of the top surface of the plank as shown, at a convenient distance beyond the elevation. **A-D** and **A-E** are the half-minor axes required, and **A-B** and **A-C** are the half-major axes required. It is not necessary to draw the centre line

9 HANDRAILING

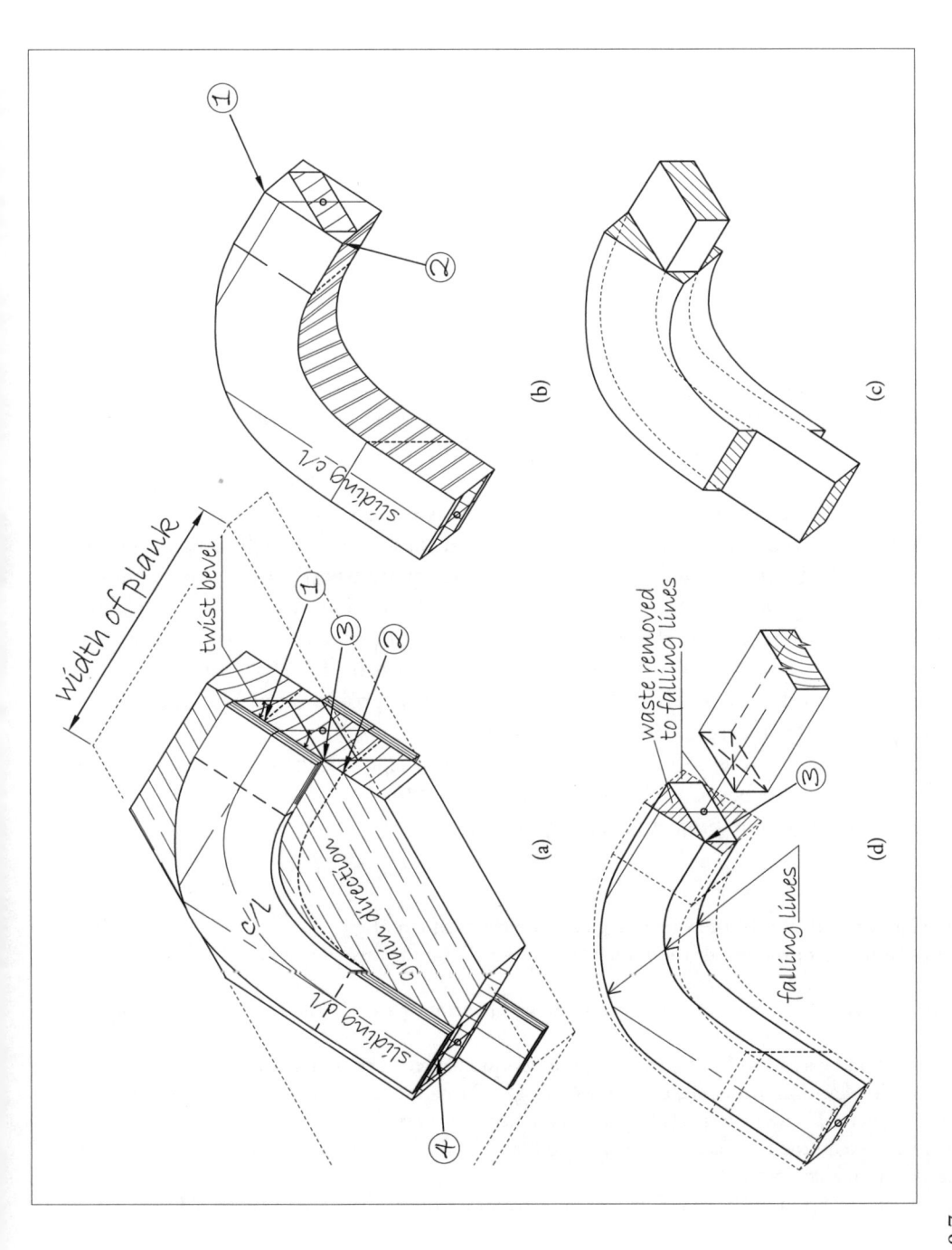

Fig. 9.7a
Fig. 9.7b
Fig. 9.7c
Fig. 9.7d

(**c/l**) of the wreath at this stage. Draw the elliptical curves using any of the methods covered in Chapter 6 on arches.

Fig. 9.7a shows the face mould on the plank of timber, with the shank ends cut at the joint line and the twist bevel and outline of straight handrail marked on the end grain. The face mould is also shown in position underneath the plank, as it is necessary to mark the mould lines on both surfaces. This is achieved by sliding the mould along the sliding centre line (**c/l**), as indicated, and setting it to positions **1, 2** and **3**. Positions **1** and **2** give the lines for the band-saw cuts, while position **3** gives a falling line.

Fig. 9.7b shows the wreath following the band-saw cuts. **Fig. 9.7c** shows the waste from the shanks removed and the wreath ready for shaping. **Fig. 9.7d** shows the wreath finished following removal of waste to the falling lines.

Example 2

Before tackling Example 2, a study of oblique planes is necessary (see Chapter 17 on constructions). The following example is of a wreath which joins lower and upper pitched handrails of equal pitch. **Fig. 9.8** shows a pictorial representation of this type of wreath. Note that the face mould and eventual wreath emerge from a hollow quarter-cylinder, which is in itself part of a square-based prism. The geometry for Example 2 below is based on these principles.

In **Fig. 9.9a** a pictorial view of the geometry involved for a wreath which joins lower and upper straight pitched rails of equal pitch is shown. The square-based prism which contains the **c/l** of the wreath, **g1-o1-f**, is shown being cut by an oblique plane **VTH**; with **f-h1** being the lower pitch tangent and **g1-h1** the upper pitch tangent. The plan of the **c/l** of the wreath is **f-o-g**. Note: This pictorial view is for visualising the process of development only. It will not reveal the true development. This is done in orthographic projection in **Fig. 9.10** below. It will be helpful to refer to **Figs. 9.9** and **9.10** together.

Fig. 9.9c shows the prism reproduced with the cylindrical aspect of the wreath evident. The method of obtaining twist bevels is also indicated in **Fig. 9.9c**.

To set up the orthographic projection in **Fig. 9.10a**:

Method

- In the plan, draw the **c/l** of the wreath **f-g** and construct square **fhga**.
- Join **a** to **h** to give **o**; **a-o** is half the minor axis of the ellipse for this wreath **c/l**.
- Draw **f-T** parallel to **a-h** and extend to **H**.
- Set off the pitch of the lower handrail from **f-h**, as shown in the plan, and pivot the resulting rise to **h1**.
- Join **T** to **h1** and extend to **V**. The traces of the oblique plane **VTH** are now drawn.
- Complete the elevation by establishing **o1**, **a1** and **g1**. The elevation of the (cut) square-based prism is now complete. The cut surface **h-h1-g1-a1** contains the centre line of the wreath.
- In the plan, extend the arc **f-o-g** to **k**; **a-k** is the plan of half the major axis of the ellipse. Project **k** to the elevation as shown to become **k1**.
- Point **p** is an extension of the major axis in plan and is found in elevation at **p1** on **VT**.

9 HANDRAILING

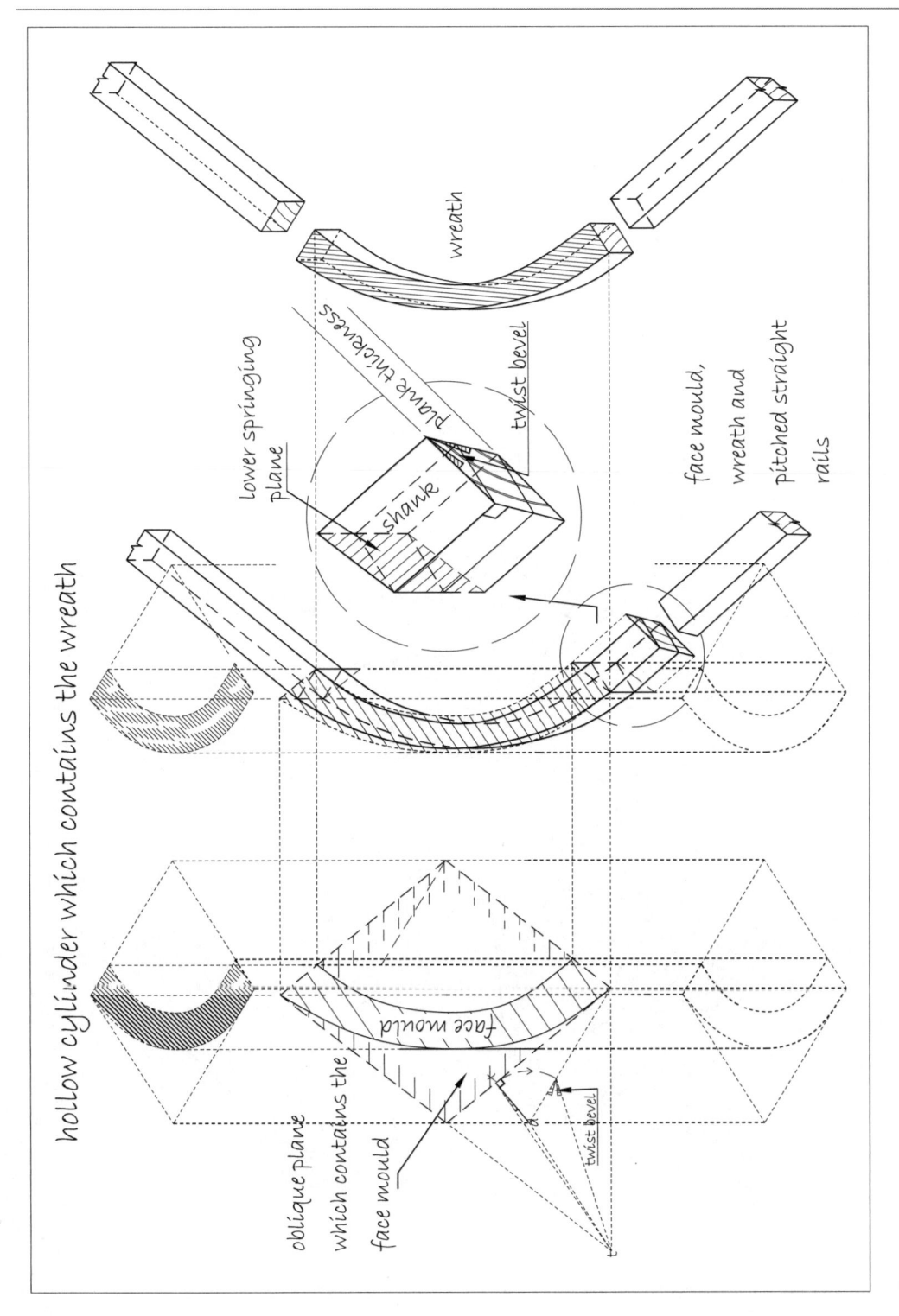

Fig. 9.8

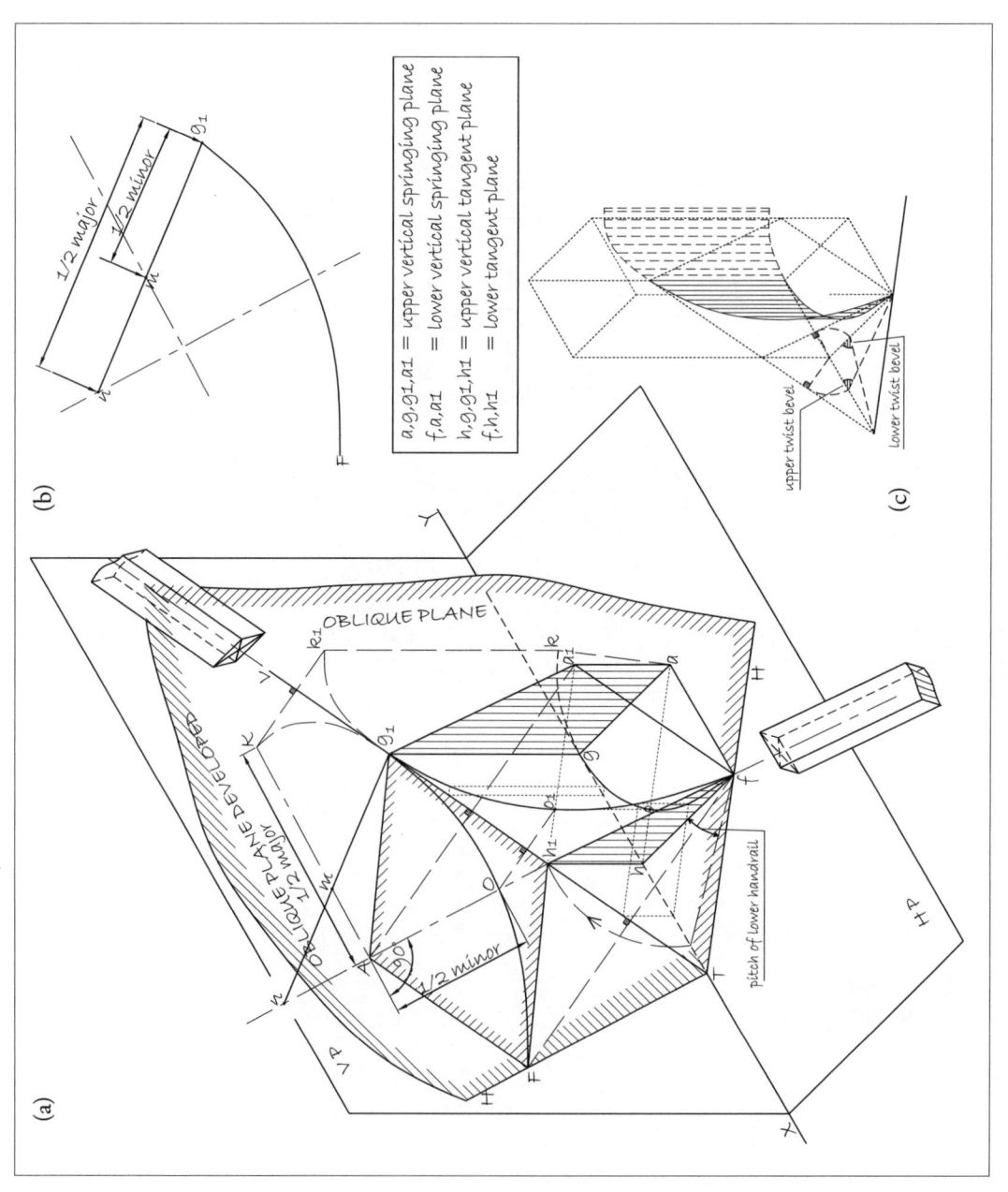

Fig. 9.9a
Fig. 9.9b
Fig. 9.9c

9 HANDRAILING

The task now is to develop the oblique plane **VTH** (as in **Fig. 9.9a**). The developed plane will contain the true shape of the centre line (**c/l**) of the wreath.

Method

- Project **f** square-off **V-T** (note: in the elevation of **Fig. 9.10**, **h** and **f** are the one point) and make **T-f** and **T-F** equal.
- Join **F** to **h1**. Project **a1** square-off **V-T** and make **F-A** and **V-T** parallel. **F-A-g1-h1** is the developed surface of the cut prism. **A-g1** is the upper springing line and **A-F** the lower springing line; **g1-h1** is the upper tangent (the **c/l** of the upper rail) and **h1-F** the lower tangent (the **c/l** of the lower rail).
- Join **h1-A** and project **o1** square-off **V-T** to **O**.
- **F**, **O**, and **g1** are points on the wreath **c/l** curve (which is part of an ellipse). **A-O** is half the minor axis of the ellipse and a line square-off at **A** gives the direction of the major axis. At this stage it is possible to draw the elliptical curve because the length of the minor axis is known, as well as some points on the ellipse.
- For clarity, **Fig. 9.9b** reproduced shows the method:
    - Strike the length of half (1/2) the minor axis from **g1** to the major at **m**.
    - Extend to the minor axis at **n** to give **g1-n** = to the half (1/2) length of the major axis of the ellipse.
    - The ellipse can be drawn using the trammel method or any other as shown in Chapter 6 on arches.

Note: On the HP (horizontal plane) the **c/l** of the wreath is extended to **k**, which results in the plan of half (1/2) the major axis as **a-k.** This is projected to the oblique plane at **k1** and, ultimately, to the development at **K**, resulting in **A-K** being half the length of the major axis.

With half the major and minor axes now known, it is possible to draw the ellipse in the normal way.

The **c/l** of the wreath has now been determined as the elliptical curve **F-O-g1**, and this is reproduced again in **Fig. 9.10b**.

To complete the face mould, refer to **Fig. 9.10b**:

Method

- Join **O-g1** and **O-F** and construct a circle at **O**, with a diameter, **r-t**, equal to the width of the handrail.
- Make **r-s** and **t-u** parallel with **O-g1** and do similarly with the left-hand side.
- The springing line on the wreath is **s-u** and defines where the elliptical curves end (similar on left-hand side at **s1-u1**).
- **A-r** and **A-t** are the two half minor axes of the inner and outer elliptical curves and **s** and **u** are, respectively, points on the inner and outer curves. Using the trammel method, the two curves may now be drawn. Note: It is not necessary to draw the **c/l** of the curve at this stage.
- Extend **h1-g1** and mark the length of the shank beyond **g1**, square-off **p-j** from **h1-g1** and draw **j-u** and **p-s** parallel to **h1-g1**.
- On the face mould, mark the lines: **h1-g1**, **h1-F**, **s-u** and **s1-u1**. The face mould is now complete and ready for use.

Fig. 9.10a shows the twist bevel for the wreath, lower and upper being the same in this case.

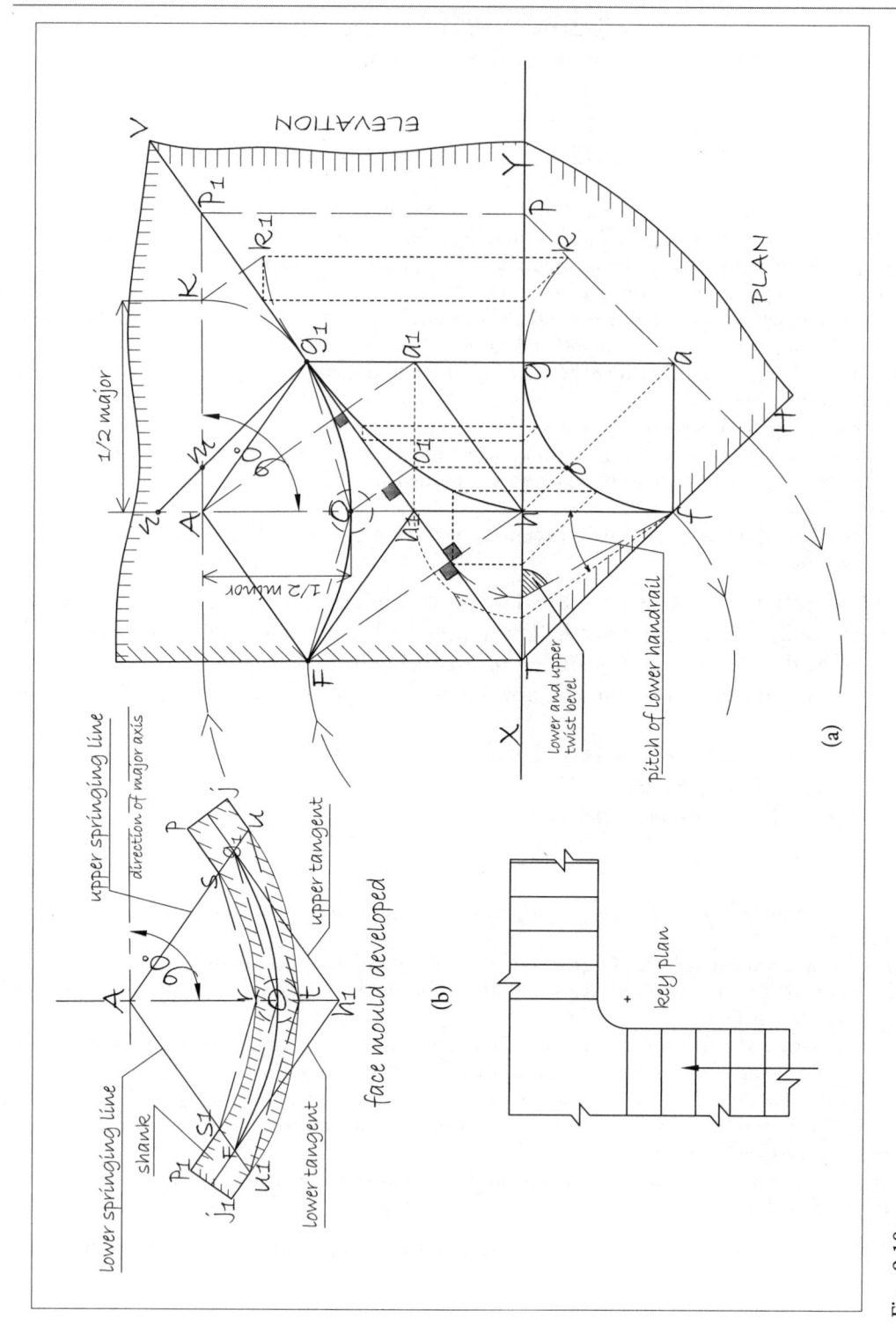

Fig. 9.10a
Fig. 9.10b

9 HANDRAILING

In **Fig. 9.11a** the twist bevel is applied to obtain plank thickness:

Method

- Draw horizontal line **a-b** and apply the twist bevel as shown from **g**.
- Set off half the width of the handrail on either side of **g-h** to find point **c**.
- Draw **c-f** square-off **g-h**.
- Apply thickness of handrail and draw **c-d** and **f-e** parallel with **g-h**.
- Draw horizontal through **e** to give plank thickness.

Note: **Fig. 9.11b** shows a more acute bevel being used for the same handrail section and it should be observed that a greater plank thickness is achieved.

In some situations one of the straight handrails may be pitched differently. When this happens the plank thickness is found by using the more acute twist bevel, however, this does not apply to the wreath given in Example 2 here. **Fig. 9.11c** shows the face mould applied to the wood plank with the twist bevels marked on the end grain. The outline of the wreath within is also indicated.

Fig. 9.12 shows an alternative method of developing the face mould. The plan of the wreath is drawn, which is the same as saying, the plan of the face mould is drawn (minus the shanks), on the oblique plane **VTH**.

The procedure is to find points on the face mould in the developed oblique plane. To find point **a**:

Method

- Draw the plan of the wreath.
- Draw the oblique plane **VTH**, the same one used for **Fig. 9.10**.
- Develop the oblique plane having **T-H** as the hinged line of the development. This is done by making **T-V** and **T-V1** equal and projecting square-off at **H**.
- Draw **a-w** parallel with **T-H** and in elevation **w1-a1** results.
- Find **W** on **T-V1** by projecting **w** square-off **T-H**. **W-A** is parallel to **T-H**.
- Any points of the face mould on line **a-w** in the plan can now be projected to the developed line **A-W**.
- Do similarly with any number of lines or points to give the face mould outline.

This method is known as the *level ordinates* method: **a-w** is a level line on the oblique plane and in reality. All lines parallel with **T-H** in the plan and parallel with **X-Y** in the elevation are level lines or level ordinates.

Note: The face mould in the development is confined to the springing lines. The shanks are added as per **Fig. 9.10b**.

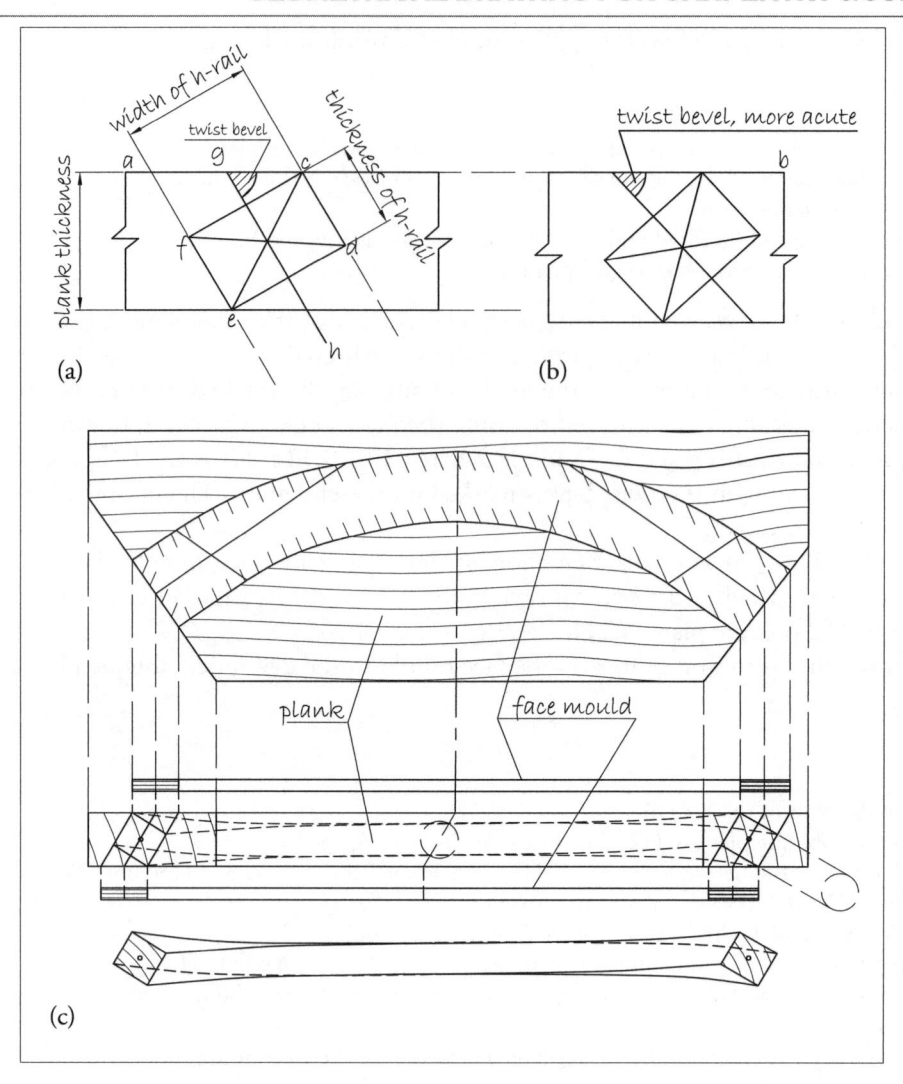

Fig. 9.11a
Fig. 9.11b
Fig. 9.11c

9 HANDRAILING

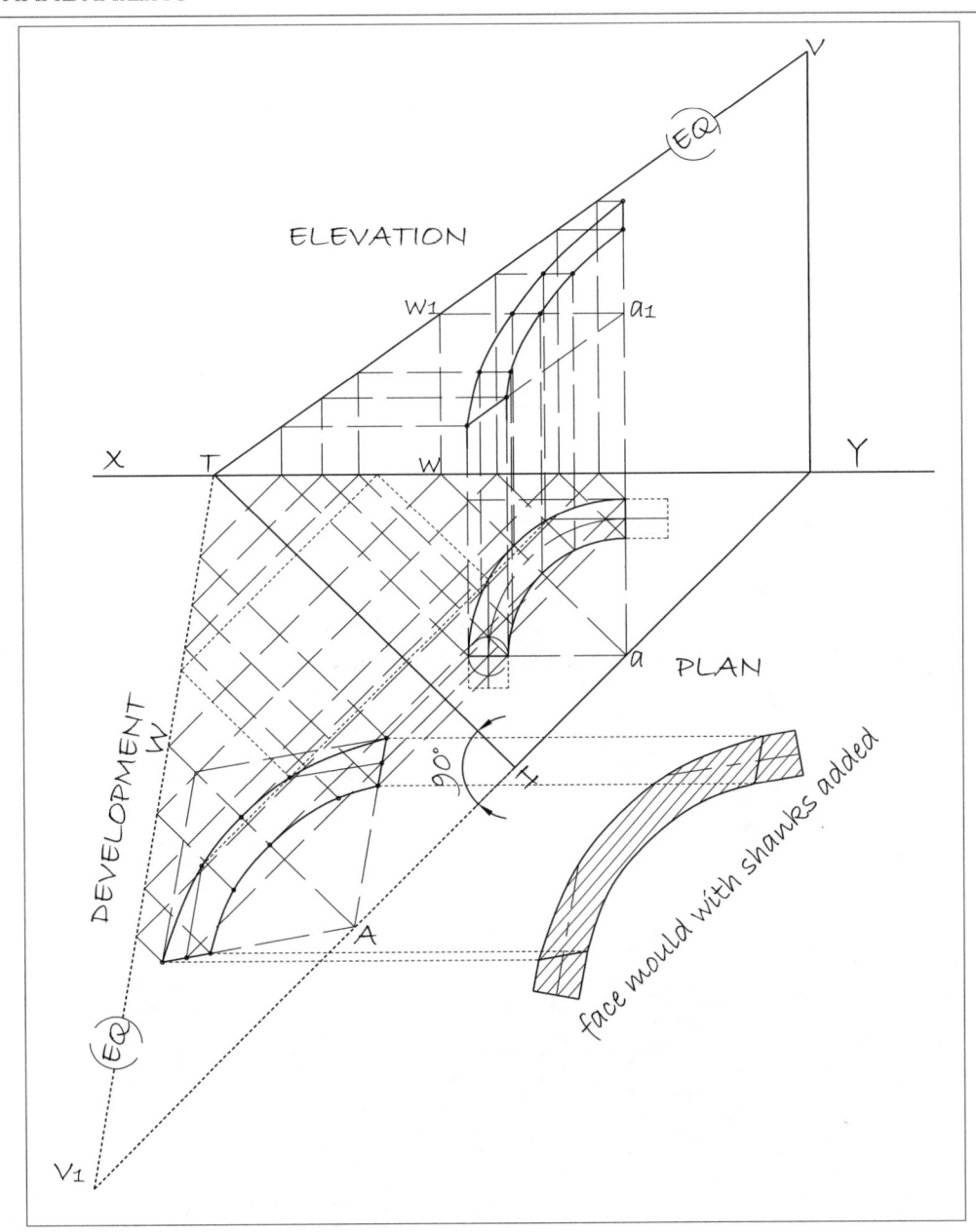

Fig. 9.12

SECTION III: SCROLLS

Handrail scrolls are used as an ornate method of terminating a handrail at the lower end. They allow the handrail to convolute (curl round). This is done using a series of diminishing arcs, or quadrants. Sometimes the scroll follows the outline of, and complements, a curtail step, which is directly underneath.

The scroll in **Fig. 9.13** is constructed using seven centre points, which draw a series of seven diminishing quadrants. The method of finding the centres and an enlarged detail of their layout are shown in the figure.

In **Fig. 9.14** a scroll with four centres is shown as well as a larger detail of the centres. The width of the handrail is **a-b** and the overall width of the scroll is four times **a-b** (**a-c**); **m** is the mid-point of **a-c**. The handrail width is divided into eight equal parts and centre points are found as follows:

- (**m-1** = one part of **a-b**)
- (**1-2** = two parts of **a-b**)
- (**2-3**, and **3-4** are also equal to two parts of **a-b**).

Attaching a scroll to a pitched handrail may be done using a ramp to ease the handrail onto the scroll or by using a wreath as shown in **Fig. 9.15a**. The scroll in **Fig. 9.15a** incorporates the wreath. The scroll shown is constructed using five centres:

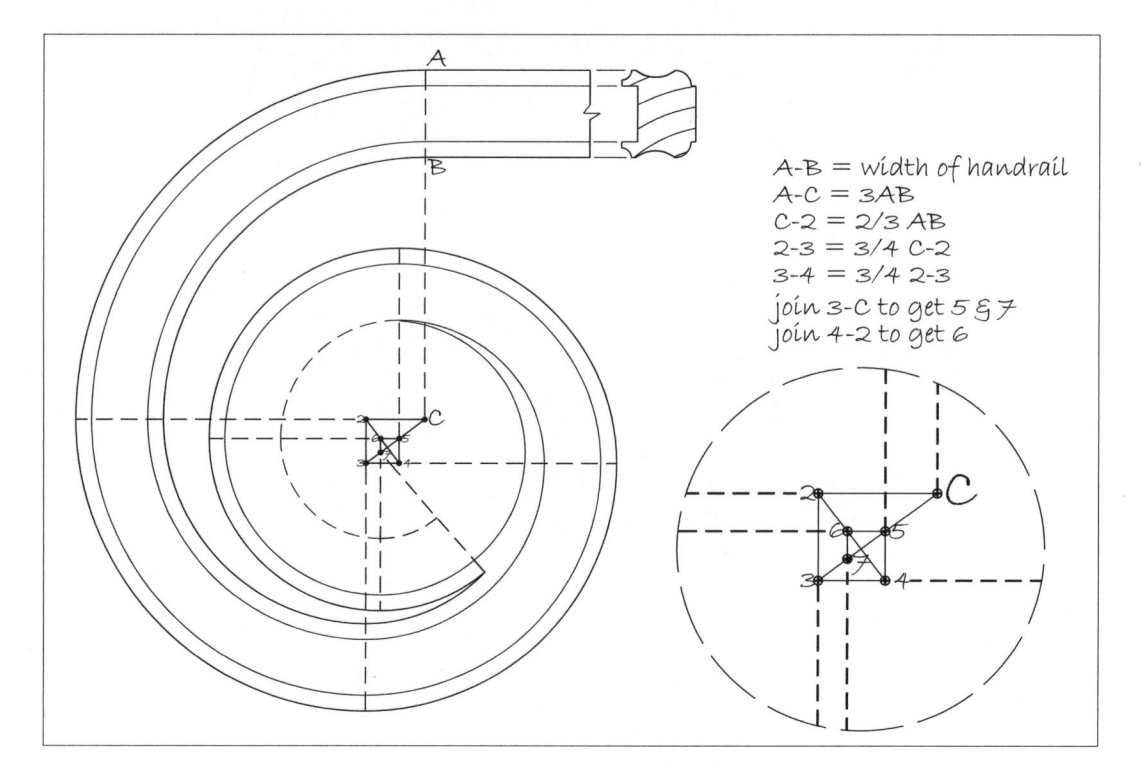

Fig. 9.13

9 HANDRAILING

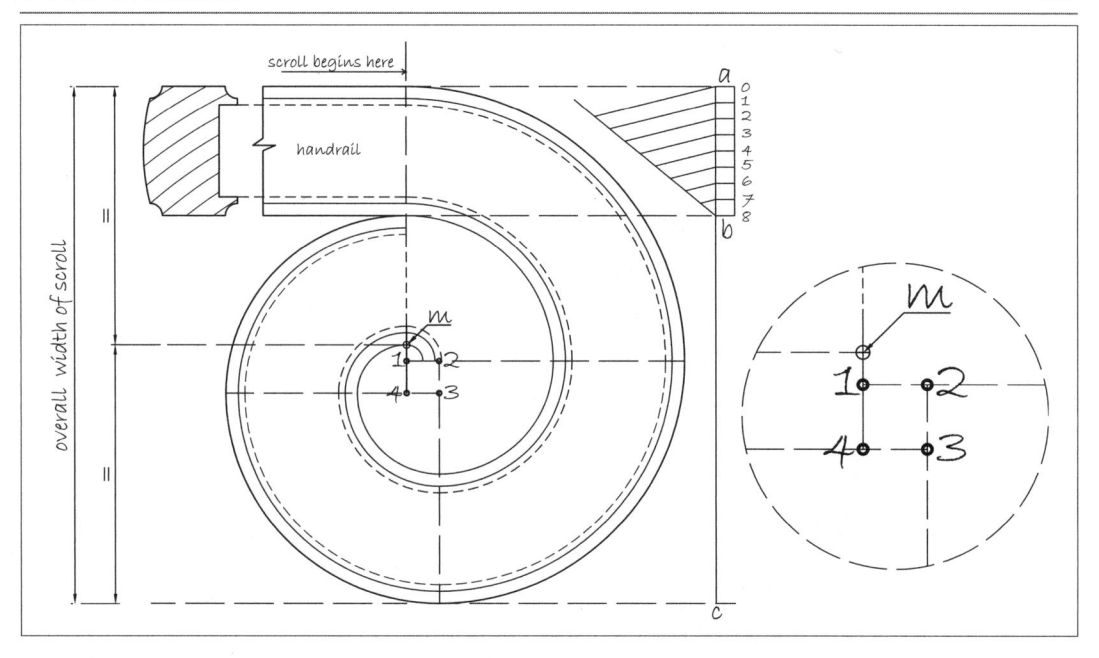

Fig. 9.14

Method

- Line **a-b** represents the width of the handrail, and **a-c** is equal to four times **a-b**.
- Point **m** is the midpoint of **a-c**.
- Divide **a-b** into eight equal parts.
- Make (**m-1** = two parts **a-b**), (**1-2** = four parts **a-b**), (**2-3** = three parts **a-b**), (**3-4** = two parts **a-b**) and (**4-5** = one part **a-b**).

The scroll quadrants can now be drawn using the centres **1**, **2**, **3**, **4** and **5**. Note: Point **m** is not a centre point for a quadrant.

The face mould for the handrail wreath is developed using methods employed in Section II above.

In **Fig. 9.16** another variation on the proportions of diminishing quadrants is given. Again, **a-b** equals the handrail width and **a-c** is four times **a-b**, with **m** as the midpoint of **a-c**.

- The line **c-d** is one half of **a-b**, **e-n** is square-off **a-d**
- The centre points, **1** to **6**, are found by using the diagonals, **1-3** and **2-4** through **n**, as shown.

Exercise

Fig. 9.17 shows a single line drawing of a pitched handrail which joins an upper straight rail and a lower scroll. Reproduce **Fig. 9.17** to the dimensions given.

- Complete upper and lower ramps.
- Complete a plan view with a scroll attached to the lower ramp.

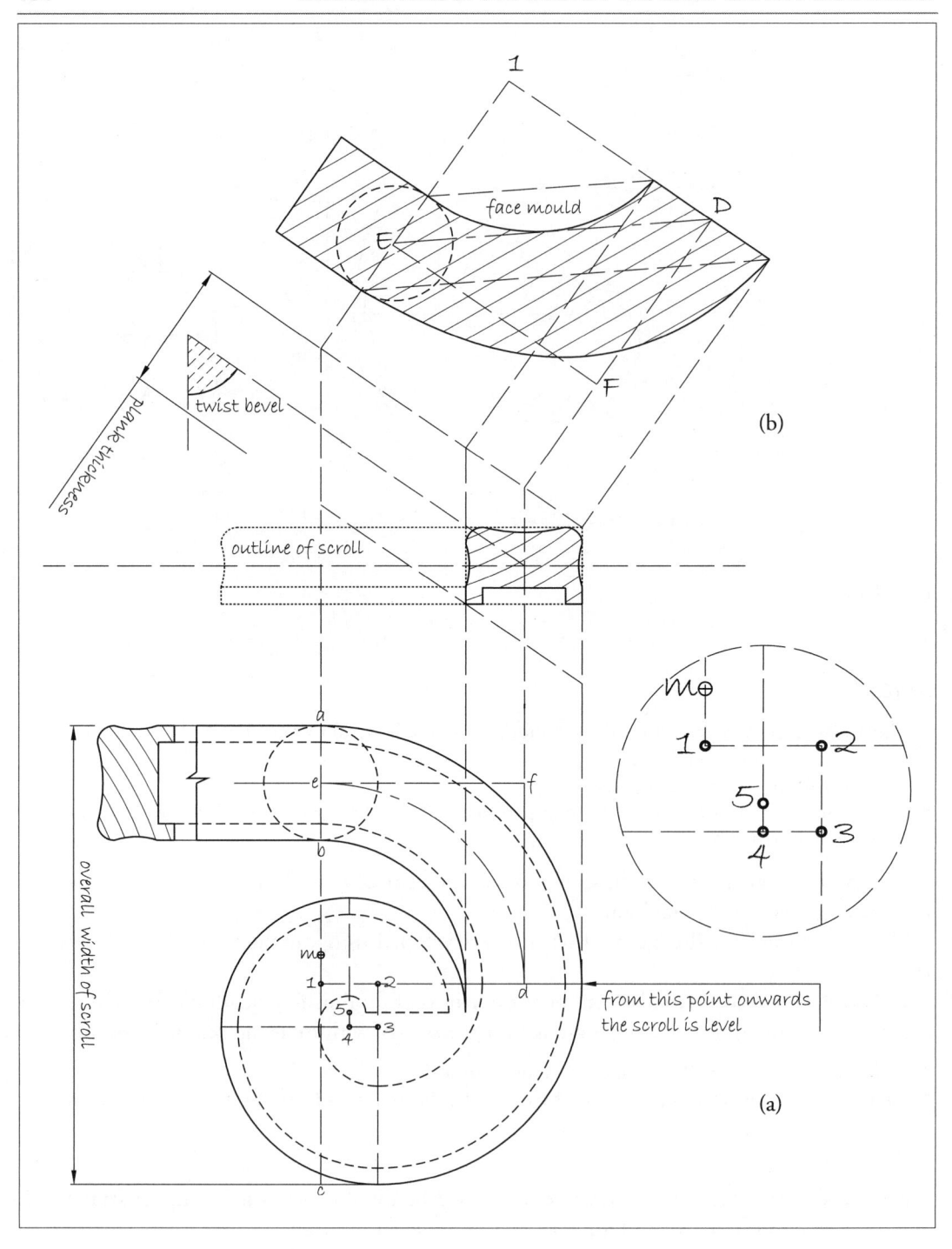

Fig. 9.15a
Fig. 9.15b

9 HANDRAILING

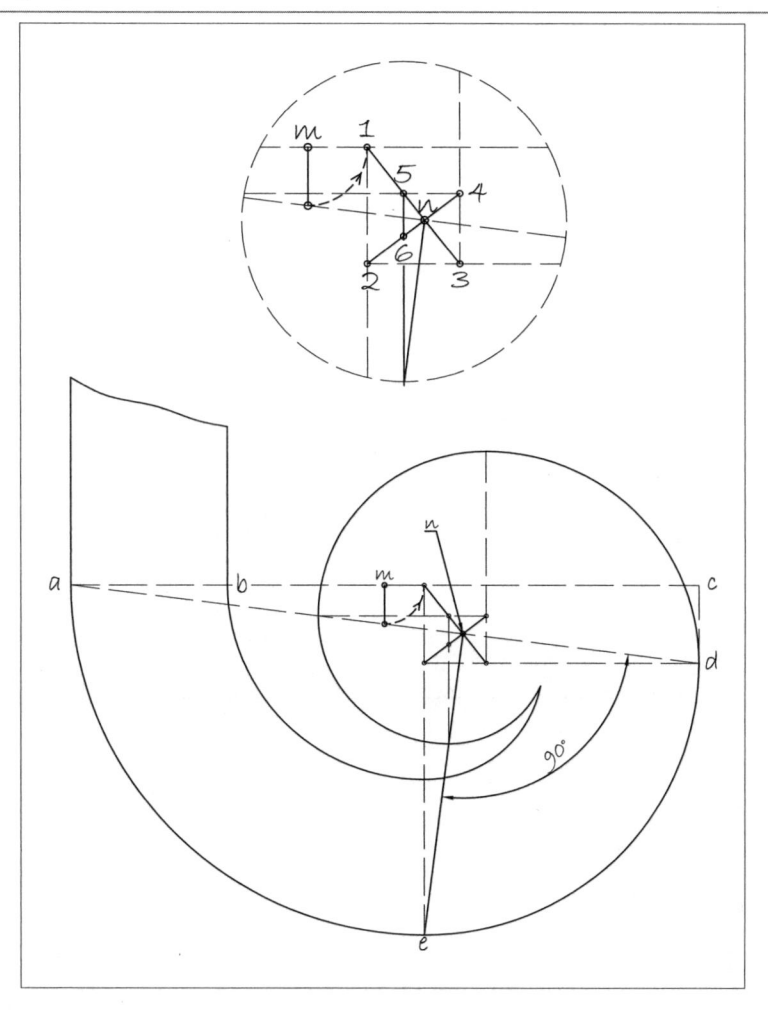

Fig. 9.16

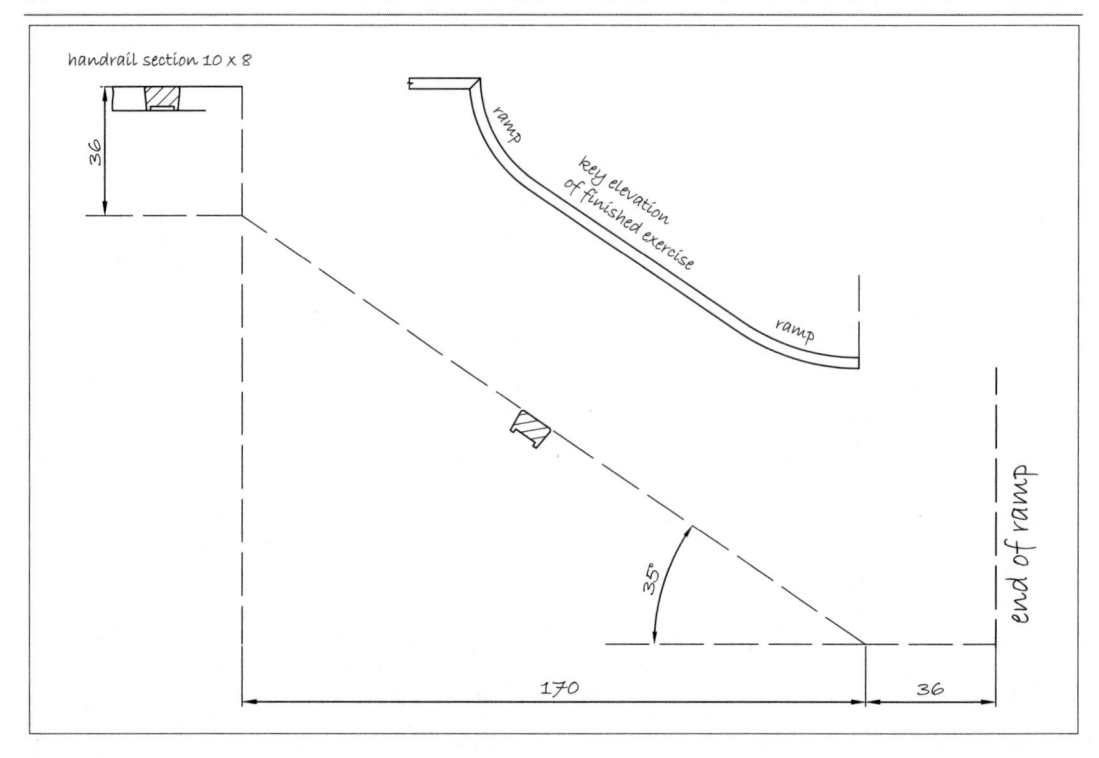

Fig. 9.17

10 GEOMETRICAL STAIRS AND WREATHED STRINGS

The geometrical stairs shown in **Fig. 10.1** is commonly called a spiral stairs, though the string of the stairs is actually *helical* in shape. In the example shown a central newel post accommodates the inner ends of the treads and risers. The overall rise is divided, in this case, into 16 step 'rises', resulting in first floor level at 16.

The development of the string for the stairs is obtained by:

Method

- Stepping out the dimension **x** (taken from the plan) along the baseline at **0** the appropriate number of times.
- Applying the margin upstand **y**, and dimension **u** above and below the relevant rise respectively each time.

The resulting development shows the riser positions and the width of the treads on the string. The drawing at **Fig. 10.1** also shows the centre post developed with the riser positions shown on it.

The true pitch of a spiral stairs occurs at the walking line of the stairs, which is shown in the plan. The walking line in a stairs with a width of less than one metre is deemed to be in the middle of the width, however, the relevant building regulations should be adhered to when determining the walking line. The pitch of the stairs on the walking line is shown in **Fig. 10.1**. The width of the step 'going' is taken from the plan of the walking line and applied to each rise. It can be seen that the resulting development looks like the conventional setting out of a straight string.

A stairs with a wreathed string to a quarter-turn landing is shown in the plan of **Fig. 10.2a**. The elevation of the lower string is drawn and attached to it, as are the developed wreath and the upper string. The plan length of the wreath, **6-7**, is expanded to its real length (using a method explained in Chapter 17 on constructions) giving **6-7₁**, which is projected to the development where the rises and margin line are set up. A width is added to the string and the developed wreath emerges.

It can be seen that where the wreath and the straight strings meet, a 'kink' is evident. In many cases, this cannot be avoided and the outcome depends on factors such as:

- the radius used for the wreath
- the going of the step
- where the risers are positioned around the turn in the plan.

It is preferable to have continuity of line in the developed wreath, which in turn ensures the same in the finished wreathed string in position.

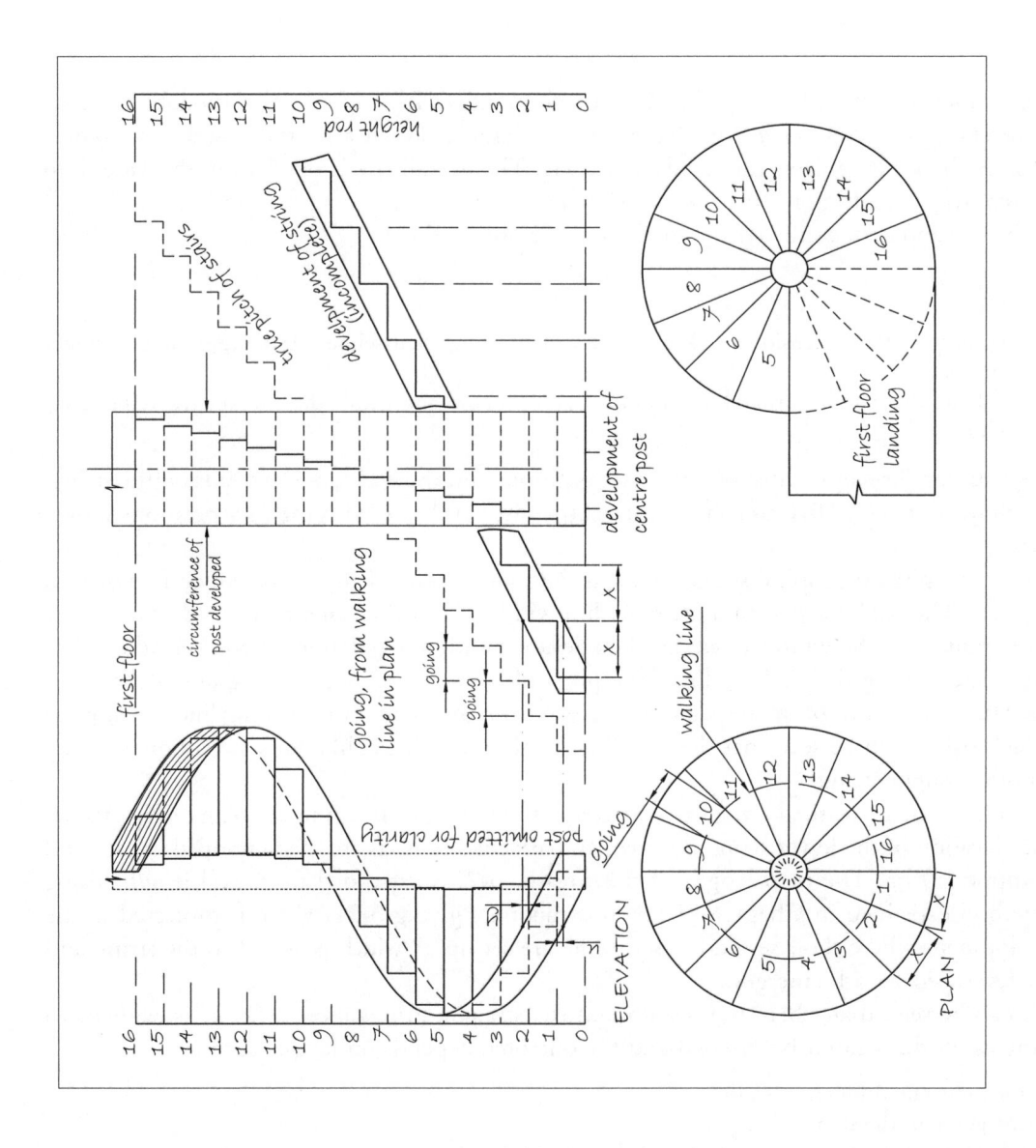

Fig. 10.1

10 GEOMETRICAL STAIRS AND WREATHED STRINGS

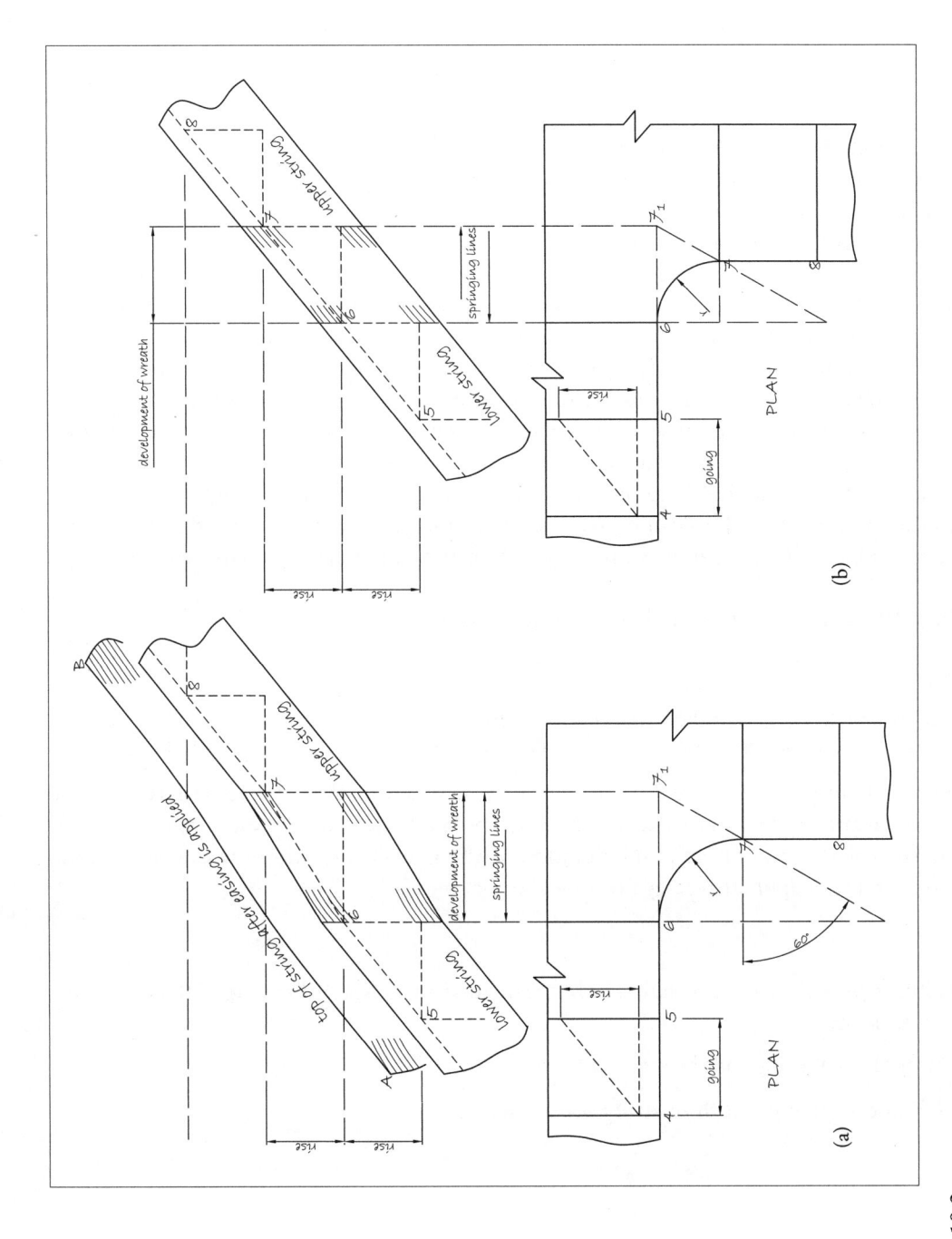

Fig. 10.2a
Fig. 10.2b

The use of *easings* at these junctions between wreath and straight string help the aesthetic continuity of line between the two items. The easings can be drawn as an arc of a circle, though in most cases the radius for the arc is so large that it runs off the development board; alternatively a *fair* curve may be drawn freehand aided by an experienced eye. The curved line **A-B** shown above the development in **Fig. 10.2a** shows the top of the string following the application of the easing.

In **Fig. 10.2b** the same stairs is shown (same going and rise), the only difference being that the radius of the wreath has been reduced so that the length of the quadrant concurs exactly with the going of the step. It can be seen in this case that the developed wreath has the same pitch as the straight strings and, consequently, has no need for an easing.

In **Fig. 10.3a** the plan of a cut-string stairs with a quarter-turn wreathed string is shown. The turn has three tapered steps. The goings at the tapered end of the steps, **6-7**, **7-8** and **8-9**, are developed in the plan. The developed goings are projected to the elevation, where a height rod of the rises is set up. The outline of the steps at the wreath of the string can now be drawn. To find the lower edge of the string an arc of dimension **x** is drawn from the bottom corner of each step as shown and a fair curve is drawn as a tangent to the arcs. This ensures continuity of material substance under each step and a suitable easing onto the straight strings.

In practice, it is not advisable to have the joint between wreath and straight string occurring on the springing line. This can be avoided by forming the joint at the next riser position, such as **5** (**a-a**) and **10** (**b-b**) in this case. **Fig. 10.3b** shows the development before easing is added.

In **Fig. 10.3c** an alternative method of forming the easing is given.

Method

- Divide a straight portion of the string edge into a number of equal parts, **o** to **p** in this case.
- Join **a-a**, **b-b** and **c-c**, then draw a fair curve tangential to the lines. This gives a parabolic curve.

A wreathed cut-string for a half-turn stairs is represented in **Figs. 10.4a** and **10.4b**. The developed wreath of the string is achieved using the method used in the example above. In **Fig. 10.4b** a plan, elevation and end elevation of the wreathed portion are drawn with the joint lines to the straight strings, again shown at **a-a,** and **b-b**.

Exercise

Fig. 10.5 shows a closed string stairs with wreathed strings. Reproduce **Fig. 10.5** and complete the exercise:

- Develop the wall string and the inner string for the stairs.

Note: Choose a suitable width of string for the stairs.

10 GEOMETRICAL STAIRS AND WREATHED STRINGS

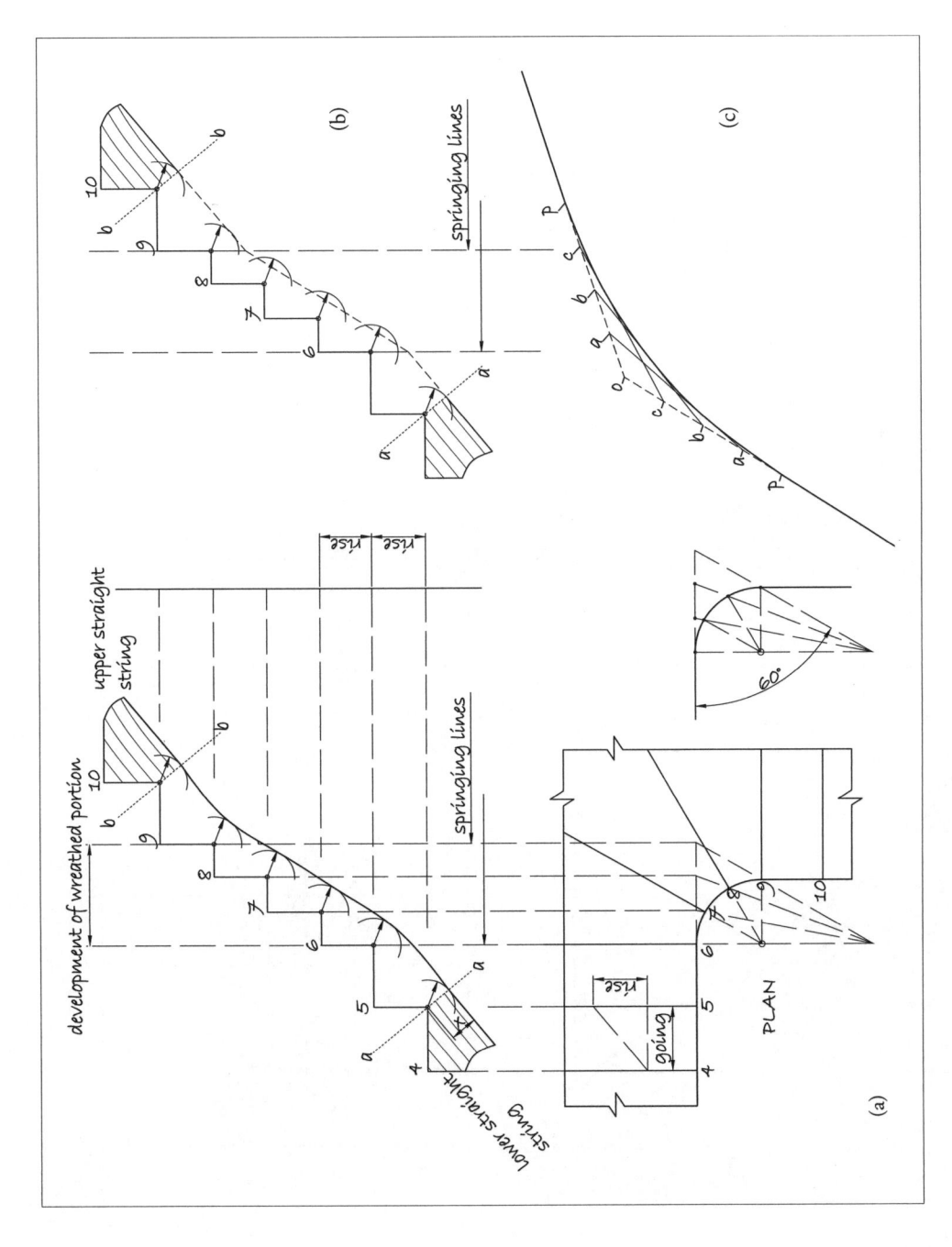

Fig. 10.3a
Fig. 10.3b
Fig. 10.3c

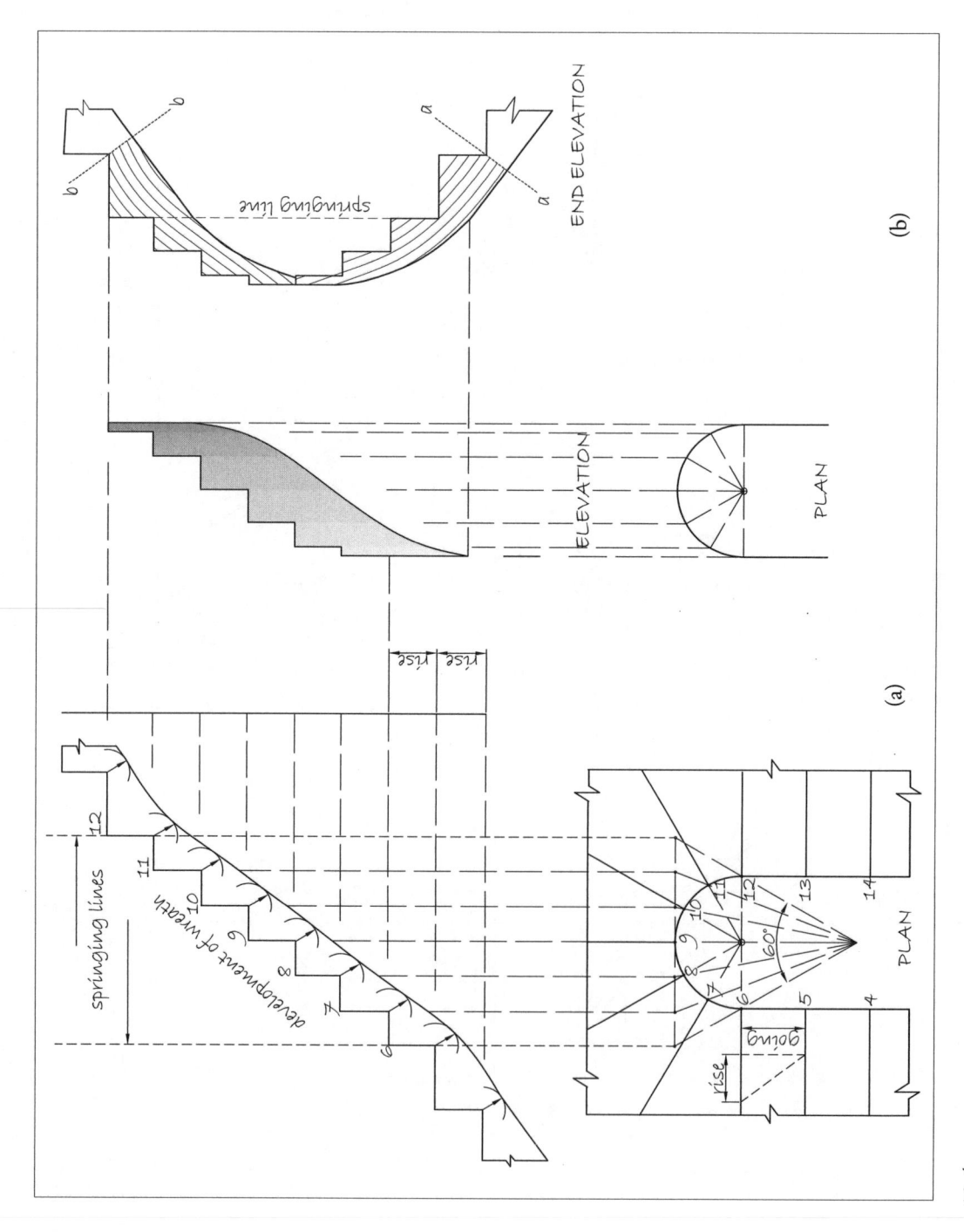

Fig. 10.4a
Fig. 10.4b

10 GEOMETRICAL STAIRS AND WREATHED STRINGS

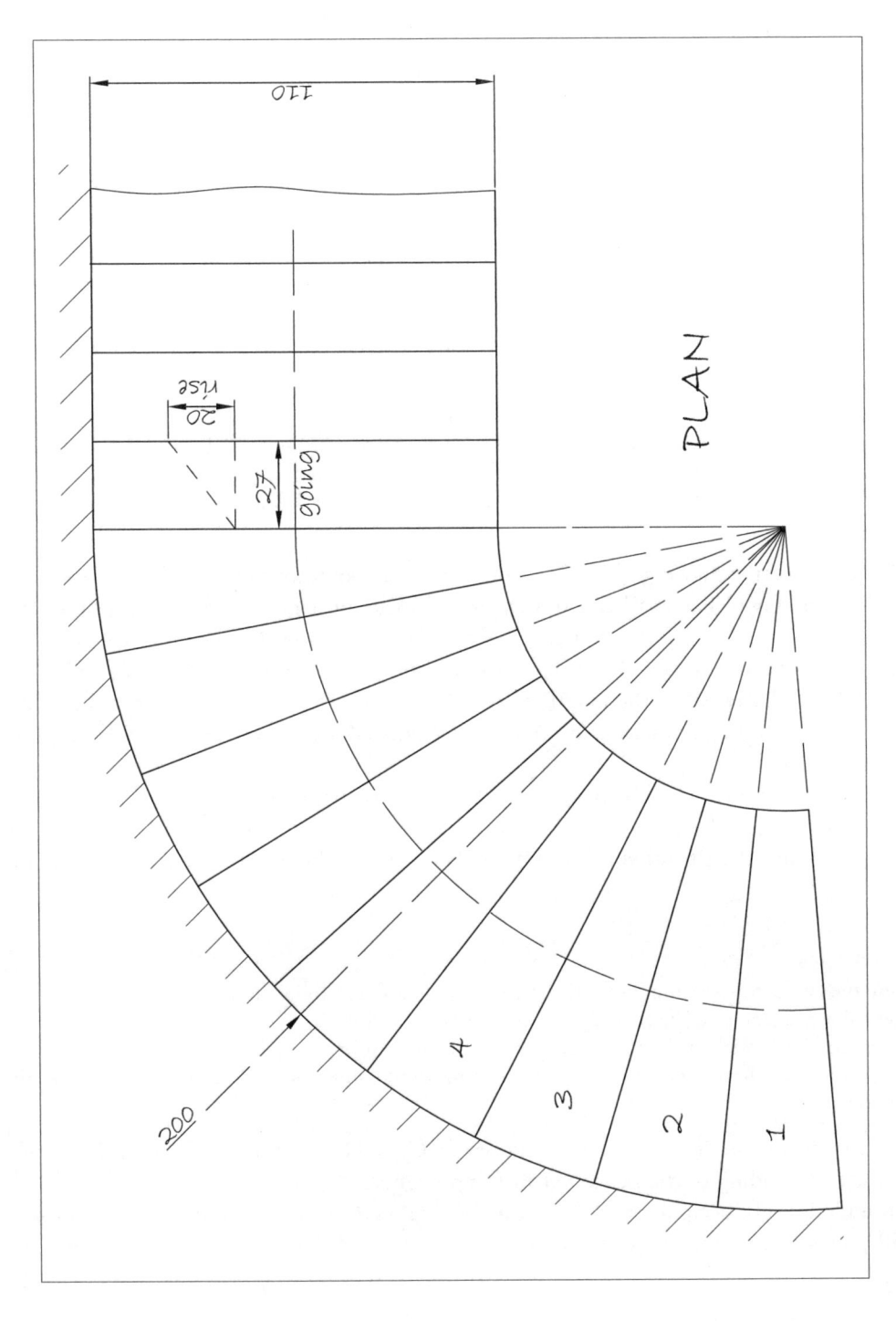

Fig. 10.5

11 LOCI

The locus (plural: loci) of a point is the path traced by a point. For example, a circle is the path traced by a point which is equidistant from the centre point (of the circle). A straight line is the path of a point which is heading in one direction only. The direction may be horizontal, vertical or follow any point of a nautical compass. So it may be said that by starting with a point and a set of predetermined conditions the locus of that point can be determined.

The first example, **Fig. 11.1**, deals with the route of a planned road which begins at **x** and continues as follows:

- The road passes within $1\frac{1}{2}$ kilometres to the east of point **A**.
- It then proceeds to a point which is two kilometres from point **B** and one kilometre from track **C-D**.
- From there, it continues to a point within half a kilometre north of **E** and terminates two kilometres from **E**.

As can be seen from the drawing, the route is determined to conform to the criteria set down.

The ellipse in **Fig. 11.2** has been drawn as the locus of a point. The focal points (see Chapter 6 on arches) **F1** and **F2** are needed for this construction. **F1-d** plus **d-F2** are equal to the length of the major axis. This may also be expressed as: **F1-d-F2**. **F1-a-F2**, **F1-b-F2** and **F1-c-F2** are also equal to the length of the major axis. The ellipse is therefore the locus of a point for which the cumulative distances to **F1** and **F2** are always equal to the length of the major axis.

The involute to a circle is shown in **Fig. 11.3**. This is the locus of a point which represents the unwinding of the circle. Imagine a piece of string wound taut around a cylinder for one revolution, and then to unwind it. The path traced by the end of the string forms the involute.

Method

- Divide the circle into a number of units, 12 is convenient, and index these.
- Draw a line tangential to the radius at **1** and mark one unit on it.
- Draw a line tangential to the radius at **2** and mark two units on it. Continue around the circle to **12**, which will have a tangential line measuring twelve units. Join all points to complete the involute.

Involutes to a pentagon and hexagon are shown in **Figs. 11.4** and **11.5**. The involutes here represent the unwinding of the perimeter of the polygons.

When a circle rolls along a straight line, as in **Fig. 11.6**, a given point on its circumference will plot a locus, which is called a *cycloid*. In the example given, two half-cycloids are drawn.

11 LOCI

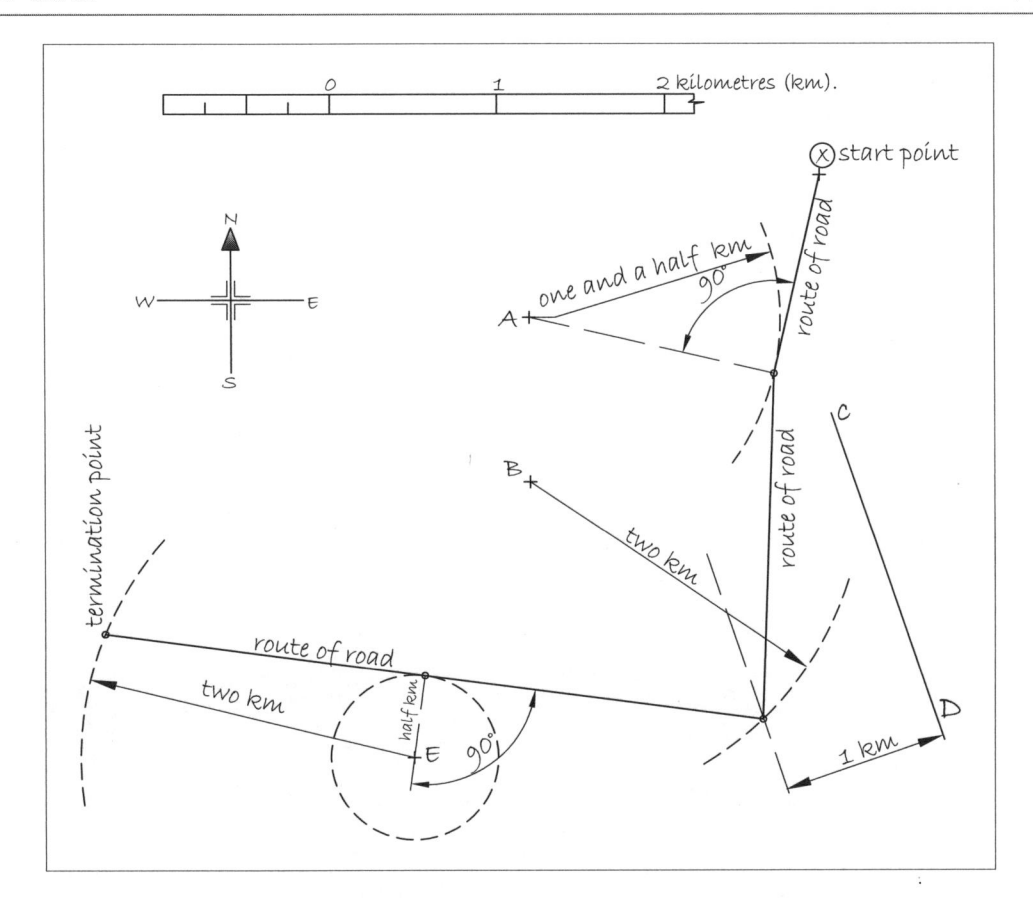

Fig. 11.1

Method

- Divide the circle circumference into a number of units (in this case, 12) and plot these along the baseline. On each of these 'stations', construct a copy of the circle in question.
- Construct a series of levels determined by the main circle as shown.
- As the circle rolls to the right, point **P** drops one level to the circle on station 1.
- As it continues to roll, point **P** drops to the circle on the next station each time.
- When one revolution is completed, a series of points will have been determined on each level. Join the points to complete the cycloid.

As the wheel of a bicycle revolves, the tyre valve plots a *cycloidal* curve for every revolution, something the cyclist is completely oblivious to.

Fig. 11.7 represents what is called a 'lazy tongs'. This principle has many applications including the extendable bathroom mirror. If it is split along the centre line (**c/l**), one half of it could represent the plan of sliding or folding doors, which are hinged at **a**. In this example, the loci of points **b** and **d** are found, as points **c** and **e** move along the centre line (**c/l**) at a constant rate, while point **a** remains a fixed pivot point.

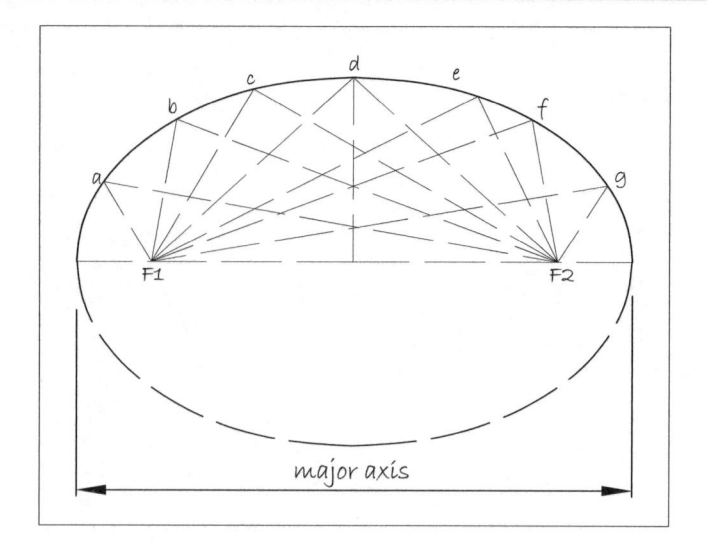

Fig. 11.2

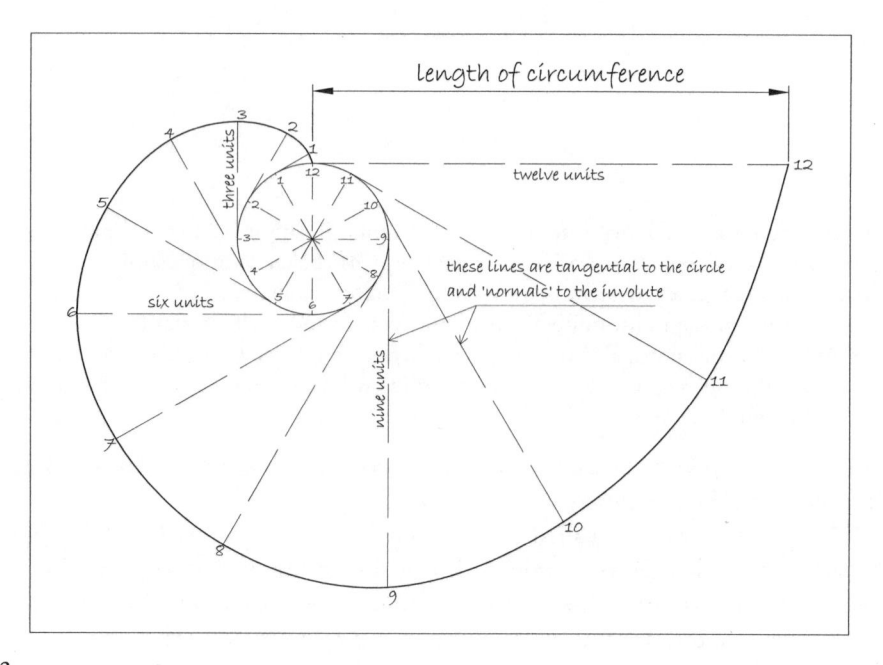

Fig. 11.3

11 LOCI

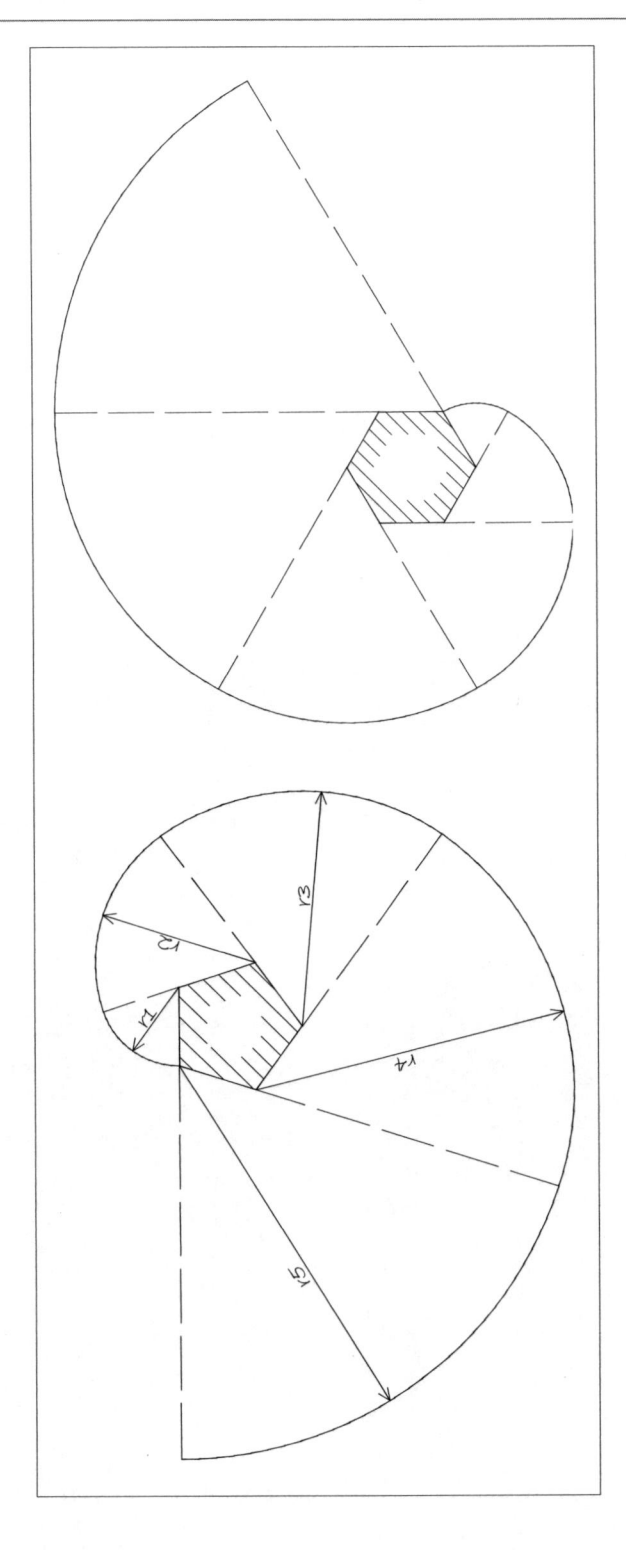

Fig. 11.4
Fig. 11.5

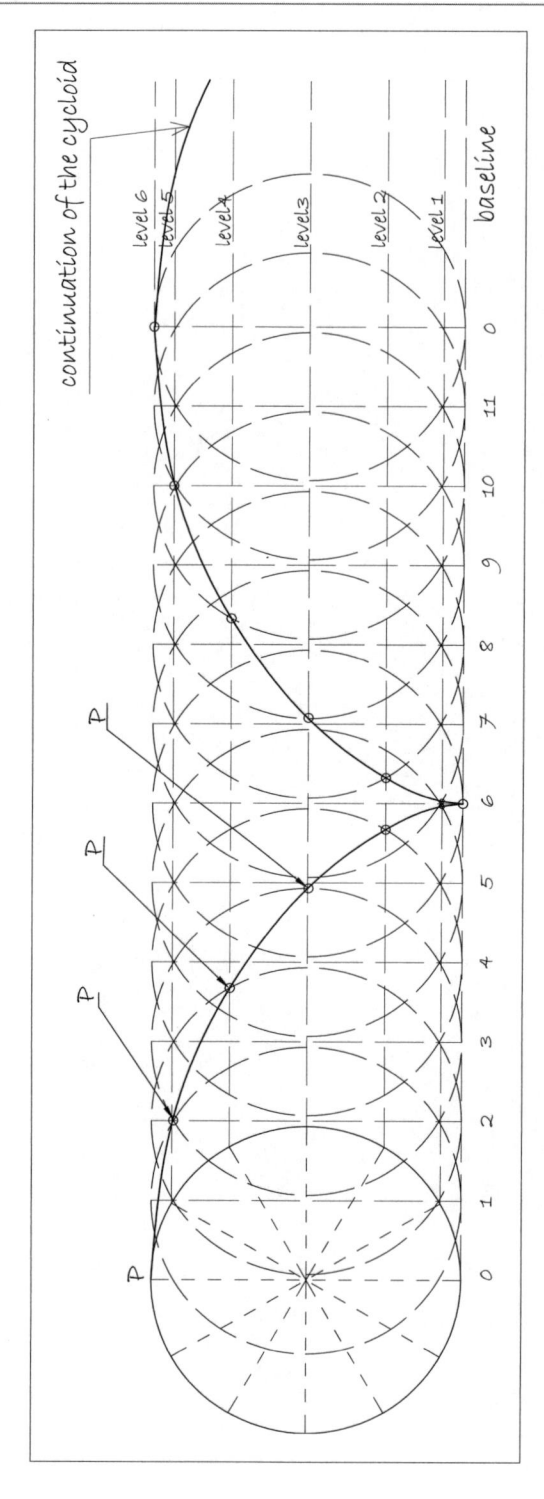

Fig. 11.6

11 LOCI

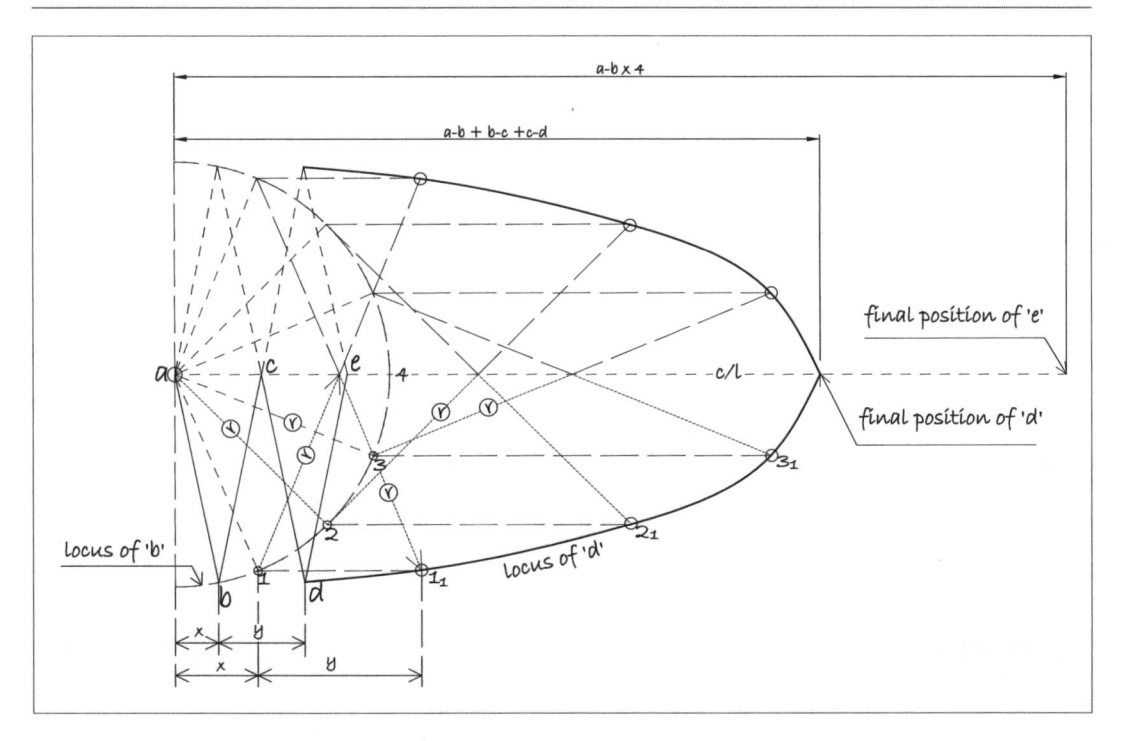

Fig. 11.7

For the lower half, below the centre line (**c/l**):

Method 1

- Pick points **1** to **4** on the locus of **b** (which is a quadrant) and construct horizontal lines from these points.
- From **1**, strike the radius **a-b** to the centre line and back to the horizontal projection from **1**. This gives a point on the curve.
- Work similarly from points **2** and **3**, giving further points on the locus as well as the final position of **d**, which is given.

If the mechanism of the movement is studied, it will be noted that dimensions **x** and **y** have a ratio of 1:2; the horizontal dimension from **a** to **b** is one unit and the horizontal dimension from **b** to **d** is two units. This will be the case for every position of the movement, and this principle is employed in method 2.

Method 2

- Draw the quadrant **a-4**.
- Divide the arc, as shown, into **1, 2, 3** and **4**.

- From **1**, **2** and **3**, draw horizontal lines.
- Measure the horizontal distance from **1** to **a** and make **1-1¹** twice that.
- Measure the horizontal distance from **2** to **a** and make **2-2¹** twice that.
- Work similarly with point **3** to give points on the locus.

The drawing in **Fig. 11.8** (on the left) represents the jib of a crane with a load **l** suspended. The jib pivots around **p** and completes a full quadrant in movement. The task is to determine the locus of **l** for the movement of the crane.

Method

- Divide the quadrant into a convenient number of parts (in this case, eight are chosen), and index **a**, **b**, **c**, etc. as shown
- From each position on the quadrant, suspend the load **l**. Join all points to give the locus.

In **Fig. 11.9** (on the right) the same jib is used, but this time as it pivots it also moves horizontally at a constant rate from **p** to **8**. The task for this example is to plot the locus of the load **l** and also the locus of the top of the jib.

Method

- As the quadrant is already divided into eight parts, also divide the distance from **p** to **8** into eight parts.
- Construct horizontal lines from all points **b** to **j** on the quadrant.
- Strike measurement **p-a** from **1**, **2**, **3**, etc., to intersect with the counterparts extended from **b**, **c**, **d** and so on.
- These intersections give points on the locus for the top of the jib.

An explanation of this: as **p** moves horizontally to **1**, the jib raises to level **b**; and when **p** is in position **2**, the jib raises to level **c** and it continues thus to level **j**. Once the locus of the jib top is determined, suspend the load **l** from it at various points, to find the locus of **l**.

    Fig. 11.10 shows a slide mechanism whereby **c** moves along the horizontal line **a-b**, as the slide moves through the fixed pin at **P**. The locus of point **d** is plotted as the movement occurs. **Fig. 11.11** shows a cabinet with a drop leaf, which is held open with a stay. The locus of the upper end of the stay is plotted as shown. For clarity, the stay is omitted from the working drawing.

    Figs. 11.12a and **11.12b** represent a stay mechanism for the sash of a window. In **Fig. 11.12a** the sash is given in the open position and the task is to plot the locus of points **6** and **7** as the sash moves to the closed position.

    The mechanism is made up of four bars: **0-1-2**, **1-3-5**, **3-4** and **6-4-2-7**, which have pivots as indicated. The finished drawing is in **Fig. 11.12b** and was done as follows:

Method

- Five positions of the movement are shown at **A**, **B**, **C**, **D** and **E**; **A** being the starting position and **E** being the final position.
- As **1** and **2** move to position **B**, the length of bar (**1-5**) is struck off from **1** on **B** to line **0-F**, giving the new position of **5**.

11 LOCI

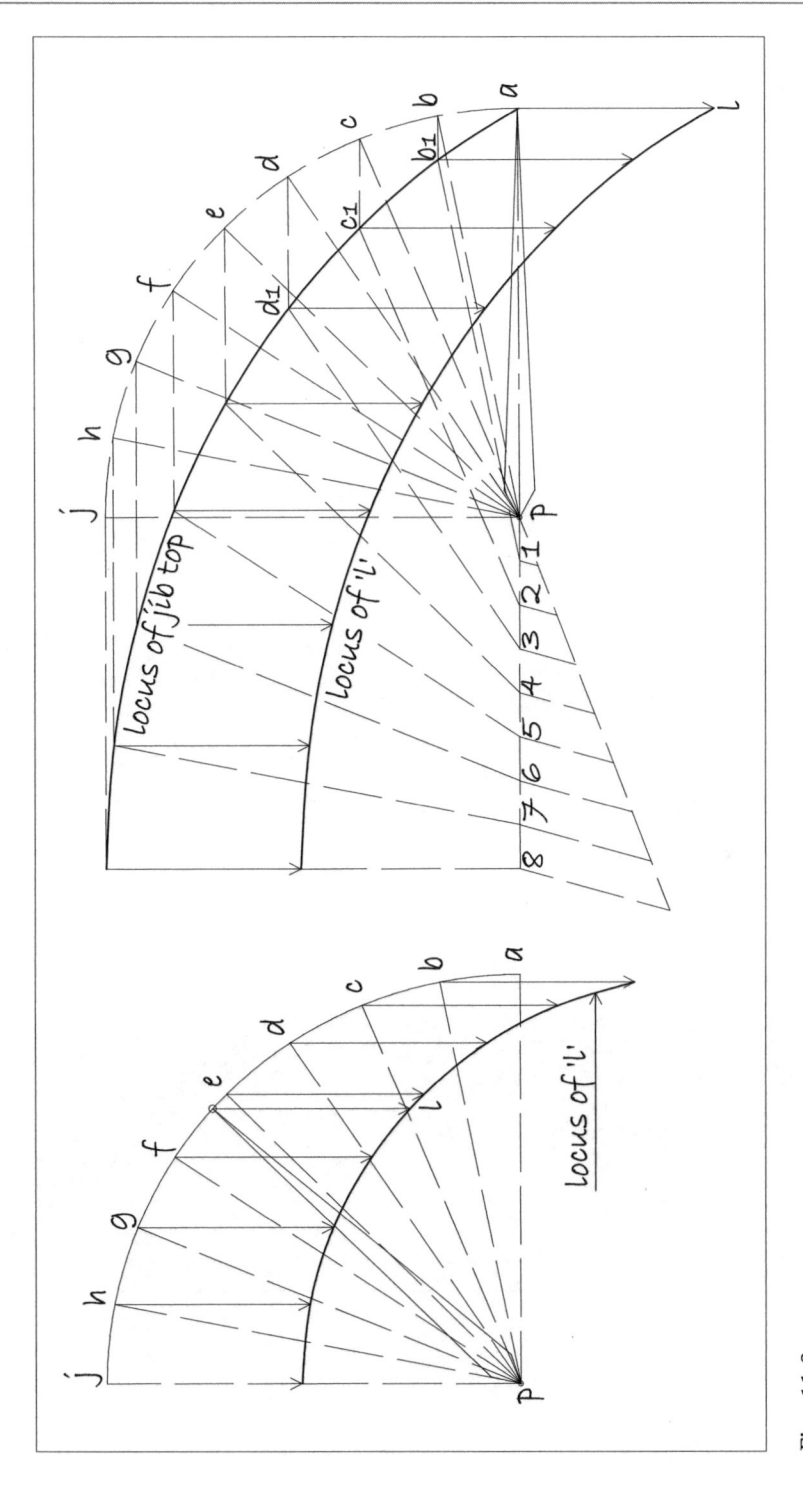

Fig. 11.8
Fig. 11.9

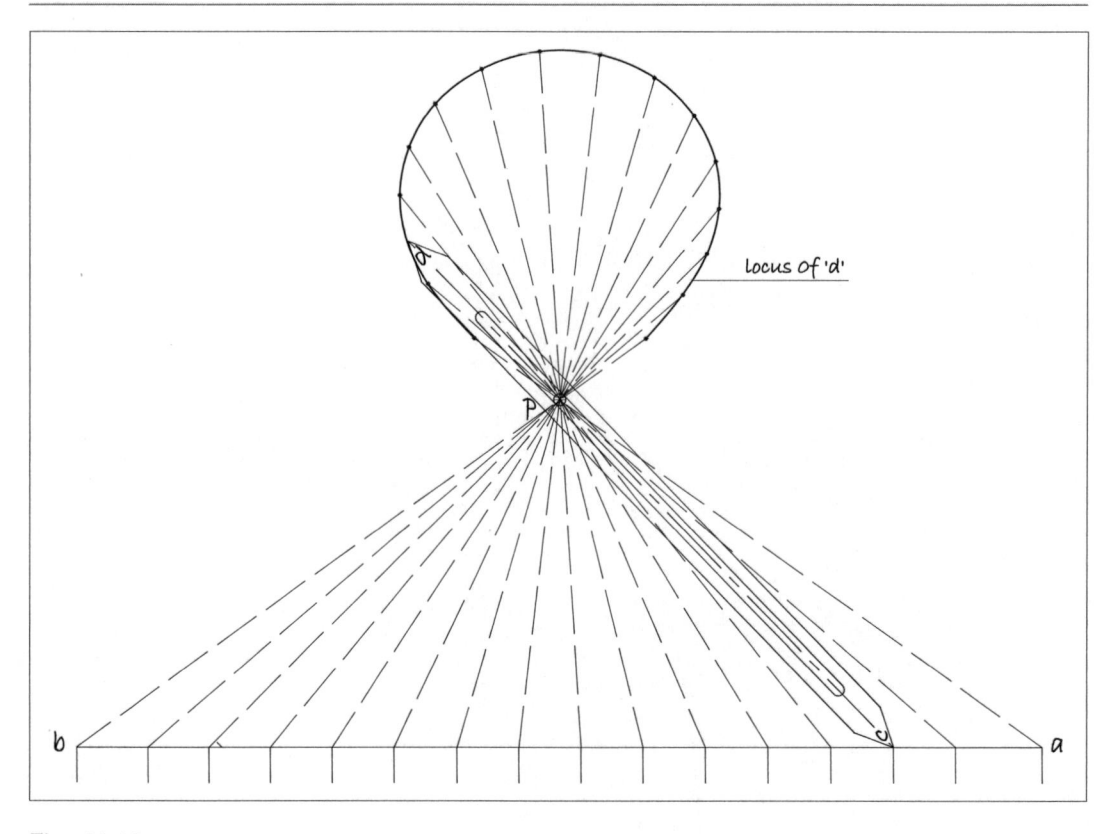

Fig. 11.10

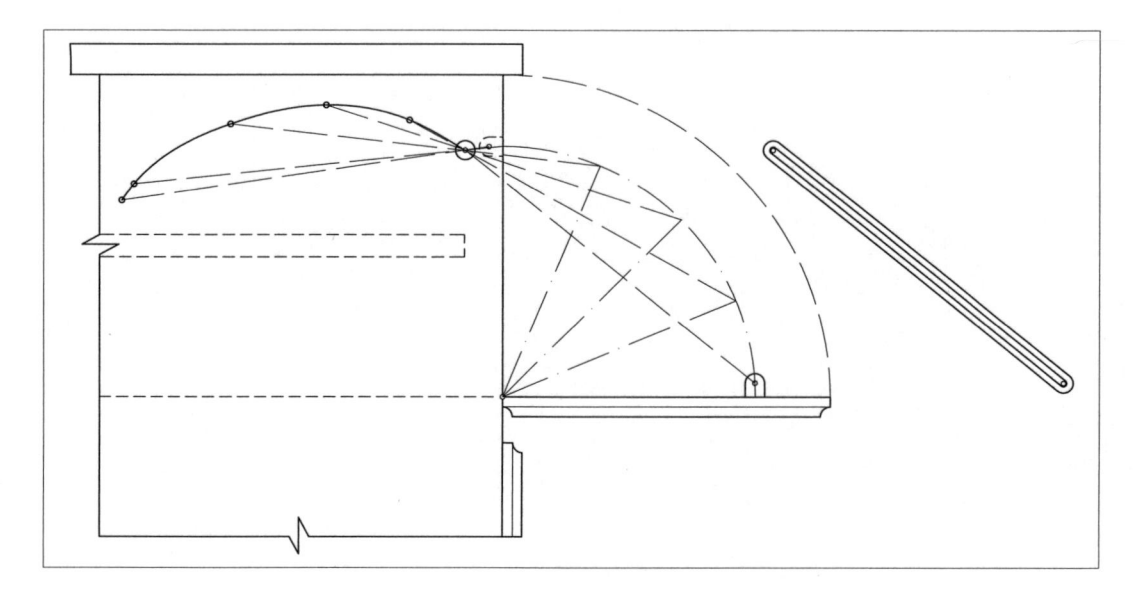

Fig. 11.11

11 LOCI

MOVEMENT CRITERIA:
- 5 is a slide which remains on line O-F at all times
- O is a fixed pivot
- pivots **1, 2, 3,** and **4** are moving pivots

TASK: plot the locus of points 6 and 7.

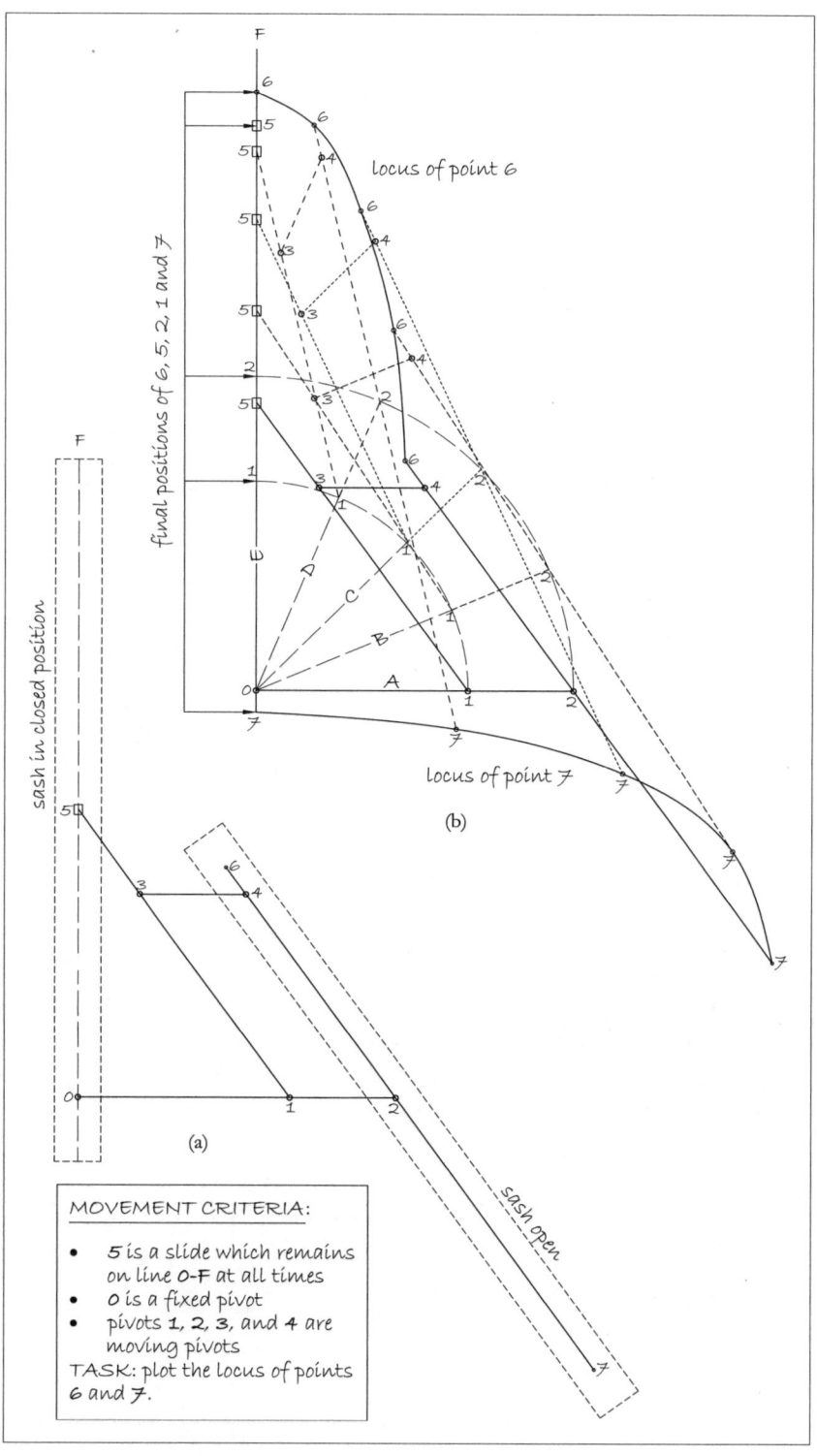

Fig. 11.12a
Fig. 11.12b

- Through the new position of **2** on **B**, draw a line parallel with **1-5** at position **B** and establish (from **2** on **B**) the constant positions of **6, 4** and **7** on this line. (Note: Bars **1-5** and **6-7** will remain parallel at all positions.)
- Continue to positions **C, D** and **E** with the same process and plot the paths of points **6** and **7**.

Exercise

1. The plan of folding doors is represented in the drawing **Fig. 11.13**.
    - At point **a,** the doors are fixed and hinged.
    - The overhead pivots move along line **a-f** at a constant rate.
    - The doors are hinged to each other as shown.
    - Reproduce **Fig. 11.13** and plot the locus of point **e** as the doors move to a closed position.

2. In **Fig. 11.14** a cabinet with a drop lid is shown. The cabinet stay, which supports the lid in the open position, is shown to the right. The pivot (**P**) on the stay collapses downwards during the movement. Reproduce **Fig. 11.14** and show the position of the stay when the lid is in the closed position.

Note: For the exercise it is not necessary to draw the stay. Use the centre line only, as shown in the drawing.

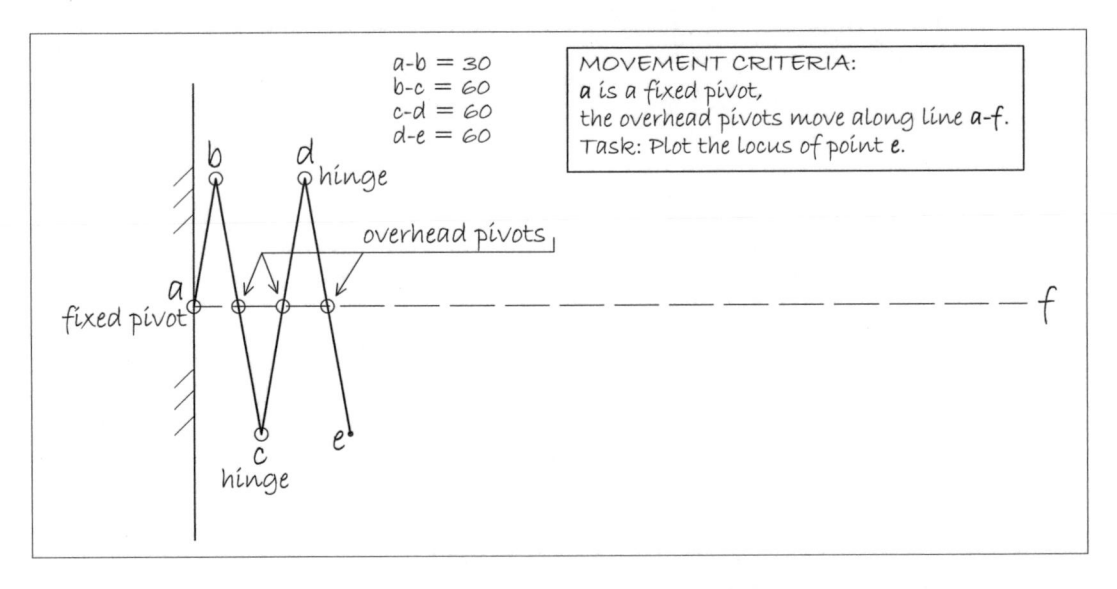

Fig. 11.13

11 LOCI

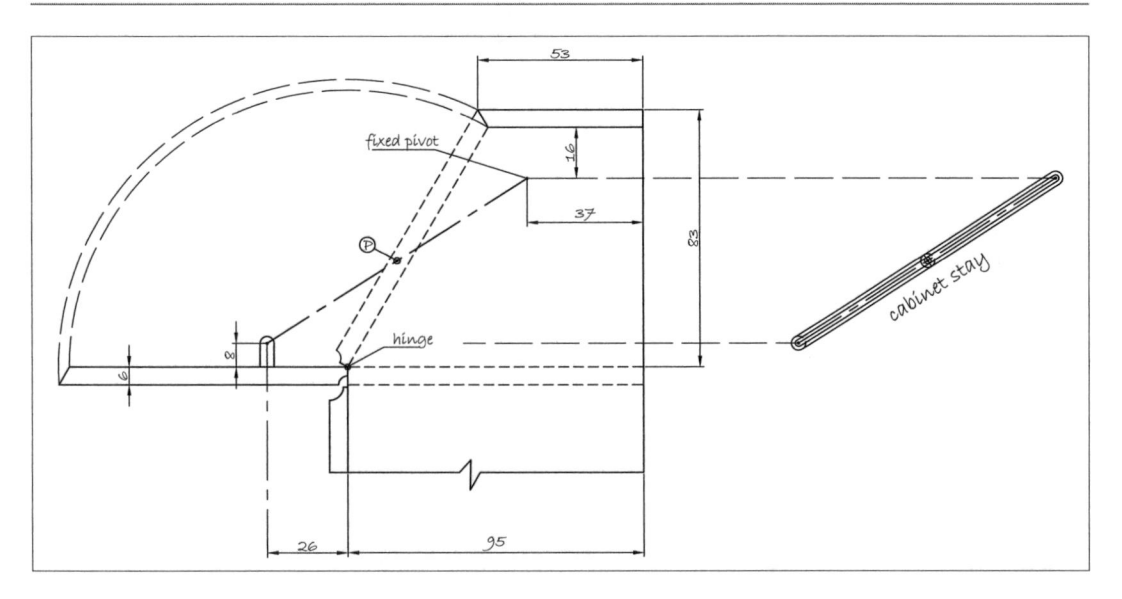

Fig. 11.14

12 FRAMED LOUVERS

SECTION I: LOUVERS IN STRAIGHT FRAMES

Louvers are thin, sloping members which are fitted into frames. They are used in turrets, doors, windows and generally in situations which require airflow and light through the frame, but prevent ingress of rain. The louvers may be made of wood, glass, metal and slate.

The frames may be:

- square-off the louvers
- simply or obliquely inclined from the louvers
- curved.

In all cases, the geometry required is to determine the true shape of the face and edge of the louver and to determine the trench bevel in the frame.

In **Fig. 12.1** a louver frame in its simplest form is shown. In the section, the louvers are pitched in this case at 45°. The true width of the louver is shown in the section, while the true length is shown in the elevation. All of the cuts required to make the louver are square-off (at 90°).

In the elevation of **Fig. 12.2** the louver frame on the left shows the louver square-off, as in the example in **Fig. 12.1**, while the frame on the right is inclined or splayed. It may be said, therefore, that the louver is simply inclined to the frame. In this situation the louver and the frame will have splayed cuts.

To find the true shape of the face of the louver, which will have the required bevel on it:

Method

- On the section, pivot the face of the louver (using **p** as the pivot) to a vertical position, resulting in **a** becoming **a1**. As this action occurs in the section, line **a** moves up to **a1** in the elevation.
- In the elevation, connect line **a1** to line **p** to reveal the true shape of the face of the louver (outlined in a dotted line).

It is important to note here that the pivot point **p** does not move in either view; the pivot is **p** in the section and it is line **p** in the elevation.

The edge of the louver is done in a similar way. It is dropped to a vertical position in the section with **b** becoming **b1**, resulting in line **b1** in the elevation. Line **b1** is connected to line **p**, as shown, to reveal the true shape and edge bevel. The trench bevel to the frame is shown to the right of the elevation. The frame is 'flattened' to reveal the trench.

12 FRAMED LOUVERS

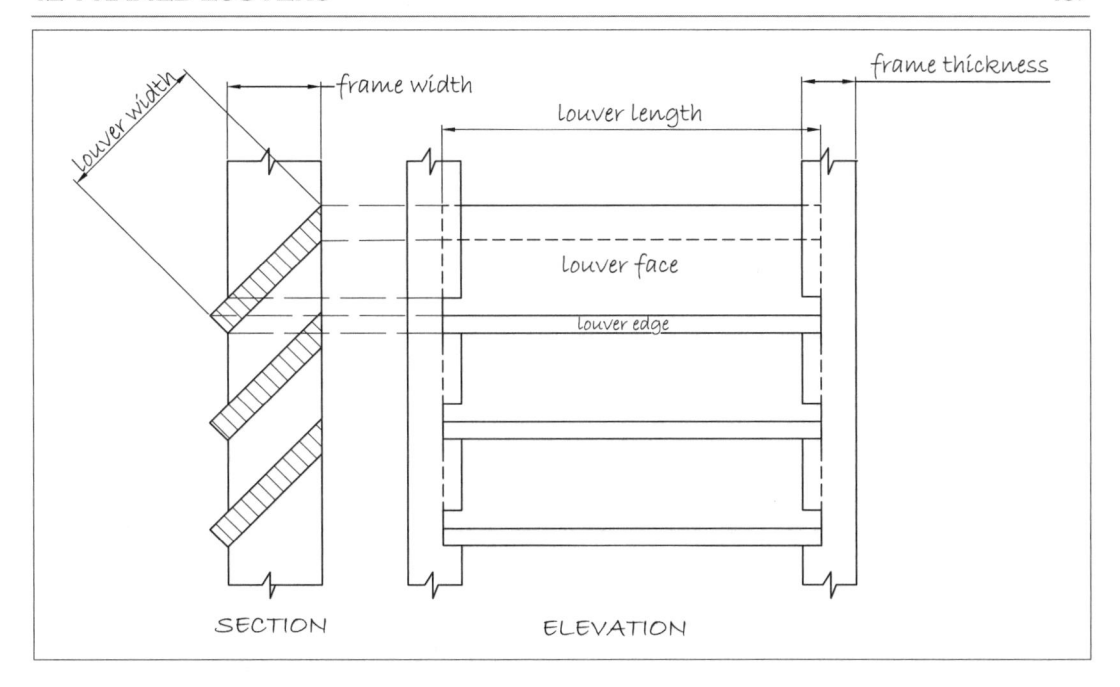

[Figure showing SECTION and ELEVATION views of a framed louver, with labels: louver width, frame width, louver length, frame thickness, louver face, louver edge]

Fig. 12.1

Method

- Set up a line parallel with the slope of the frame and set off the frame width (**w**) as shown.
- In the elevation, project the trench points at 90° to the frame onto the 'flattened' frame, ensuring that the relevant points are projected to the relevant edges.
- Join the points to get the trench bevel.

In **Fig. 12.3** the frame section into which the louver is fitted is bevelled as shown. The elevation of the trench and louver intersection is more difficult to draw than in the previous examples. The dotted trench line **a-b** is extended to meet the projection of line **p** from the end section to find the extreme left corner of the louver. The method of obtaining the bevels is the same as for the previous example.

Note: The dimension **w** is used when determining the trench bevel.

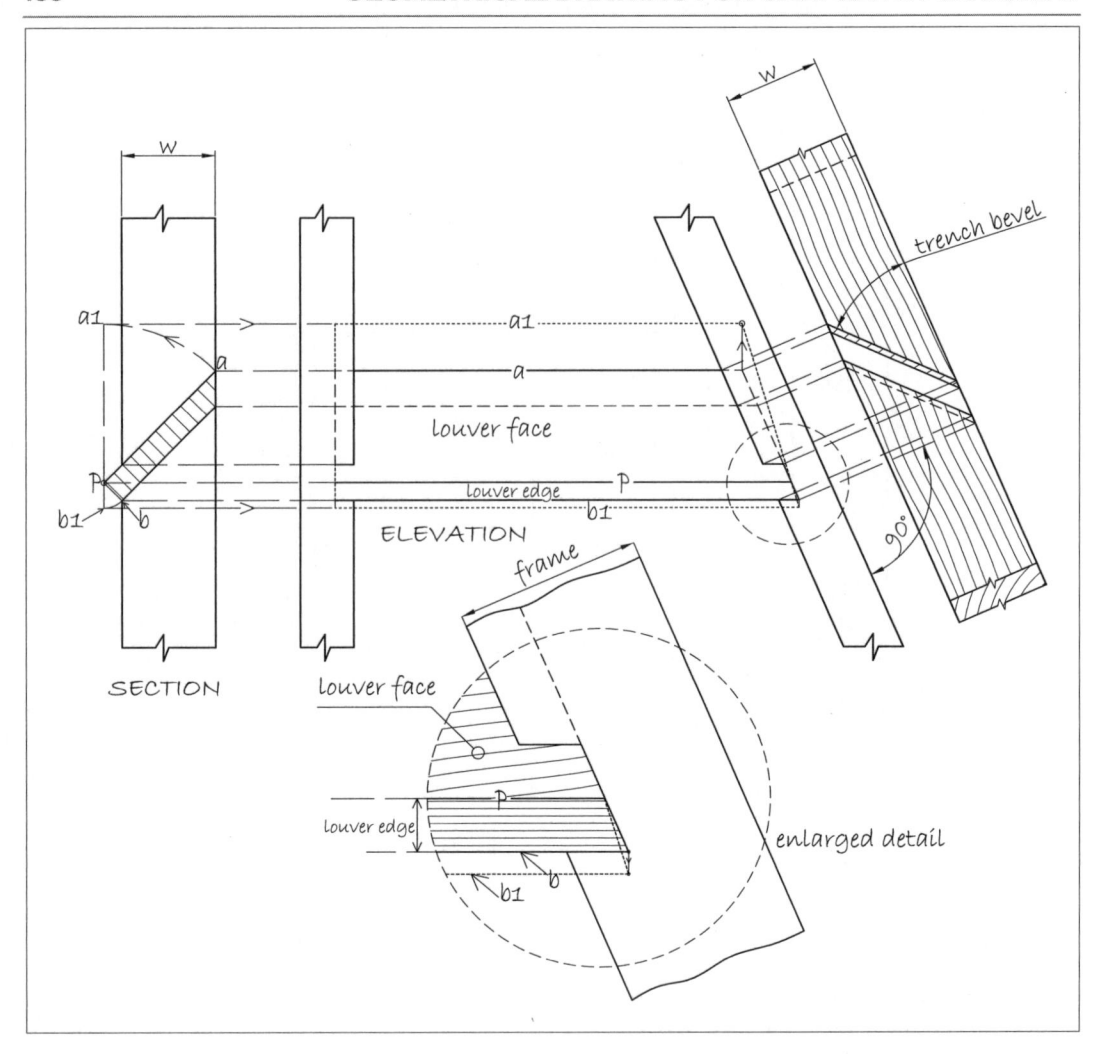

Fig. 12.2

Fig. 12.4 shows a louver, which is obliquely inclined to the frame. The structure is a pyramid turret where louvers are trenched into the hips as shown. The development of the louver is done using two methods.

Method 1
- Off the section **c-c**, lines are drawn at 90° to the louver face and dimensions **a**, **b**, and **d** (taken from the plan) are set off to find the true shape of the louver face and edge.

12 FRAMED LOUVERS

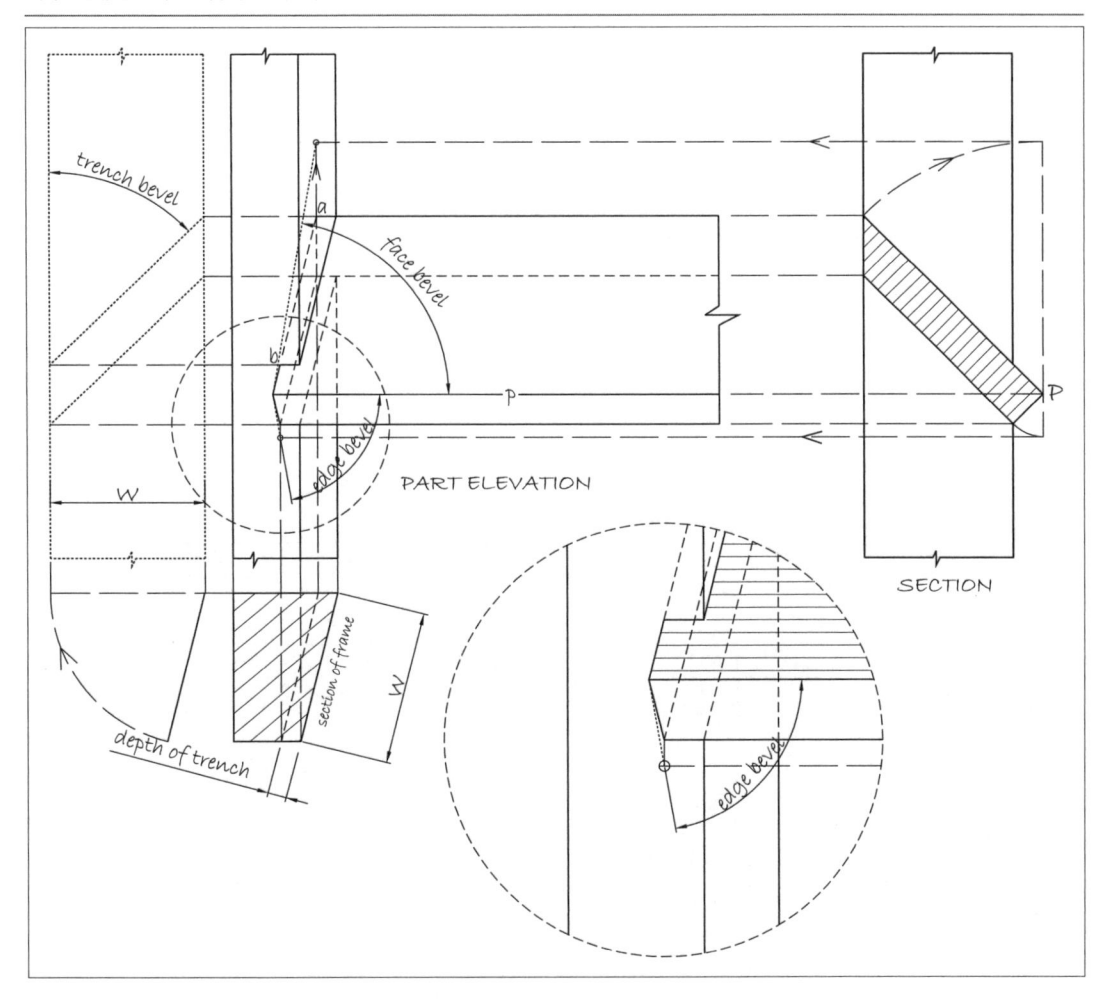

Fig. 12.3

Method 2

- The louver face is pivoted to a horizontal position in the section and projected to the plan, where the pivoting edge stays put and the outer edge moves out to the projection line.

This method is similar to the previous examples, however, in this case the louver is being brought to a horizontal plane. In the previous examples the louver was pivoted to a vertical plane. Note that the edge of the louver was swung to the same plane as the louver face, thus enabling the edge bevel development in both methods. To see the true trench bevel, an auxiliary elevation of the hip is drawn, using the relevant heights indicated.

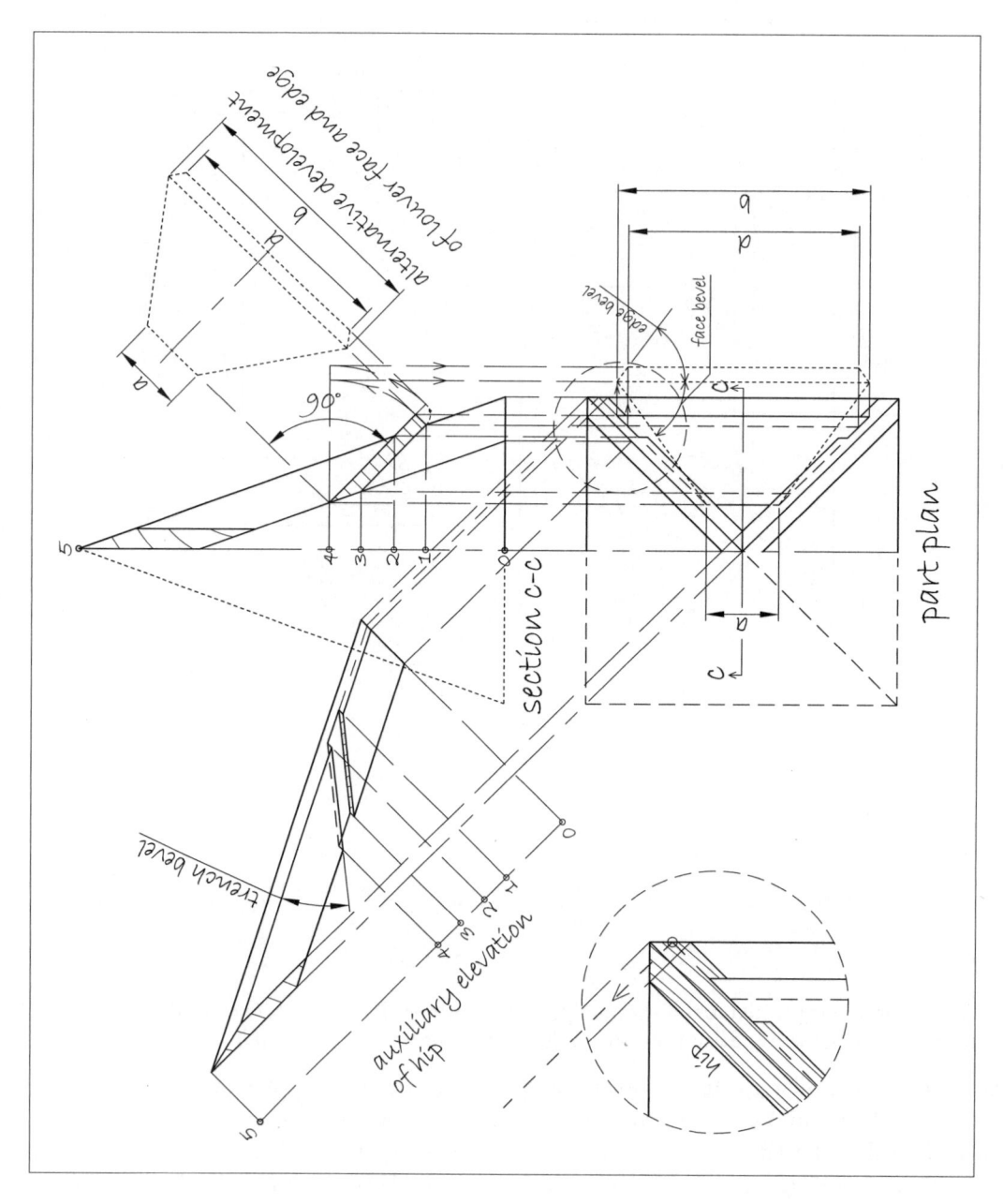

Fig. 12.4

12 FRAMED LOUVERS

SECTION II: LOUVERS IN A CURVED FRAME

Where louvers are fitted to a curved frame, the face of the louver, where it enters the trench, will have a curved outline, consequently, the trench will have to be undercut to accommodate the louver correctly. Note that in cases where the curvature of the frame is not extreme it may be possible to get away with a straight seating to the trench.

The geometry for obtaining the curve on the louver is shown in **Fig. 12.5**; remember that the principle of pivoting the louver used in the previous examples in Section I still applies.

Method

- In the section, divide the louver face into a number of parts as **p, a, b, c** and **d** and project to the elevation.
- The louver face is pivoted upwards in the end section, and the 'new' heights are projected to the elevation, where they intersect with their counterparts, which move vertically to meet them (see the enlarged detail on the bottom drawing).

The edge bevel is found in the same way as described in Section I.

Marking the trench lines on the frame may be done from a full-size drawing, although this can be laborious if many louvers are involved. **Fig. 12.6** shows a drawing of a device which speeds up the marking of the trenches and the method illustrated is very accurate.

Method

- A piece of wood is prepared with the pitch of the louver in its section. This is called a spile batten.
- The batten is positioned on the frame to the louver positions, and a scriber draws the trench line along the inner edge of the frame.

This method is simplicity itself and you may wonder after seeing it: Why didn't I think of that? The assembled frame may also be placed in a rectangular box where the louver positions are marked along the side walls, from where the spile batten picks up the louver positions. This is convenient when several frames have to be made up.

Exercise

Reproduce the frame in **Fig. 12.7** and determine the face and edge bevels for both ends of the louver.

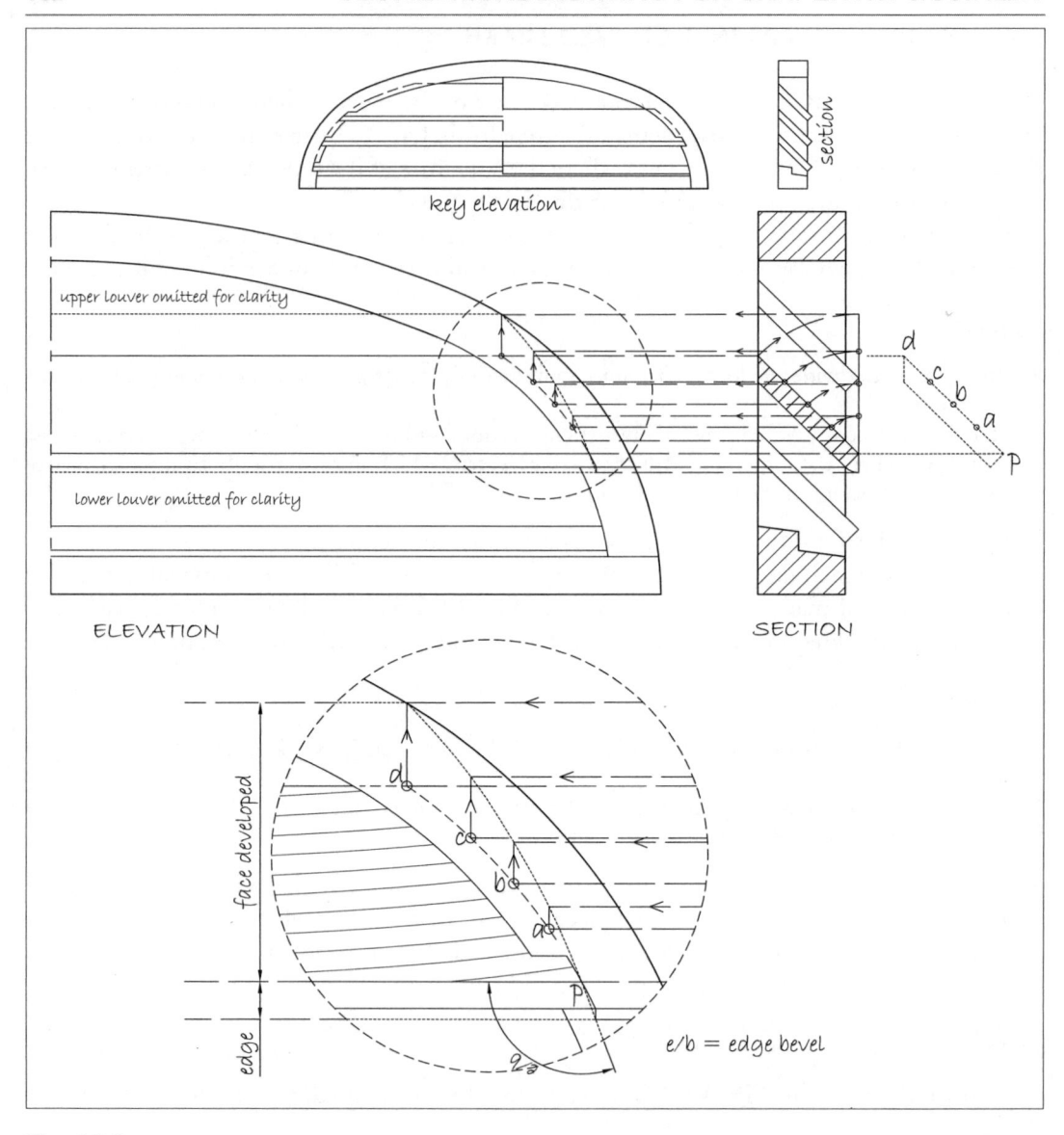

Fig. 12.5

12 FRAMED LOUVERS

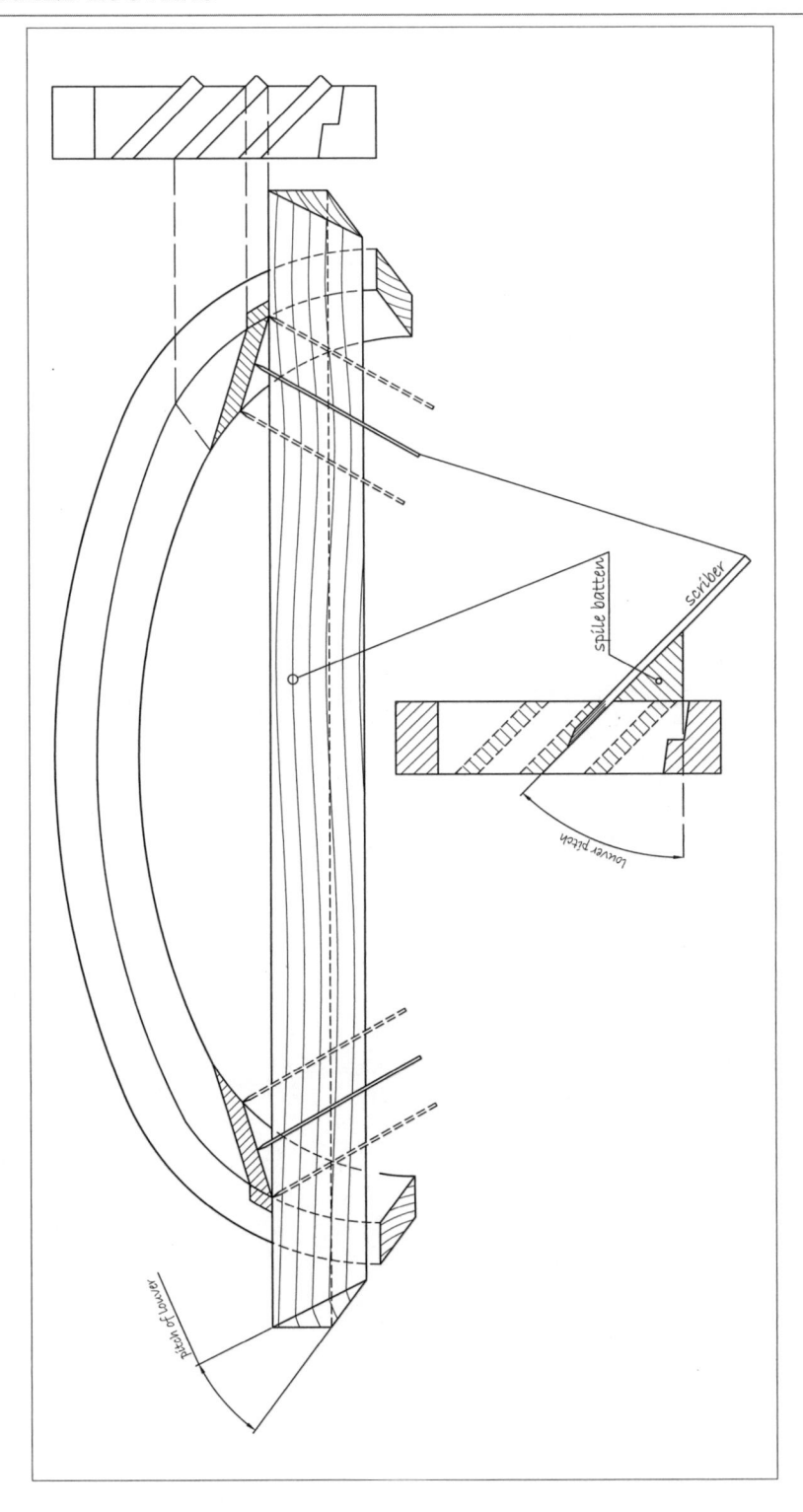

Fig. 12.6

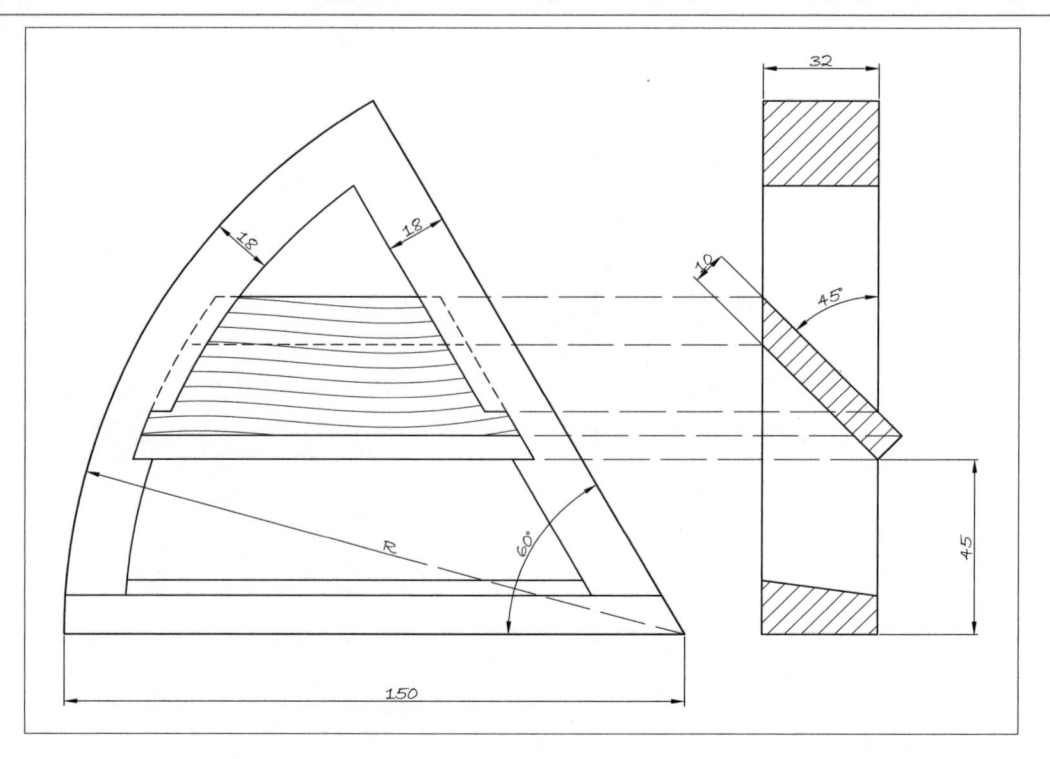

Fig. 12.7

13 MOULDINGS

SECTION I: INTERSECTION OF MOULDINGS

This section deals with mouldings on the same plane (the same flat surface) and how they intersect with each other on different mitre lines.

In **Fig. 13.1a** four mouldings, **F**, **G**, **H** and **J** are shown, all a continuation of the given moulding. The mitre line **1-2**, between **F** and **G**, is at 45°, bisecting the corner angle of 90°; the section of both mouldings is therefore identical.

The mitre line **3-4**, between **G** and **H** does not bisect the corner angle, therefore, moulding **H** will have a different section. This section is found as follows:

Method

- Set off the thickness, **a**, **b**, **c**, **d** and **e** of the given moulding at 90° to moulding **H** as shown.
- Project the lines of moulding **G** from the mitre line **3-4** to intersect with the thicknesses and join the relevant points.
- The true section of **H** is shown superimposed and hatched.

The mitre line **5-6** bisects the corner angle, therefore, moulding **J** is identical to **H**.

In **Fig. 13.1b** the true section of the mitre on **5-6** is found by setting off the thickness, **a**, **b**, **c**, **d** and **e** at 90° to the mitre line as shown. It may be deduced, therefore, that when the mitre line bisects the corner angle of the return moulding, both moulding sections remain the same. Note: To avoid confusion over the width and thickness of mouldings, it will be assumed that the elevational dimensions show the width, while the dimension that protrudes from the wall surface represents the thickness.

In **Fig. 13.2** a section of architrave is given at **A**. The section is mitred at 45° to a horizontal return moulding (which will be identical in section), which in turn is mitred to an obviously wider moulding. The resulting section of the wider moulding is given at **B**. The development of the mitre is shown at **C**.

In all cases, the thickness of the mouldings is constant and represented by the dimensions **a** to **f**.

In **Fig. 13.3** a curved moulding is shown intersecting with straight mouldings. Mouldings **G** and **J** are identical in section; this results in a curved mitre line at **c-d**. The true section of **J** is determined on any radial line from the centre mark **c/m**. To determine the mitre line **c-d**:

Method

- Set up the section of moulding **G** on the radial line from the centre mark **c/m**.
- From **c/m**, draw arcs through the points **a** to **b** to intersect with their counterparts on moulding **G**.
- Join the intersections to give the curved mitre.

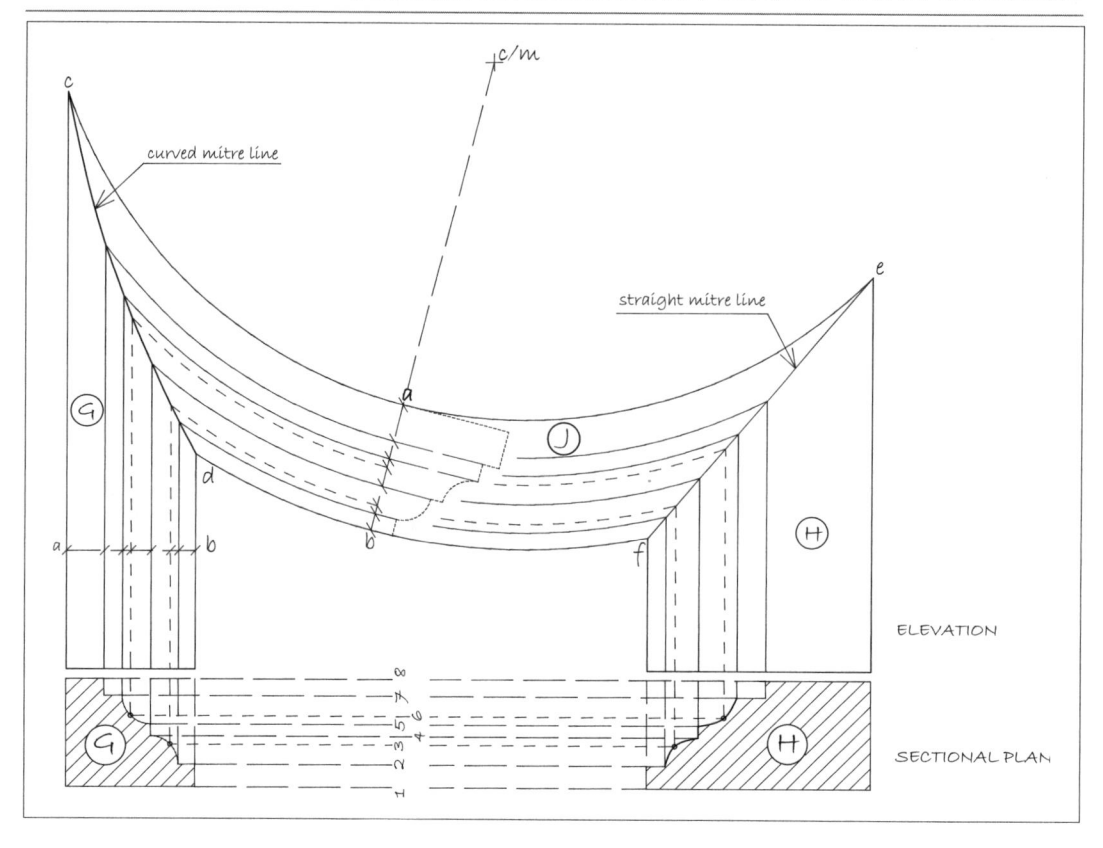

Fig. 13.3

In **Fig. 13.5** a further exploration of how curved and straight mouldings intersect is given. Here are some rules governing the intersection of straight and curved mouldings:

- A straight and a curved moulding of identical sections intersect at a curved mitre line.
- Curved mouldings of equal section but of different radii intersect at a curved mitre line.
- Curved mouldings of equal section and equal radii intersect at a straight mitre line.

Fig. 13.6 shows a gothic tracery window, in which intersecting curved mouldings are typical of this style of work. The enlarged detail shows the moulded parts divided equally, thus determining the curved mitre line. This of course means that all mouldings have the same sectional profile.

13 MOULDINGS

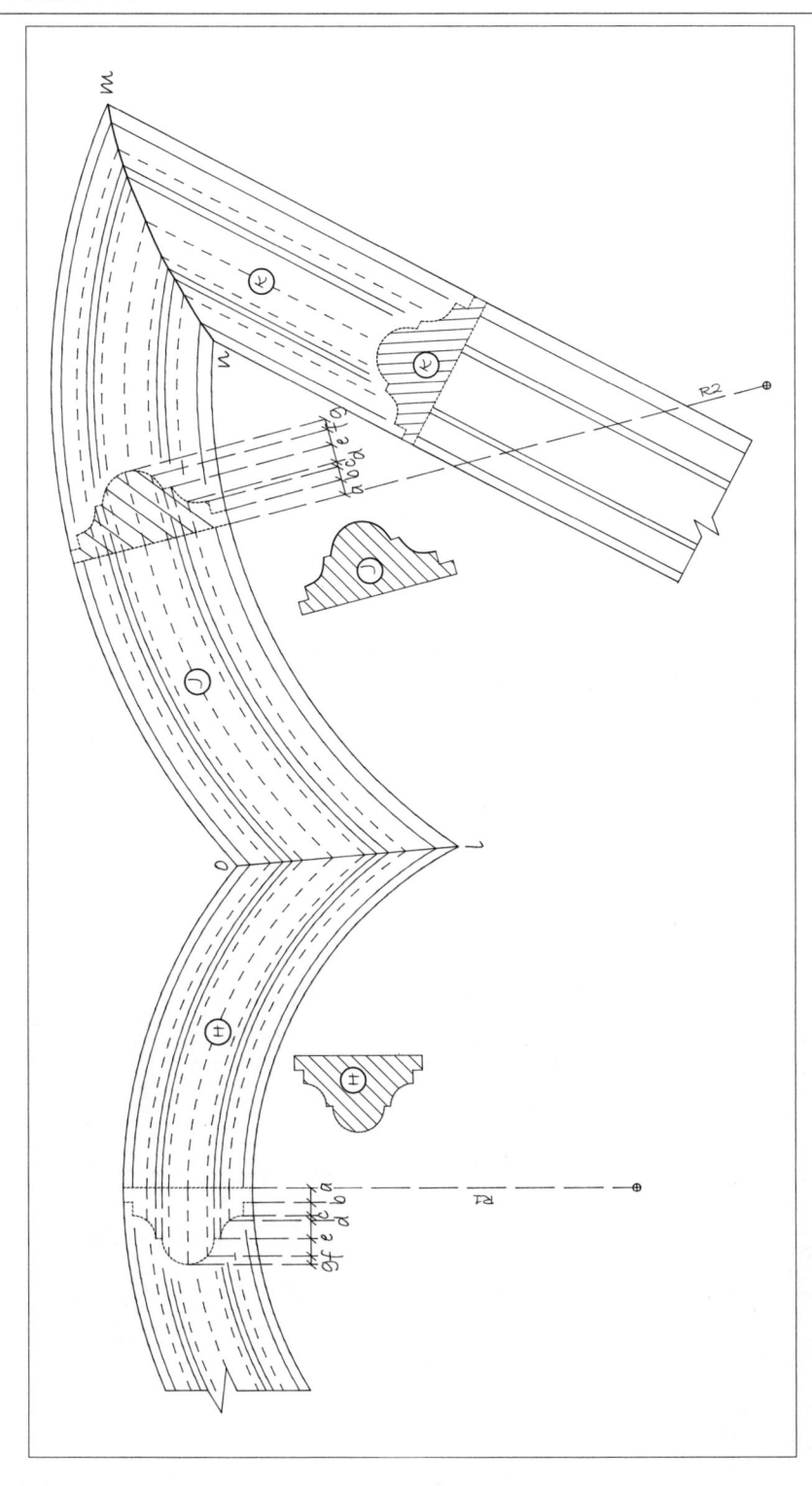

Fig. 13.4

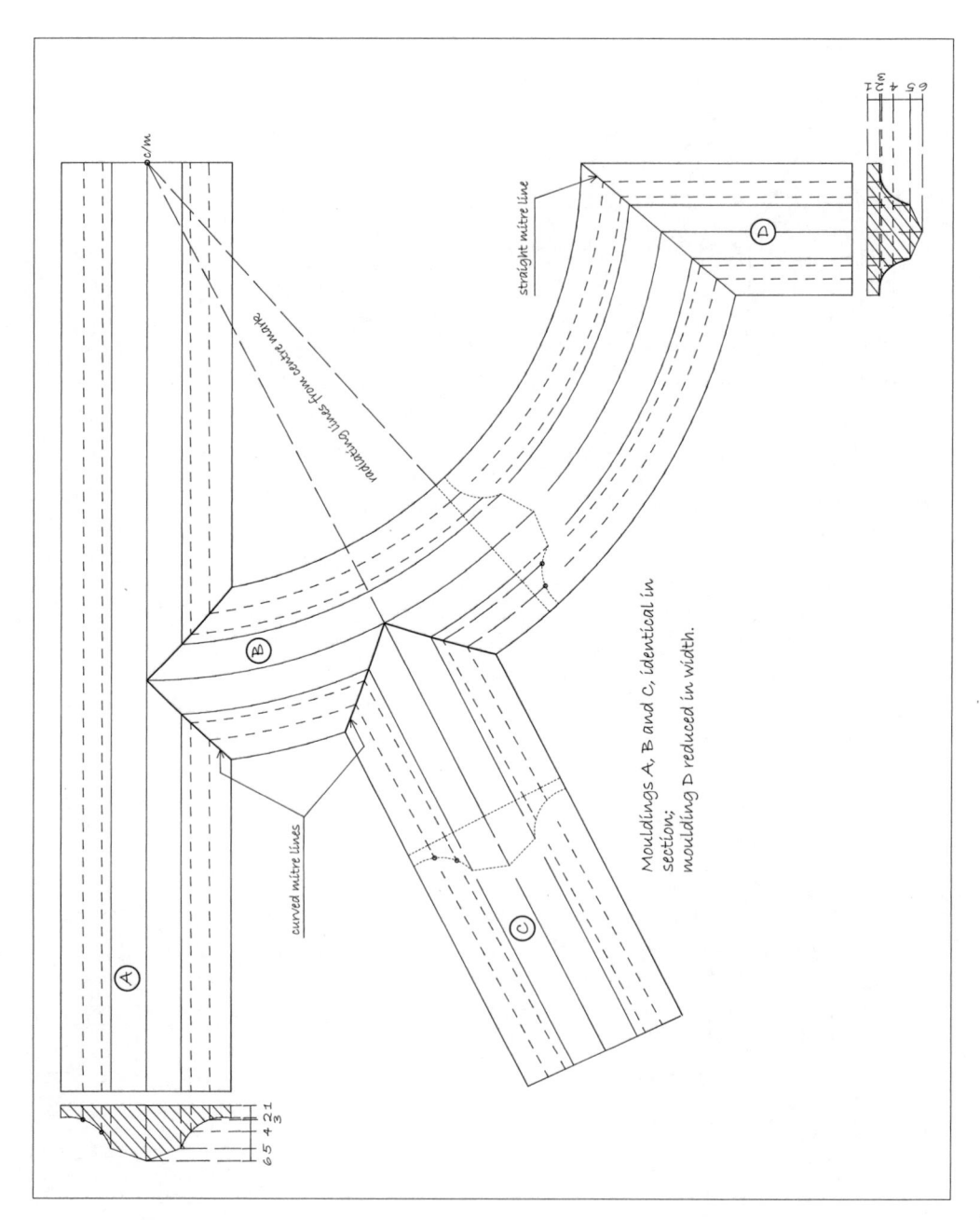

Fig. 13.5

Mouldings A, B and C, identical in section; moulding D reduced in width.

13 MOULDINGS

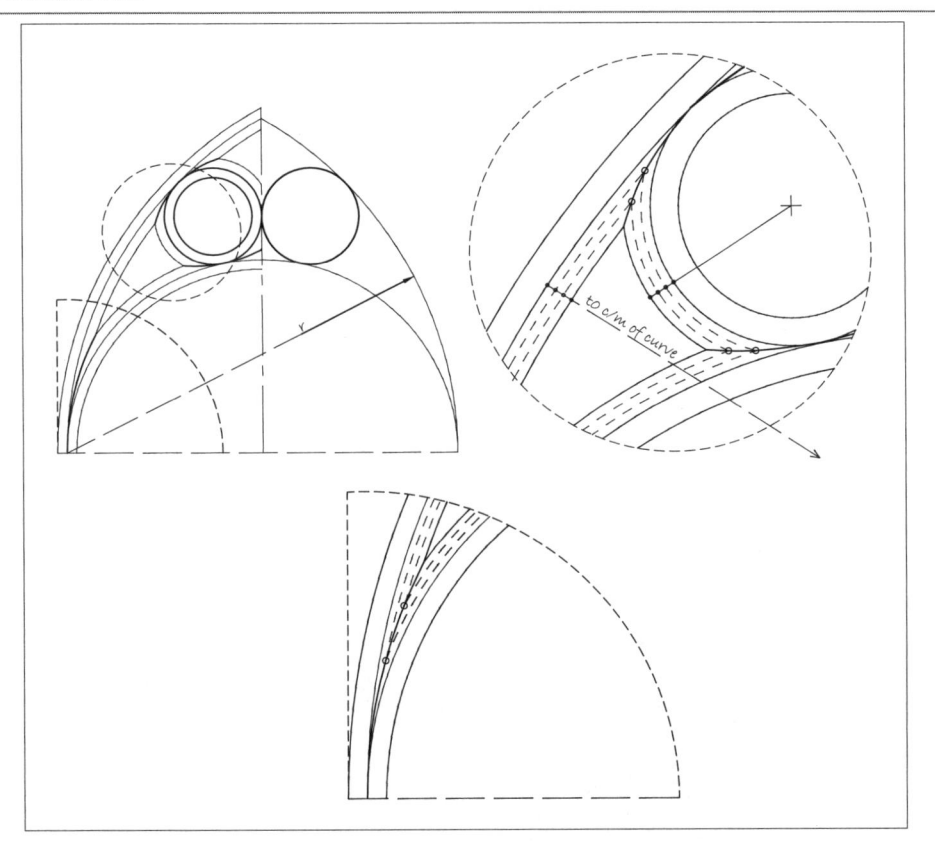

Fig. 13.6

SECTION II: RAKING MOULDINGS

Fig. 13.7a shows a pictorial view of a moulding, which runs horizontal and then rounds a corner where it changes pitch, that is to say, it becomes a raking moulding. The return moulding in this case goes onto a different plane, and in the given view the 'new' plane is at 90° to the one that holds the level moulding.

This is the fundamental point about raking mouldings, that mouldings not only change direction but, while doing so, continue on a different plane.

In **Fig. 13.7b** a plan and elevation of the mouldings in **Fig. 13.7a** are drawn, looking in the direction of arrow **A** (as in **Fig. 13.7a**). The task is to determine the true section of the raking moulding.

172 GEOMETRICAL DRAWING FOR CARPENTRY & JOINERY

(a)

30 degrees

raking moulding

level moulding

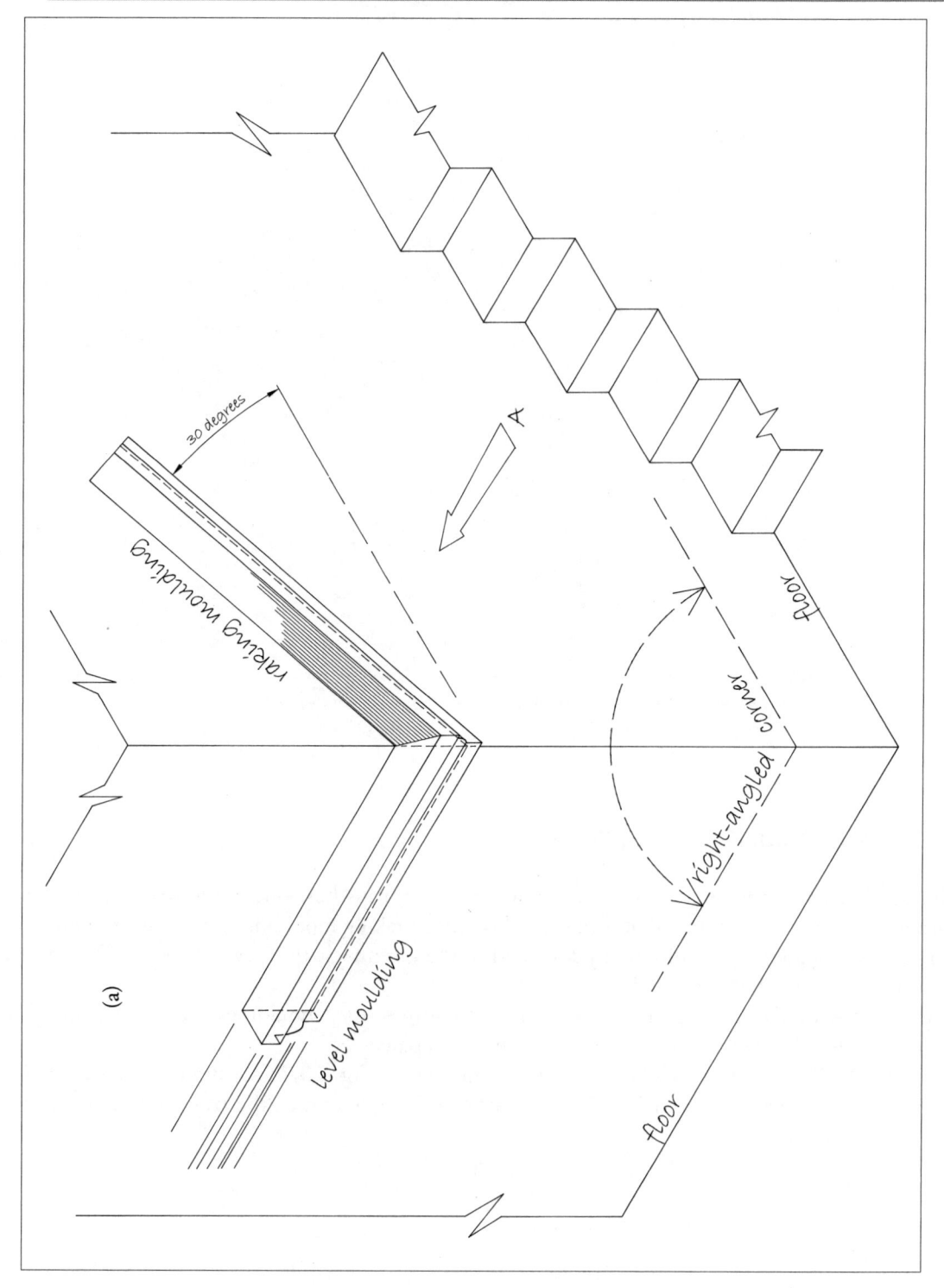

A

right-angled corner

floor

floor

Fig. 13.7a

13 MOULDINGS

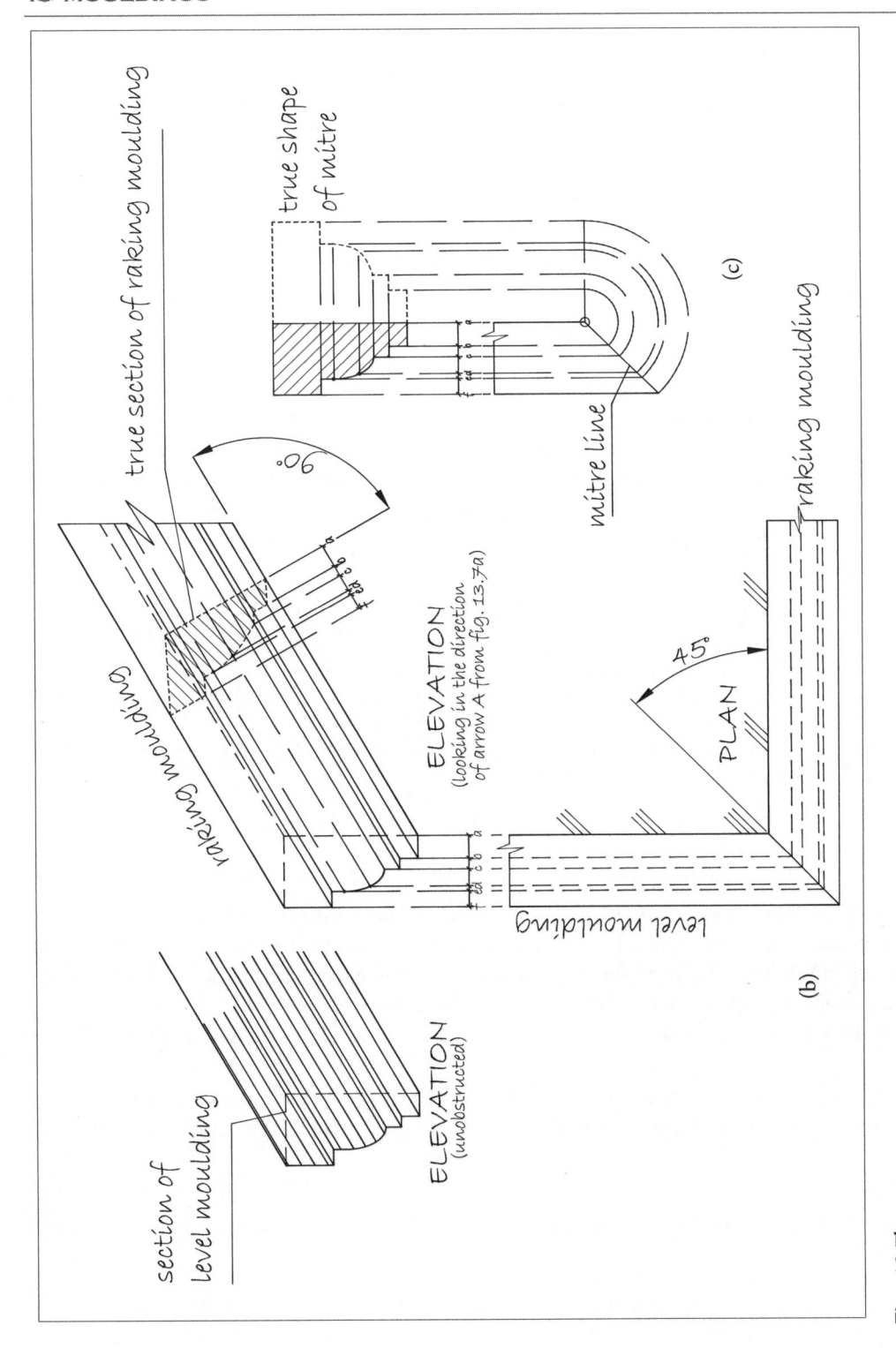

Fig. 13.7b
Fig. 13.7c

Method

- Set out dimensions **a** to **f** at 90° to the raking moulding in the elevation.
- Find the intersections for the relevant moulding lines in elevation, with their relevant dimension offset from **a** (let **a** represent the moulding surface against the wall).
- Join the intersections to reveal the true section of the raking moulding.

Note: When dealing with a moulding section which is curved (in profile), pick points on the curve (in this case, **e** and **d**), to facilitate finding them in the required section.

In the raking moulding section, the dimensions projecting from the wall are identical to those of the given level moulding; in other words, the thickness of the moulding does not change, only the moulding width changes. This is a rule worth remembering.

In **Fig. 13.7c** the level moulding section and the mitre line from the plan are reproduced. The mitre line and all points on it are pivoted to a horizontal position. These points are then projected to the elevation, where they intersect with horizontal projections from the level moulding section. The resulting section in the grid is the true shape of the mitre.

In this situation the moulding widths (or heights, depending on how you view them) do not change, but the thickness does because it is elongated by the mitre line.

Fig. 13.8 in another example, shows an alternative method of finding the section of the raking moulding.

Method

- In the elevation, the raking moulding is cut at an angle of 90° as shown.
- The cut surface, and all points on it is pivoted to a horizontal line. This procedure is known as *rebatment*.
- All points are projected to the plan where they intersect with their counterparts, which are projected horizontally (from the plan).

Note: This method works in this case because the raking moulding is horizontal in the plan and shows its true length in the elevation.

In **Fig. 13.9** an arrangement of mouldings is shown in which a moulding running along a sloped ceiling intersects with a level moulding, which is on a different plane (a wall). In this case, the raking moulding is the 'given' moulding and the section to be determined is that of the level moulding. The method employed is self-explanatory from the drawing. In the example shown, mouldings **B** and **C** are identical in section.

In **Fig. 13.10** a level moulding intersects with a curved moulding, which is on a different plane.

This example is treated in the same way as the previous raking mouldings, except that the true shape of the curved moulding section is determined on a radial line to the curve's centre mark.

Method

- From the section of the level moulding in the elevation, arcs are drawn (form the centre mark) through the relevant points.
- On any radial line, the dimensions **a** to **h** are set off square-off (the radial line).
- Where the moulding lines intersect at the radial line, they are squared-off (the radial line) to intersect with their counterparts on **a** to **h**.

13 MOULDINGS

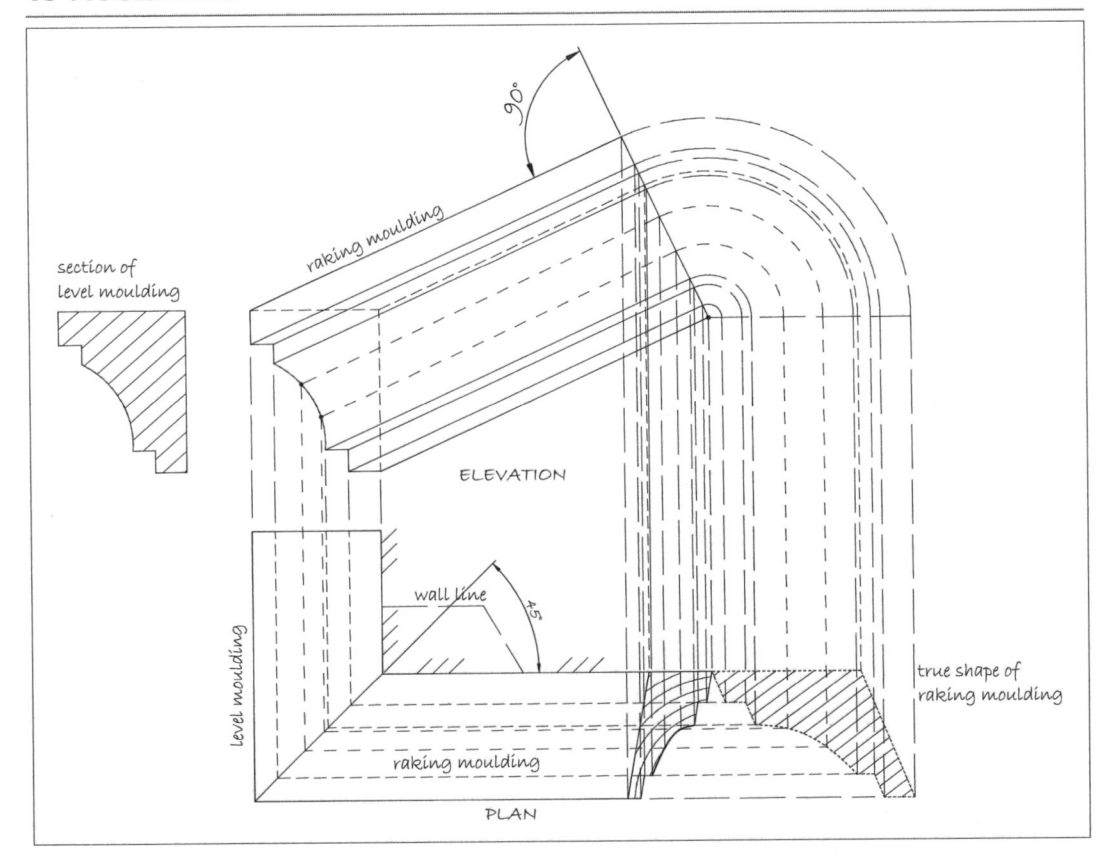

Fig. 13.8

In the examples above, the raking mouldings covered were all returning on a plane which was square-off the given plane; in other words, the plans of the examples given showed a 90° wall corner. In **Fig. 13.11** an example of a raking moulding which is not square-off in the plan is shown.

Method

- The plan is drawn from the section of level moulding given.
- The elevation of the level moulding is then drawn by transferring the moulding widths, **a** to **h**, as shown.
- The elevation of the mitre is determined and all points are projected (from the mitre) at the given rake.
- Square-off the raking moulding in the elevation the thicknesses **a** to 7 are set off and the section of the raking moulding is determined.

The top edge of the raking moulding is developed as shown.

The most difficult example of raking mouldings is the one in **Fig. 13.12**. In this case, both mouldings are inclined at an angle of 30° to the horizontal plane, as distinct from other examples shown where only one moulding was inclined.

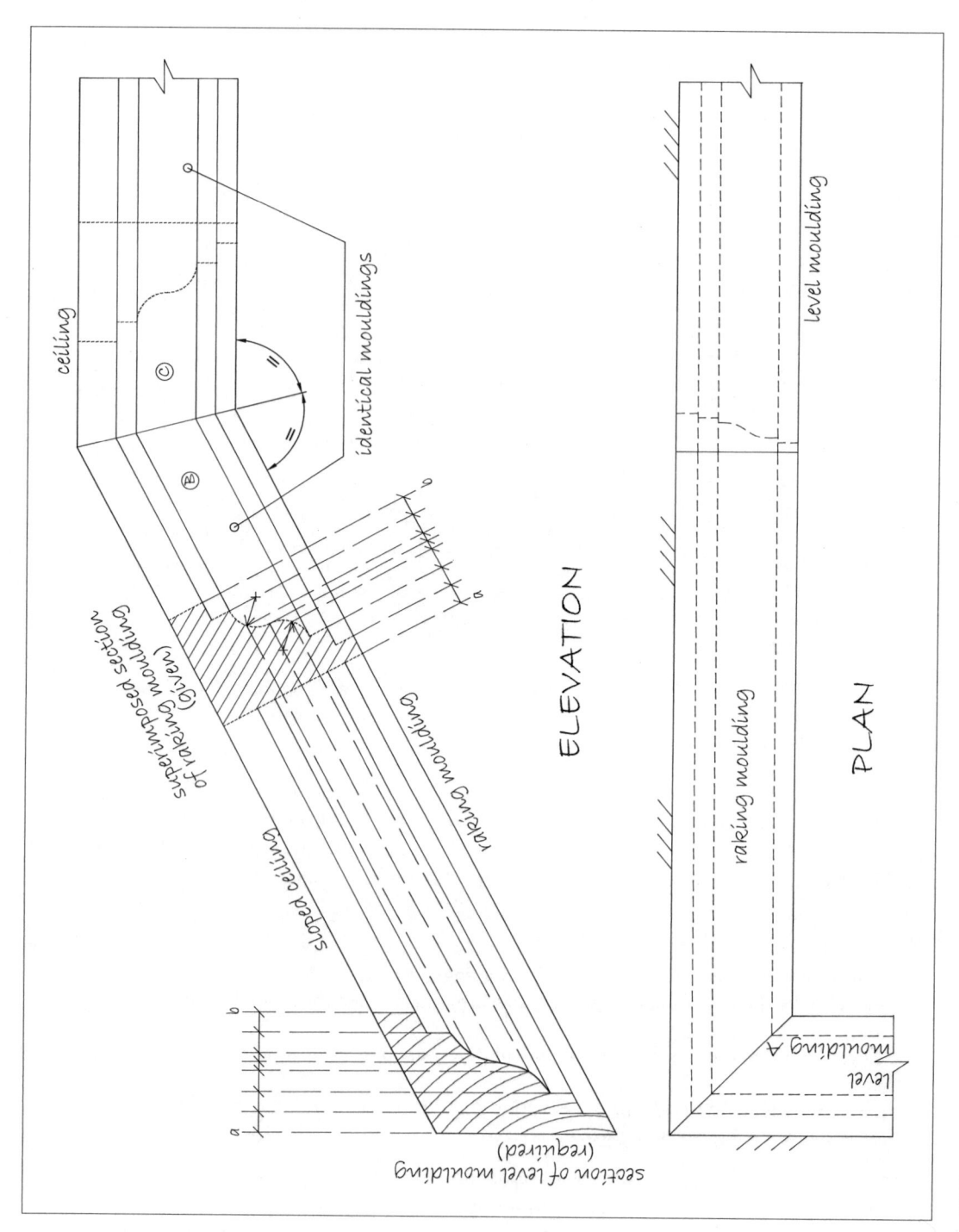

176 GEOMETRICAL DRAWING FOR CARPENTRY & JOINERY

Fig. 13.9

13 MOULDINGS

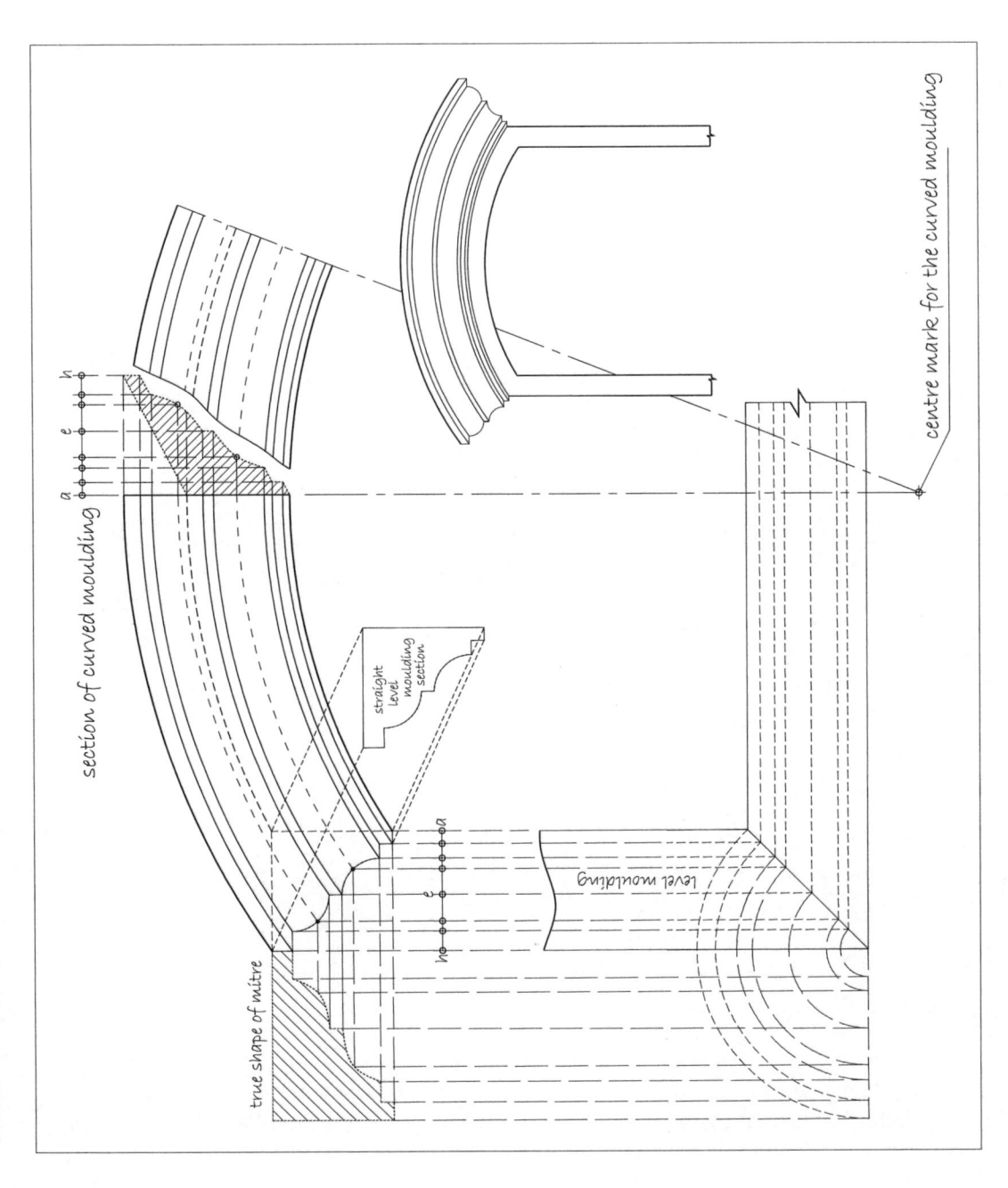

Fig. 13.10

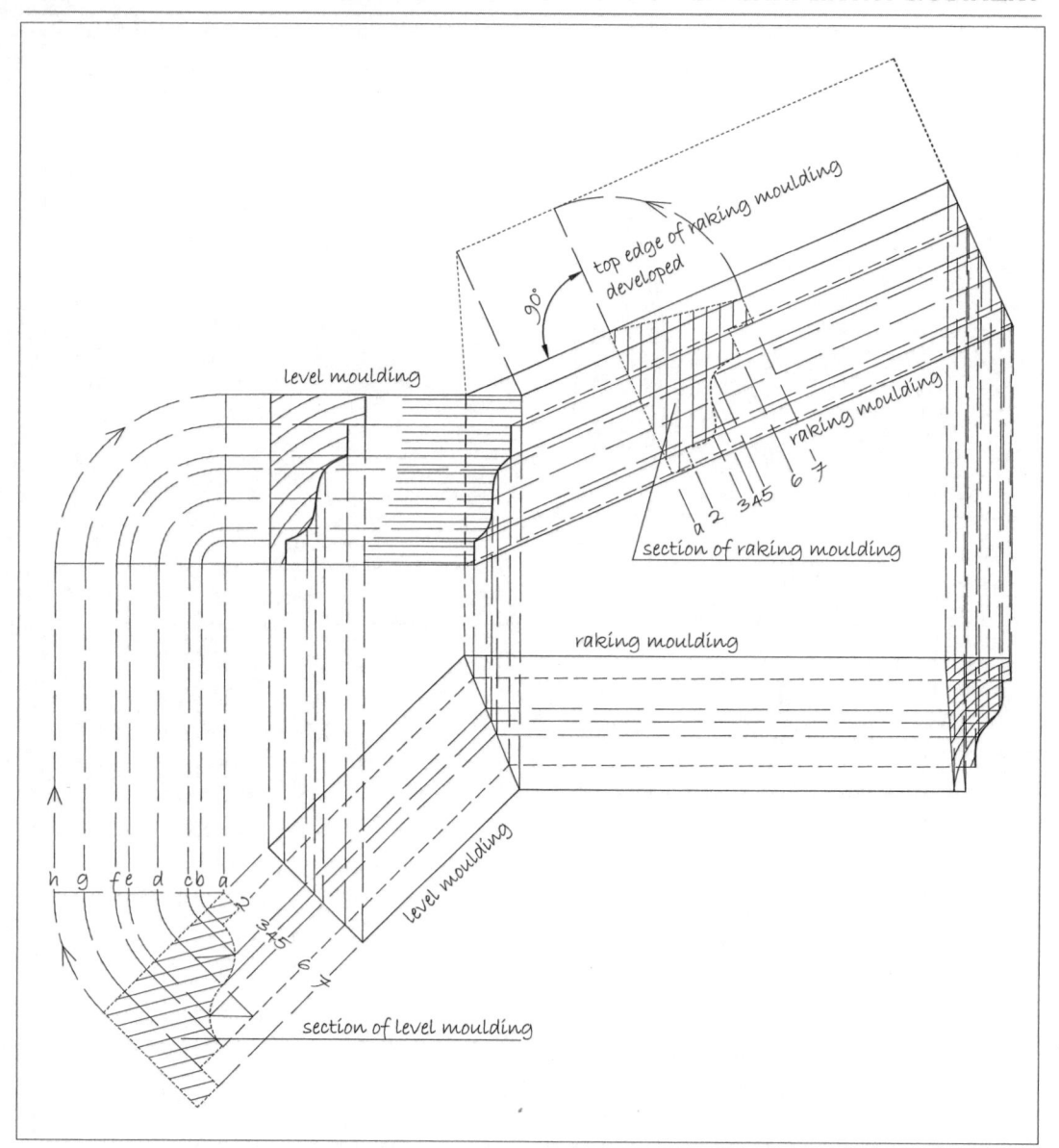

Fig. 13.11

The mouldings are presented in the plan, as shown, and they are mitred on a vertical cutting plane, represented by the mitre line in the plan. The completed **Fig. 13.12** has a lot of geometrical detail and it may seem difficult to know where to start. Below is the sequence for producing the drawing. The section of the given moulding **A** is drawn, and the plan of mouldings **A** and **B** (the right end grain of **B** comes later). An auxiliary elevation of moulding **A** is drawn.

13 MOULDINGS

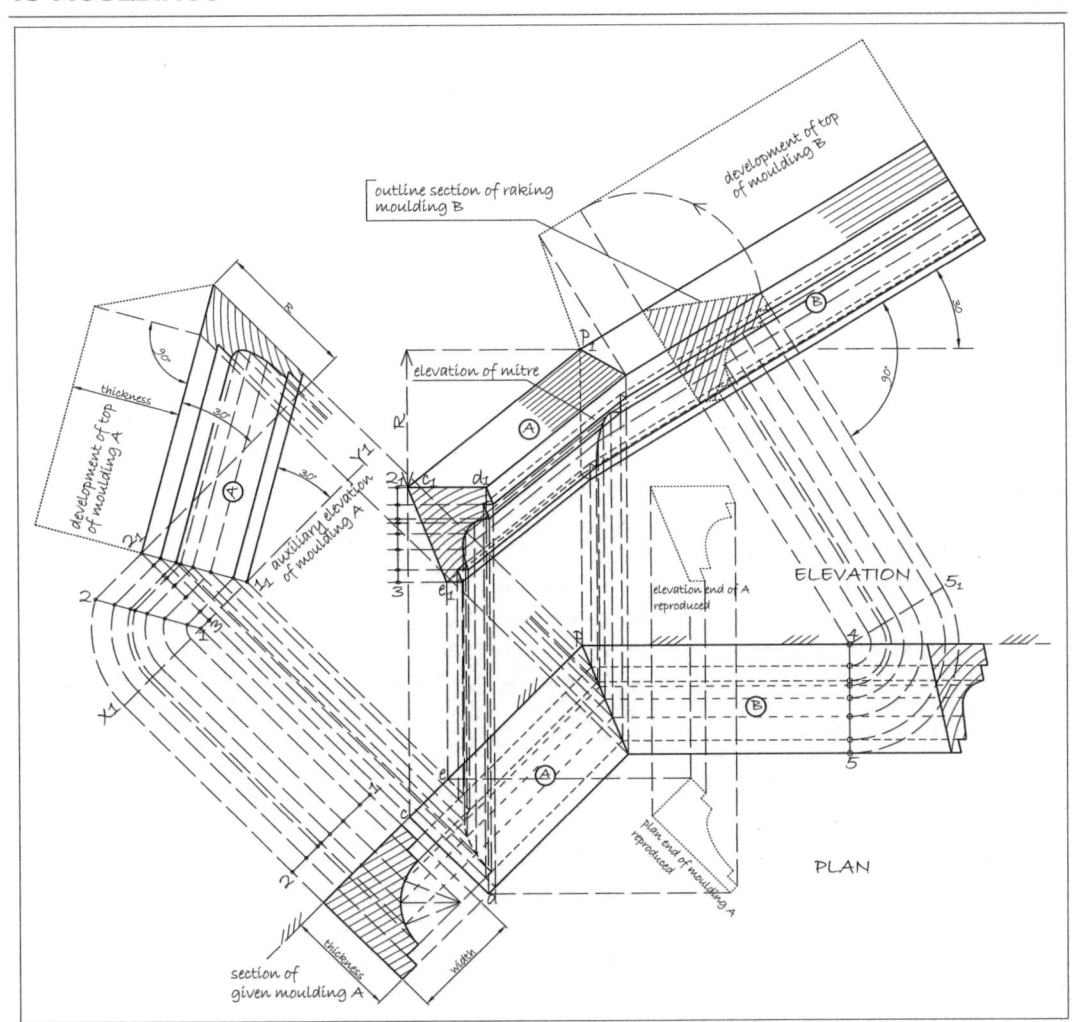

Fig. 13.12

Method

- Construct the **X1-Y1** line parallel with the plan of moulding **A**.
- Project the moulding width **1-2** and all points on it to the **X1-Y1** line, and pivot these to an angle of 60° to **X1-Y1**. (Note: A cut square-off of moulding **A** in auxiliary elevation will be at 60° to the **X1-Y1** line.)
- Move **1-2** in auxiliary to **11-21**. Point **21** is on a projection line from **c-d** in the plan.
- Project all points on **11-21** to the plan to complete the dotted outline of the end of moulding **A** (in the plan), project all points to the mitre line in the plan as shown.
- Project all points on **11-21** at 30° to **X1-Y1**, and project all points from the plan mitre line to intersect to complete the auxiliary elevation.

To draw the elevation of mouldings **A** and **B**:

Method

- Transfer the heights **21-3** from the auxiliary elevation to the elevation, as shown, and project from the dotted end outline of moulding **A** in the plan to complete its elevation.
- From **c1** in the elevation set up height **R**, which is found in the auxiliary view, and draw a line horizontally to **p1**, which is projected from **p** in the plan. (Note: The elevational rake of moulding **A** is **c1-p1**, and the true rake of moulding **A** is seen in the auxiliary view only.)
- Project all points from the (end grain) end of moulding **A** in the elevation, parallel with **c1-p1**, to intersect with the projections from the mitre line in the plan to complete the elevation of the mitre.
- Project all points from the elevation of the mitre, at 30°, to give the elevation lines of moulding **B**.

To find the true section of moulding **B**:

Method

- In the plan, draw the moulding lines from the mitre parallel with moulding **B**.
- Pivot the thickness **4-5** and all points to become **4-51**. **4-51** is parallel with the elevation of **B**.
- Project all points from **4-51** square-off the elevation of moulding **B**.
- Join the intersections with their counterparts in the elevation of **B** to give the sectional outline required.

The top surfaces of mouldings **A** and **B** are developed using methods covered in previous examples.

SECTION III: MOULDING CUTTERS

When a moulding cutter strikes the wood as it is being fed along the fence of a machine, the cutter cuts on an arc, the centre of which is the cutter block (centre). The moulding profile produced on the wood will be different from the cutter itself. The consequence of this is that for any required moulding profile, a cutter shape will have to be determined.

Traditionally, when moulding planes were used, the cutter, being sloped within the body of the plane, produced a moulding profile which was different from the cutter (because of the angle by which it struck the wood). If the cutter and the required moulding profile are

13 MOULDINGS

to be identical, the cutter must be used square-off the wood. This is the case when using, for example, a *scratch stock*, a tool used by cabinetmakers traditionally and by the author to this day for producing delicate mouldings. With a scratch stock, the cutter is held in the stock and used square-off the edge of the wood to 'scrape' rather than cut the moulding shape. It is used like a marking gauge, but instead of a spur there is a cutter inserted.

Nowadays, there is a wide variety of moulding cutters ready to go, and there is no need to wonder if the cutter in the machine is identical to the resulting moulding profile. If, however, work in Section II on raking mouldings has any relevance to this section, it is because for every new section of raking moulding that we determined, a cutter shape will have to be determined to suit.

Fig. 13.13a shows a required moulding section in line with the fence of the moulding machine.

Method

- From the cutter block centre, a line is drawn square-off the fence.
- The moulding dimensions are projected to this line.
- The dimensions are drawn using arcs from the centre of the cutter block to touch the face of the cutter.
- From the face of the cutter the 'new' dimensions are projected vertically (or square-off the cutter plan) to give points on the required cutter.
- Copy dimensions **o** to **h** to complete the required cutter shape.

Note: The dimensions **o** to **h** (which in this case are heights) do not change.

In **Fig. 13.13b** the moulding section and the moulding cutter are reproduced, and it can be seen that they are not identical. **Fig. 13.14a** and **13.14b** give another example of this.

Exercise

1. A level moulding and a raking moulding are shown in **Fig. 13.15**. The raking moulding is raking downwards at 35°. Reproduce **Fig. 13.15** and complete the exercise.
    - Draw an end elevation on the right-hand side.
    - Determine the section of the raking moulding.
    - Develop the top surface of the raking moulding.

2. In **Fig. 13.16** three mouldings intersecting are shown. Mouldings **A** and **B** are identical in section. Moulding **C** is reduced in width by the mitre line shown. Reproduce **Fig. 13.16** and complete the exercise.
    - Complete the elevation of the mouldings showing all relevant lines.
    - Determine the section of moulding **C**.

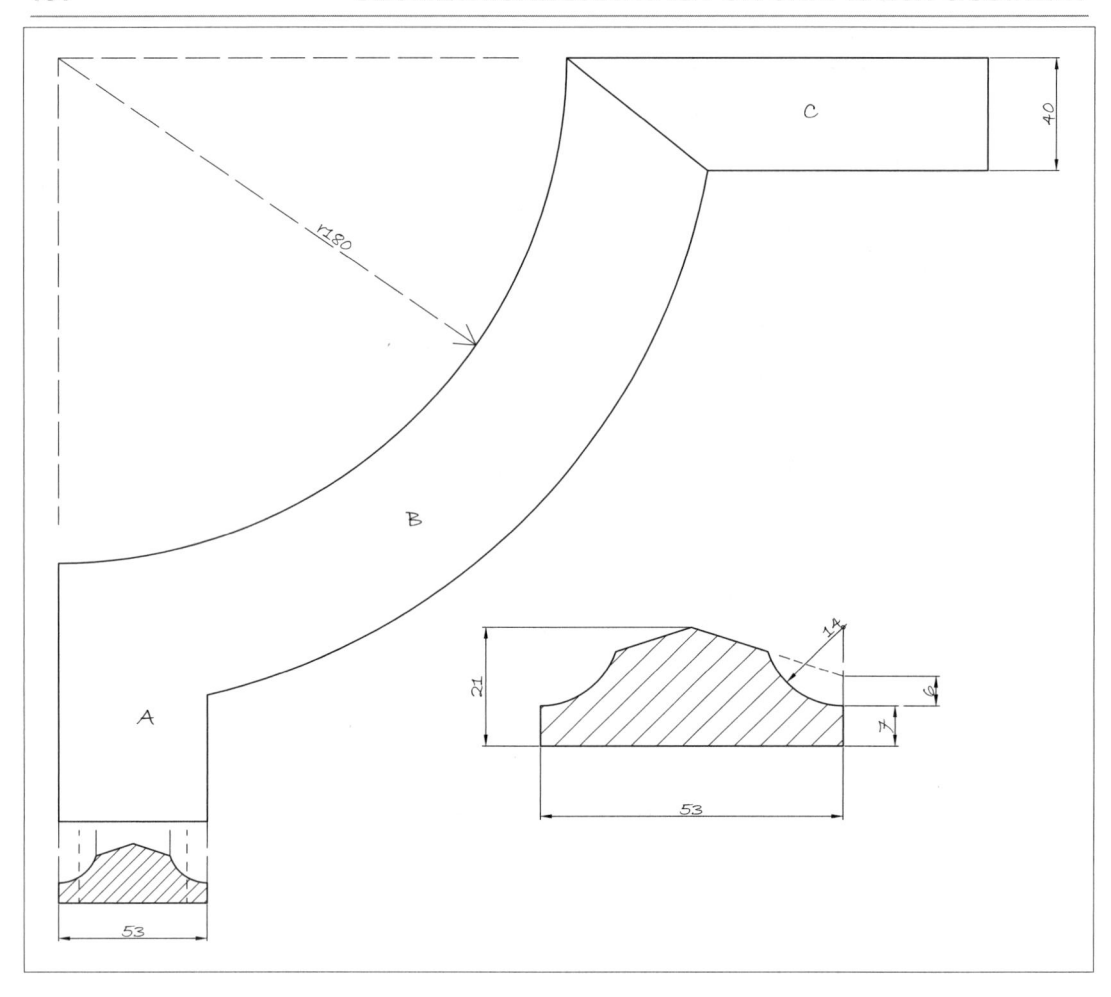

Fig. 13.16

14 INTERPENETRATION OF SOLIDS

Solid geometry deals with geometric solids, such as prisms, cylinders, pyramids, cones and spheres. All of the solid items we observe and come into contact with every day are in one way or another based on these solids. A pitched roof is a triangular-based prism, a food can is a cylinder, most pencils are hexagonal prisms, a ball is a sphere, and a traffic cone is just that, and the list goes on. This chapter explores situations where the geometric solids come into contact with each other and *interpenetrate*.

Figs. 14.1a and **14.1b** show, respectively, a triangular prism interpenetrate with a square prism and a square-based pyramid. The lines of interpenetration are relatively easy to determine, as shown.

When dealing with curved solids, the geometry involved is more advanced. In the example in **Fig. 14.2**, a cylinder is standing vertically, with, on the left, a triangular prism interpenetrating the cylinder and, on the right, a square prism interpenetrating. In this case, the lines of intersection between the solids are determined using a method called horizontal sections. For the left-hand side of the cylinder:

Method

- Construct a series of levels, **1** to **5**, on the elevation.
- Find the plans of the various levels as shown.
- Where the plan of level **4** on the prism strikes the cylinder, project the points to the elevation of level **4**, giving the points of interpenetration at that level.
- Work similarly with all other levels and join the relevant points to give the curved line of interpenetration in the elevation.

A horizontal section of the prism at level **4** is shaded in the plan. Where this section intersects with the horizontal section of the cylinder (which is the circle in plan), points of intersection are found at that level.

The square prism on the right is treated in the same way. Levels are assumed at **b₁**, **a₁ c₁**, and **d₁** in the elevation and their counterparts are found in the plan. Intermediary levels are also shown on the square prism.

The surface development of one side of the triangular prism (**Fig. 14.2**) is determined by pivoting upwards the true width of the surface and all points on it, and projecting the relevant points of intersection upwards, as shown. The square prism is done in a similar way.

Fig. 14.3 represents the orthographic projection of a pitched roof intersecting with a conical roof. As in the example in **Fig. 14.2**, the prism (the pitched roof) is divided into a series of levels, **1**, **2**, **3**, **4**, **5**, and these are found and indexed in the plan as shown. These levels represent horizontal sections of pyramid and prism at various heights.

GEOMETRICAL DRAWING FOR CARPENTRY & JOINERY

square prism

superimposed section of triangular prism

ELEVATIONS

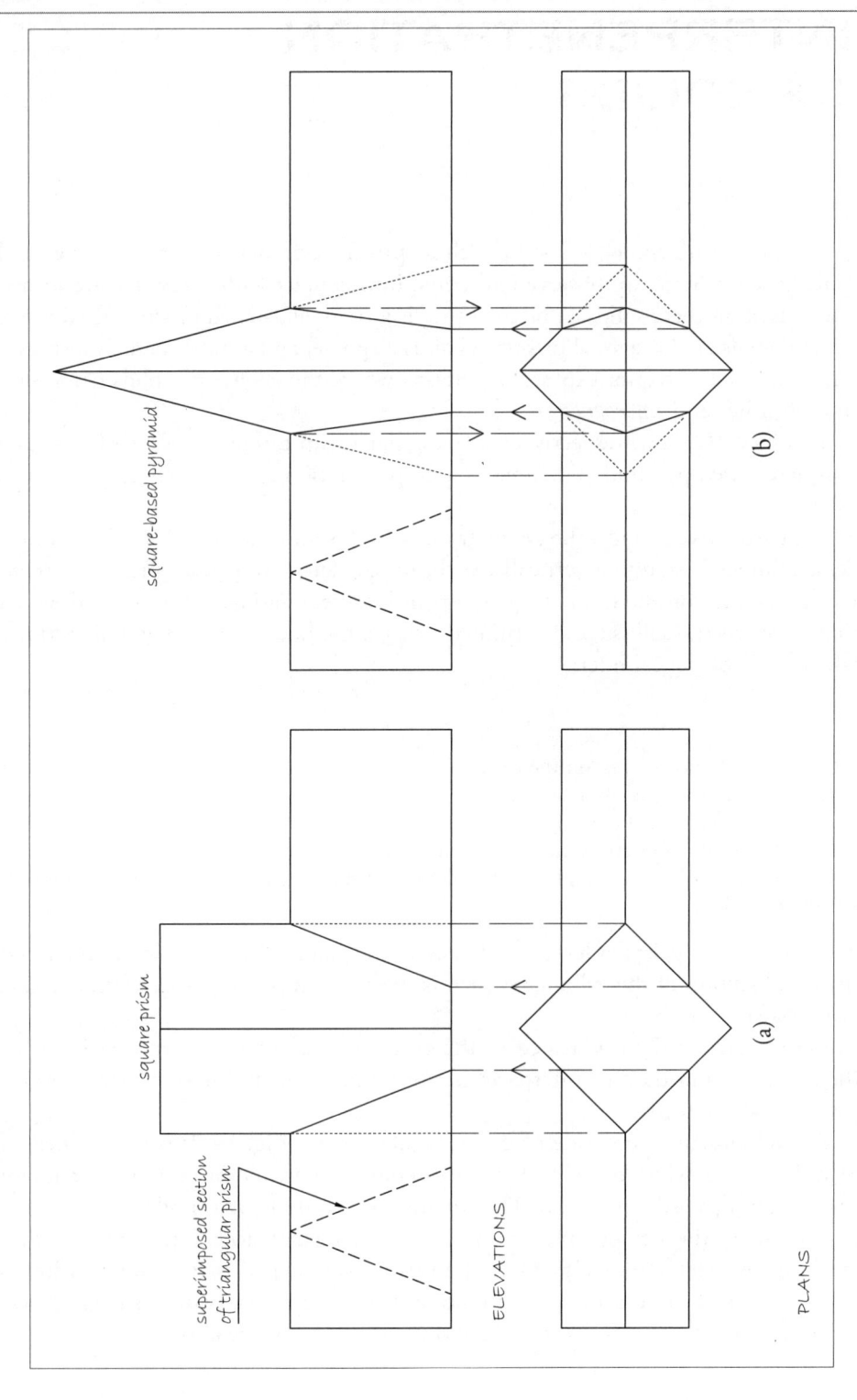

(a)

PLANS

square-based pyramid

(b)

Fig. 14.1a
Fig. 14.1b

14 INTERPENETRATION OF SOLIDS

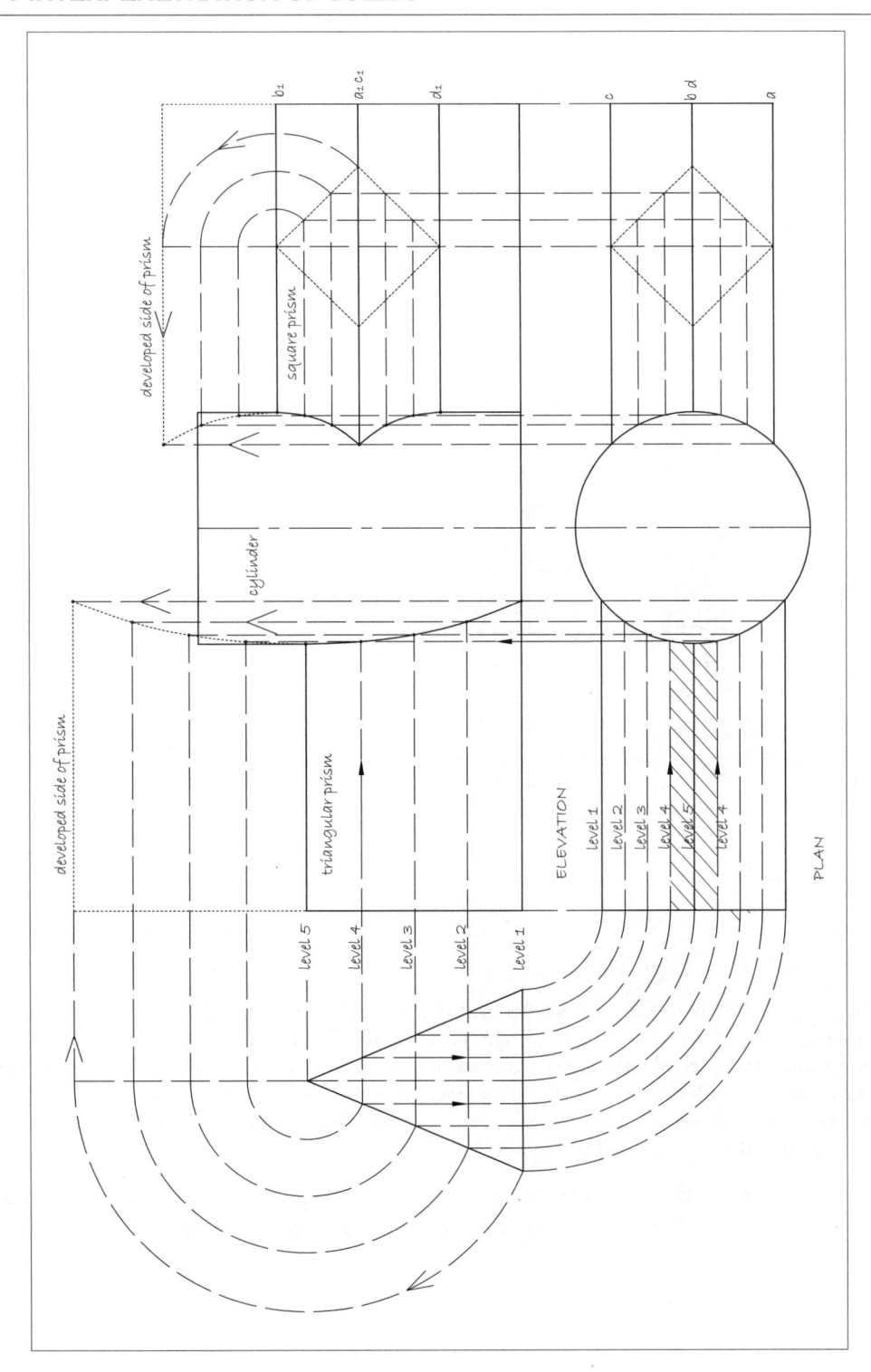

Fig. 14.2

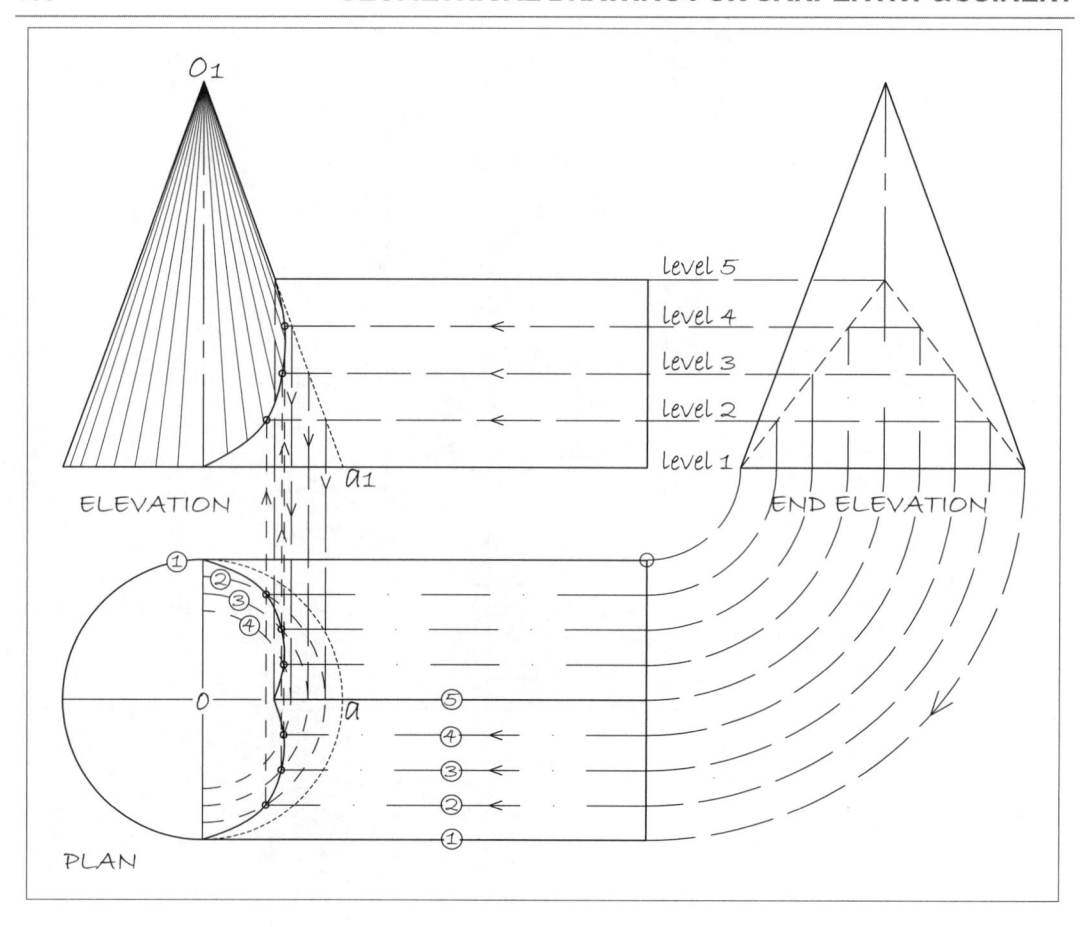

Fig. 14.3

Method

- The horizontal sections of the cone are found by dropping verticals from the outside generator, **o1-a1** to **o-a**, in the plan and constructing semi-circles at the different levels (in the plan).
- The horizontal sections on the prism are projected from the end elevation to the plan, as shown, and where they intersect with the semi-circular levels, points of intersection are found in the plan.
- Having found the intersection points in the plan, they are then projected vertically to the relevant levels in the elevation. Join all points to complete.

A dome (semi-sphere) and a triangular prism are shown in **Fig. 14.4**. The same principle of using horizontal sections is used to determine the lines of interpenetration.

14 INTERPENETRATION OF SOLIDS

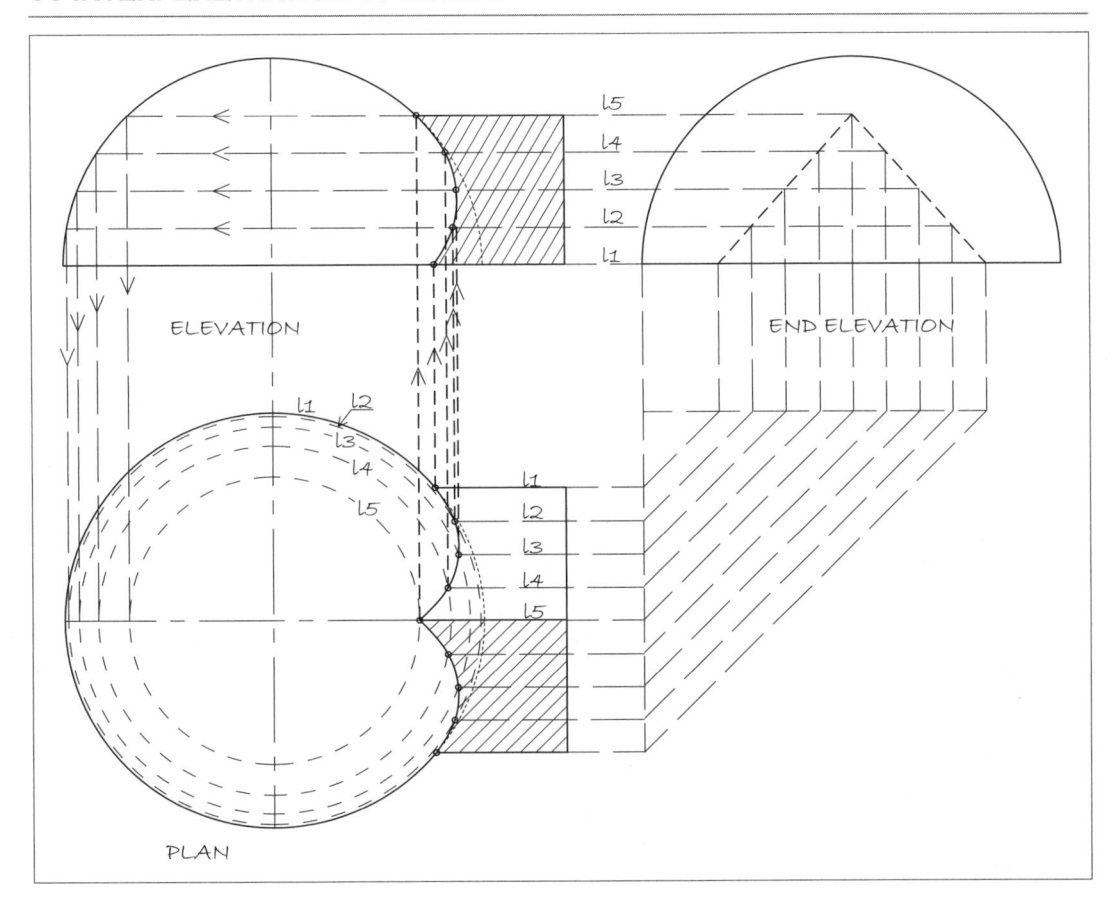

Fig. 14.4

In **Fig. 14.5** an octagonal prism is shown emerging from three sloping planes. In practical terms, this could represent a chimney emerging from a roof surface. The points of interpenetration are found by using lines on the surfaces of the sloping planes.

The method for the right-hand side is as follows:

Method

- Lines are drawn through the corners of the prism in the plan. These lines are parallel with the wall plate line **m-n** and end at **p-n**, where their positions are projected to **p1-n1**.
- Horizontal lines are drawn from the intersections on **p1-n1** to where they intersect with the vertical projections of the prism corners from the plan, thus giving the points of interpenetration.

The middle example in **Fig. 14.5** is similar to the previous one, but with the addition of an alternative method.

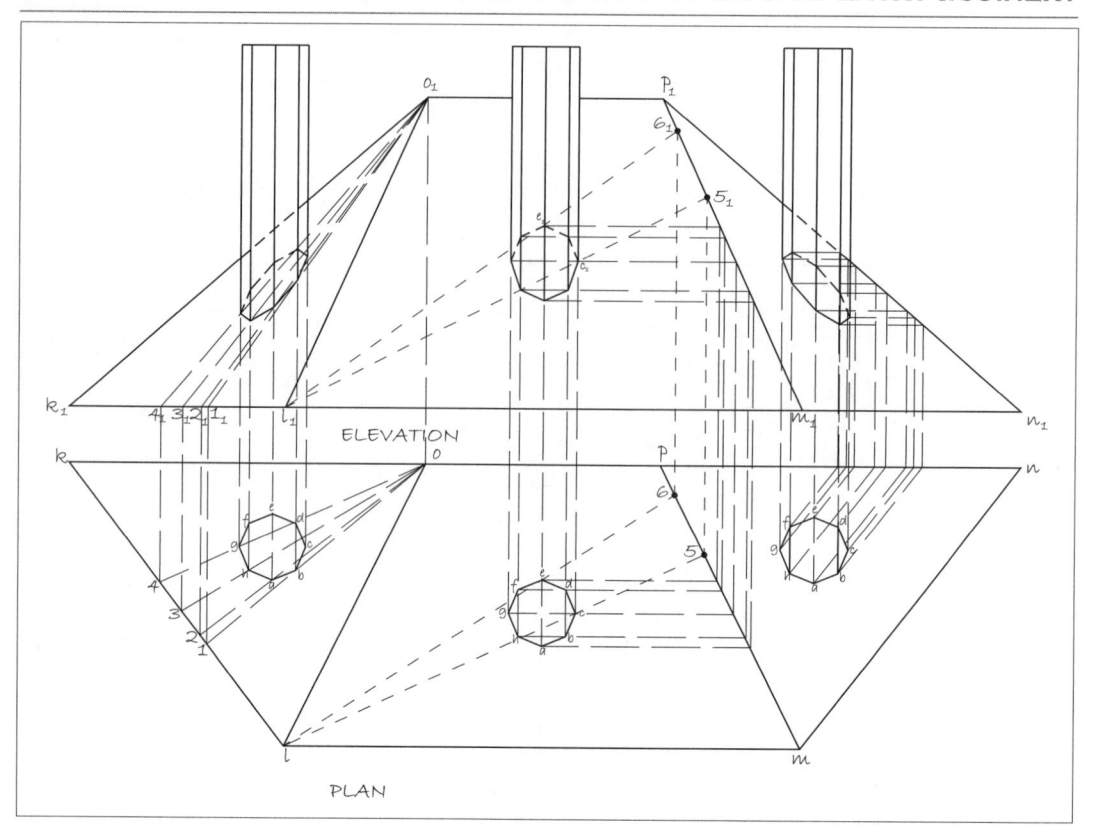

Fig. 14.5

Method

- In the plan, lines **1-5** and **1-6** are drawn through corners **c** and **e**, respectively, of the prism (for clarity, only two corners are being used).
- The elevation of these lines is found to be **l1-51** and **l1-61**.
- Where these lines intersect with the vertical projections of **c** and **e** from the plan, points of interpenetration are found (at **c1** and **e1**).

On the left-hand side, four lines that take in the corners of the prism, **o-1**, **o-2**, **o-3** and **o-4**, are found in plan and elevation. Where these lines intersect with the vertically projected prism corners from the plan, the points of interpenetration are found. The latter method may be used in many different ways to solve problems in plane and solid geometry. The concept is also referred to as using lines in space.

Note: Only four lines are used for clarity; all corners of the prism would be treated similarly.

14 INTERPENETRATION OF SOLIDS

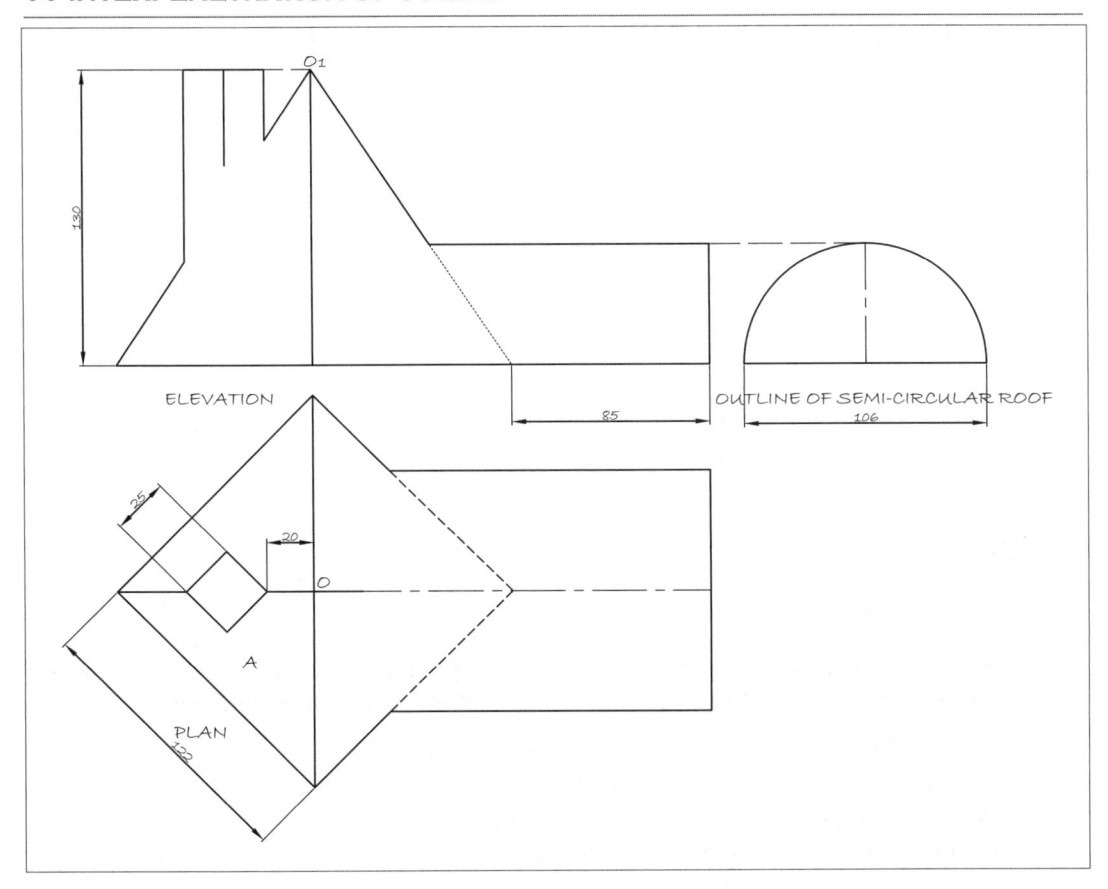

Fig. 14.6

Exercise

In **Fig. 14.6** a square-based pyramid with a half-cylinder and square prism intersecting is shown. Applied to roofing, it could represent a pyramid roof and a semi-circular roof, with a chimney emerging. Reproduce **Fig. 14.6** and complete the exercise.

- Complete the plan and elevation showing the lines of intersection between all three items.
- Develop one half of the cylindrical roof surface.
- Develop surface **A** of the pyramid roof.

15 INTERSECTING VAULTS

Vaults are arched structures forming roofs and ceilings, or tunnel-like chambers, and have varying profiles and equally varying plan outlines. Where they intersect with each other, the line of intersection is called a *groin*. They may intersect at the same level or at varying levels depending on their individual profiles. The profile of a vault is its arch outline (revision of Chapter 6 on arches may be necessary at this stage).

In **Fig. 15.1** views of a cross vault (the plan is cruciform, or shaped like a cross) are shown. The profile of the vault is semi-circular, while the plan shows the groin/s as straight lines. This occurs when the vaults intersecting have the same profile. To find the true shape of the groin:

Method

- Divide the semi-circle into equal parts, eight in this case, and project the points to the plan.
- Where the projection lines intersect with the groin, project them at 90° as shown.
- Set up a height rod in the elevation for the relevant heights and transfer them to the plan as shown.
- Draw a curve through the relevant intersection points to give the shape of the groin.

In this case, the developed groin is semi-elliptical in shape.

Fig. 15.1 also shows the development of the surface of vault **A** (revision of cylindrical development in Chapter 4 will be helpful at this stage). In **Fig. 15.2** the profile of the vaults is segmental in shape and the groins intersect at angles of 120° in the plan. In this example the groin is developed as an auxiliary elevation, clear of the given drawing, whereas in the example in **Fig. 15.1** the development was tied onto the plan of the groin.

Fig. 15.3a shows two intersecting vaults which are gothic (also called equilateral or pointed) in profile but do not have the same rise. While they both have the same springing line, their ridges are at different levels. In this case, the plan of the groin is not a straight line and consequently the true shape of the groin will have double curvature, meaning that it is curved in plan and in elevation.

To find the auxiliary elevation of the groin:

Method

- Join **c-d** and project the intersection points at 90° as shown (see enlarged detail).
- Take the heights from the elevation and set them off at 90° to the line **c-d**.
- Draw a curve through the intersection points to give the auxiliary elevation of the groin. This is the true shape of the groin in auxiliary elevation.

Note: Its curvature in plan is found between the straight line **c-d** and the plan of the groin itself (see enlarged detail).

In **Fig. 15.3b** the same vault profiles as in **Fig. 15.3a** are used, however, in this case the ceiling ridges of the vaults are on the same level and consequently the springing lines are at

15 INTERSECTING VAULTS

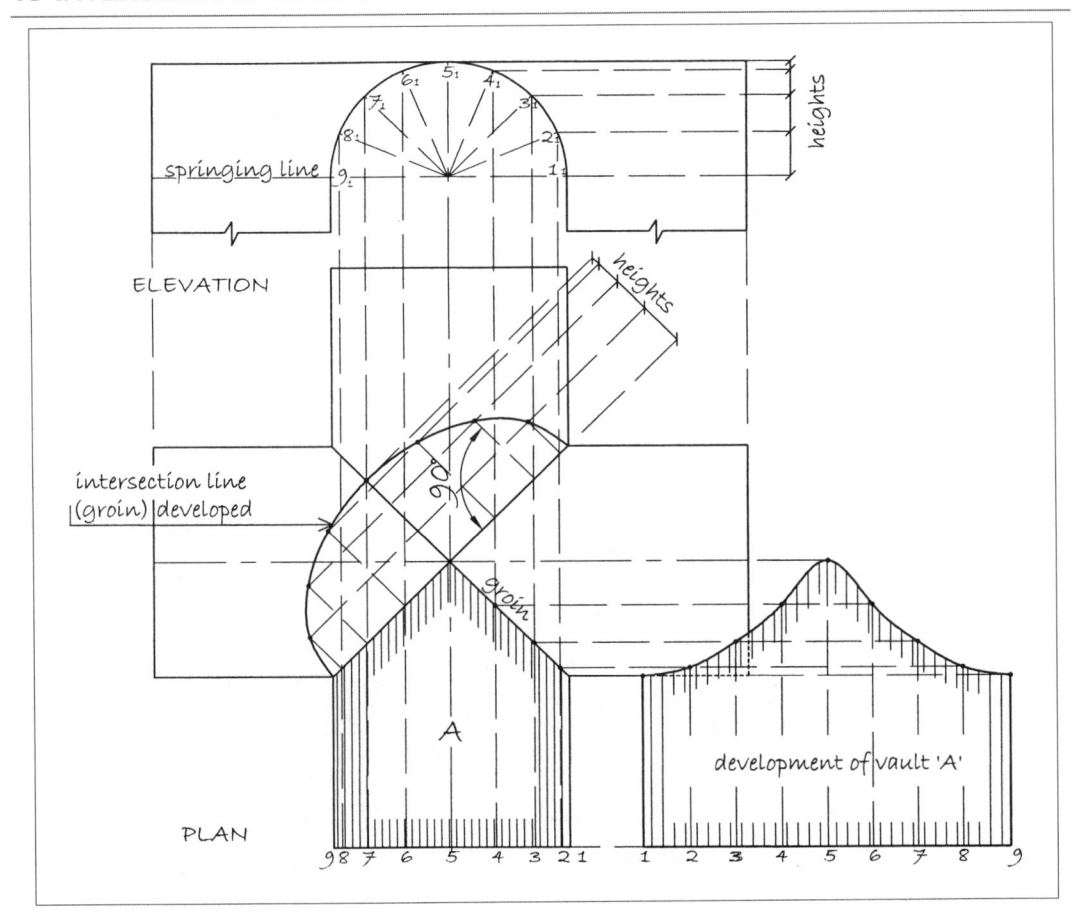

Fig. 15.1

different levels. With this arrangement, vault **B** is *stilted,* meaning it is 'propped' so that the *impost* for both vaults is the same.

The impost is the line of bearing of an arch on its underneath structure and, in most cases, the impost and the springing line occur at the same level, but the springing line is an abstract term and the impost is a concrete one.

In all of the figures above, the examples given are more accurately described as *barrel* vaults, meaning that they are continuous, or chamber-like, in construction.

The vaults shown in **Figs. 15.4** and **15.5** are individual and more specific to the description of a vault. They are called ribbed vaults.

In **Fig. 15.4** a semicircular vault is shown in pictorial form as well as in a plan view. It is important to note that in the plan the groin is shown as a straight line. This is because both vault profiles are the same.

In **Fig. 15.5a** the vault shown has two different profiles. In this case, the plan view shows the groin as a curved line, which is described as a 'waving groin'.

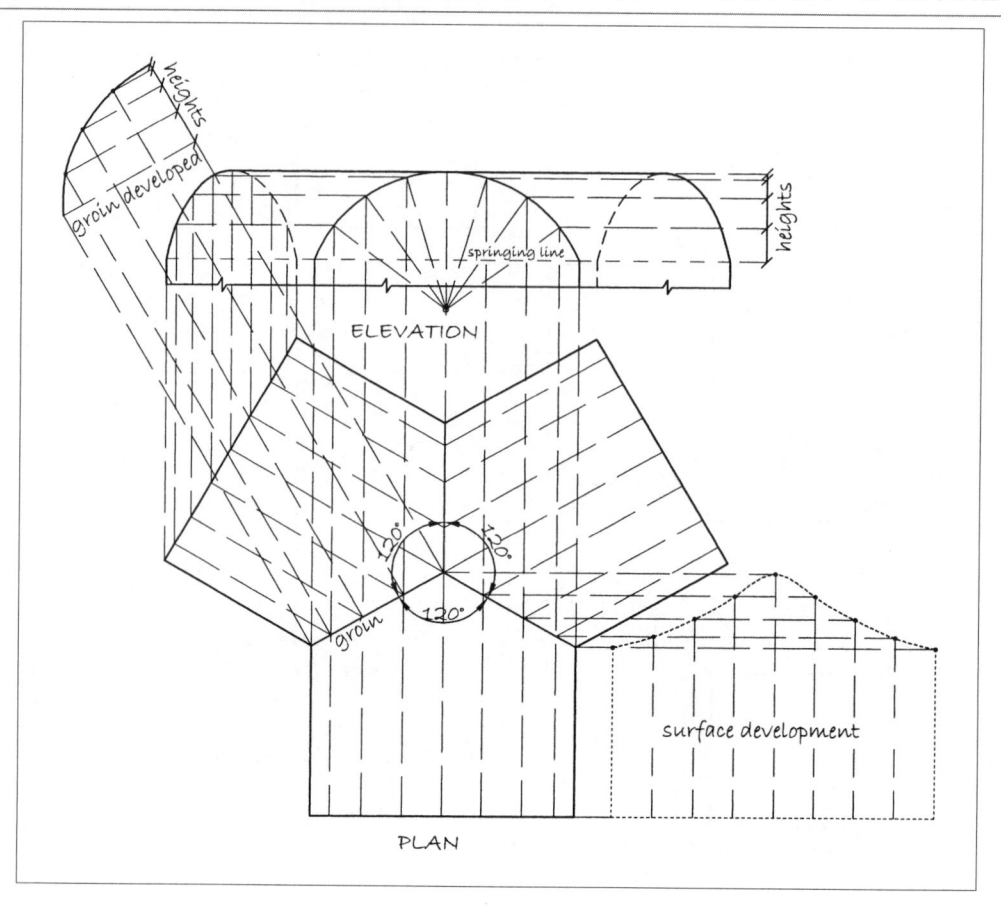

Fig. 15.2

The auxiliary elevation of the groin **a-b** is drawn:

Method

- Square-off **a-b** in the plan, lines are projected from points on the plan of the groin.
- The heights from the elevation are set off and the auxiliary shape of the groin is determined. It should be emphasised at this stage that the auxiliary shape is exactly that which is viewed from **S**. The actual groin rib will have double curvature, which is determined in conjunction with the plan curve.

Fig. 15.5b shows the plan of the groin reproduced with thickness added. An *axonometric* projection is employed in an attempt to show the double curvature of the actual solid member, which is a curved rib similar to those used in turrets (see Chapter 8 on roofs).

A cross barrel-vault was explained in **Fig. 15.1** with accompanying text. **Figs. 15.6a** and **15.6b** deal with a similar kind of vault but with an additional construction feature. Orthographic views are shown at **Fig. 15.6a**.

15 INTERSECTING VAULTS

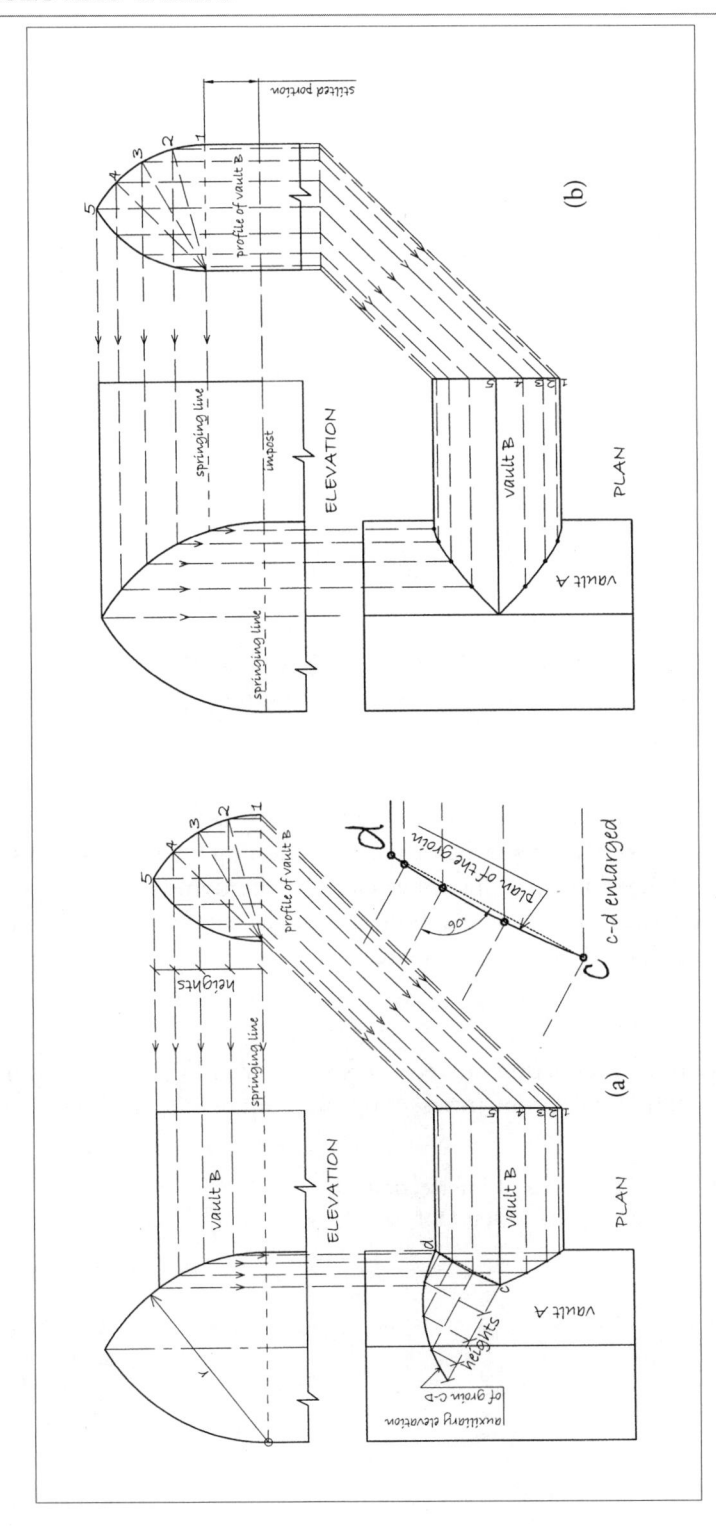

Fig. 15.3a
Fig. 15.3b

SEMI-CIRCULAR VAULT
(RIBBED VAULT)

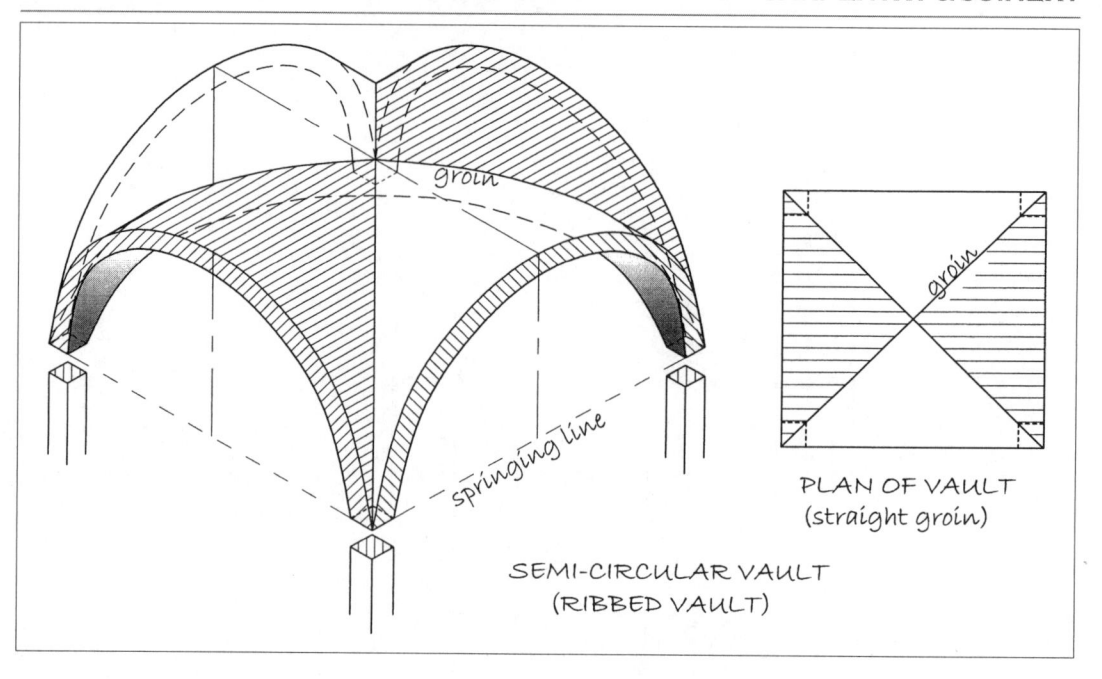

PLAN OF VAULT
(straight groin)

Fig. 15.4

At the junction of the two vaults a dome is mounted. This is facilitated by the addition of *pendentives*, which are *spherical triangles* that act as transition members from the square at the springing lines of the barrel vaults to the upper circle of the dome. The complete structure here is known as a *pendentive dome*. The spherical triangles are complex in form and their surfaces cannot be developed in the normal sense.

Exercise

The plan and elevation of intersecting curved roofs is given in **Fig. 15.7**. The main roof is segmental and has a semi-circular roof intersecting. Reproduce **Fig. 15.7** and complete the exercise.

- Complete the plan showing all lines of intersection.
- Develop the groin between the segmental roofs.
- Develop roof surface **A**.
- Develop roof surface **B**.

15 INTERSECTING VAULTS

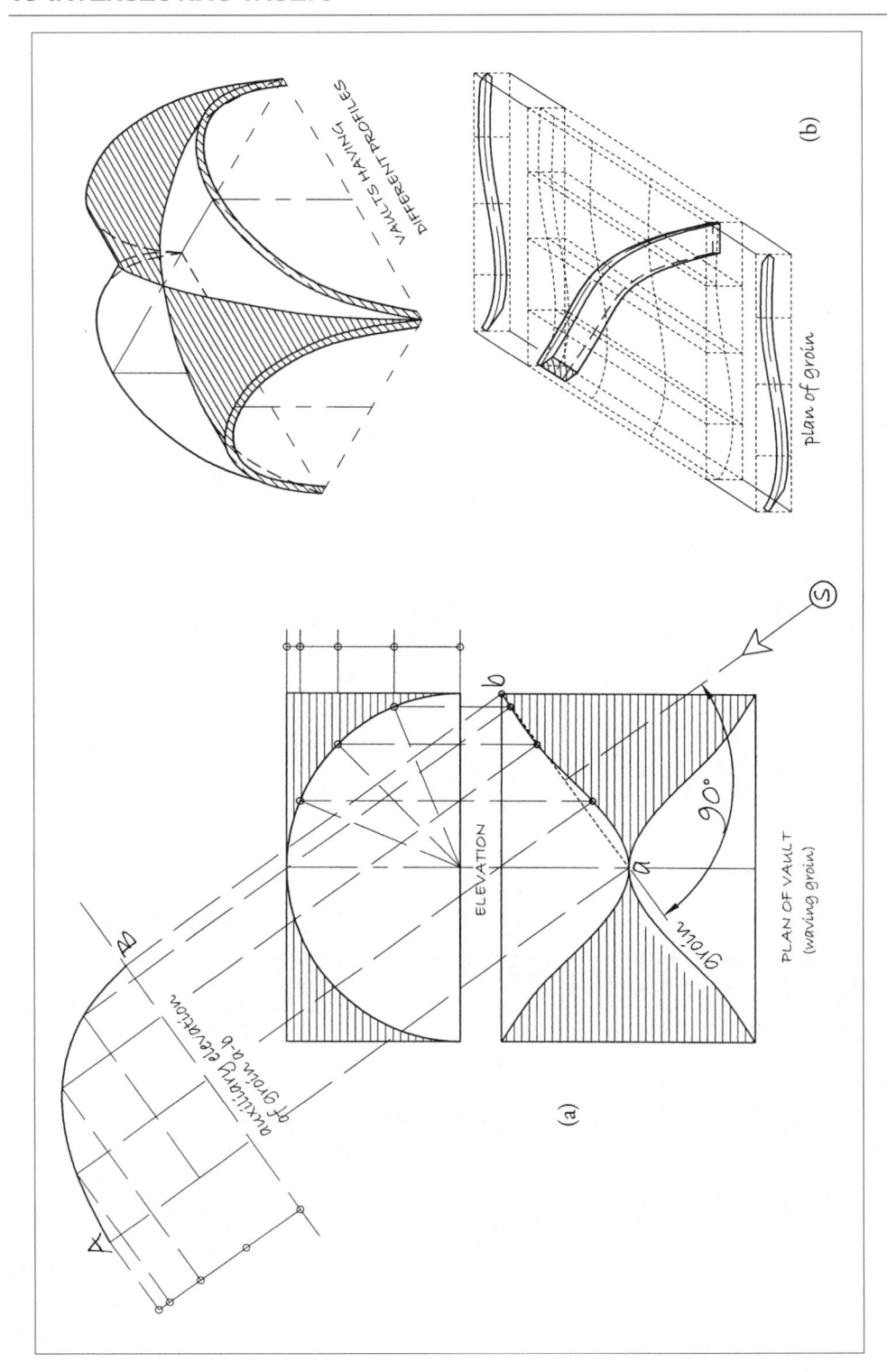

Fig. 15.5a
Fig. 15.5b

Fig. 15.6a
Fig. 15.6b

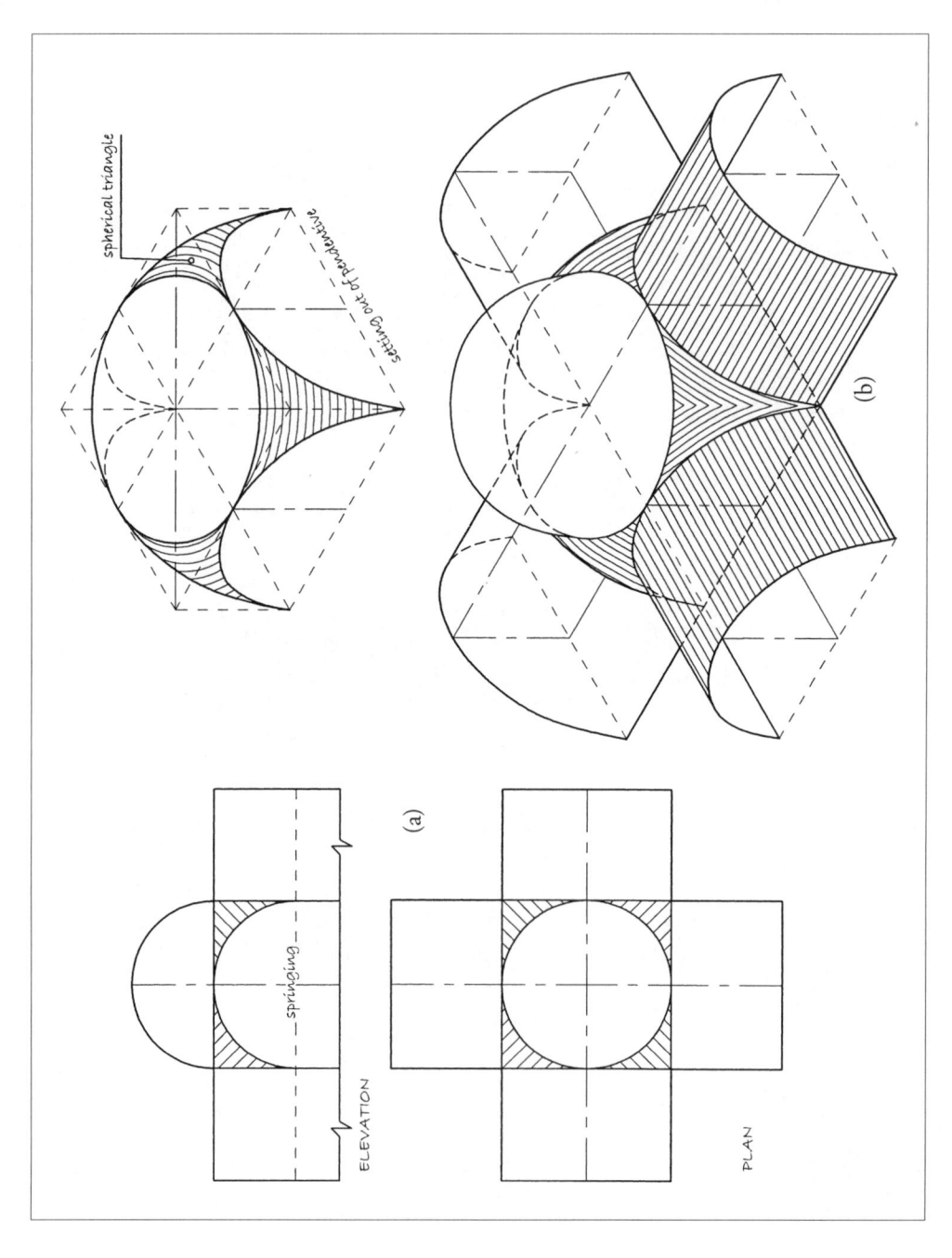

15 INTERSECTING VAULTS

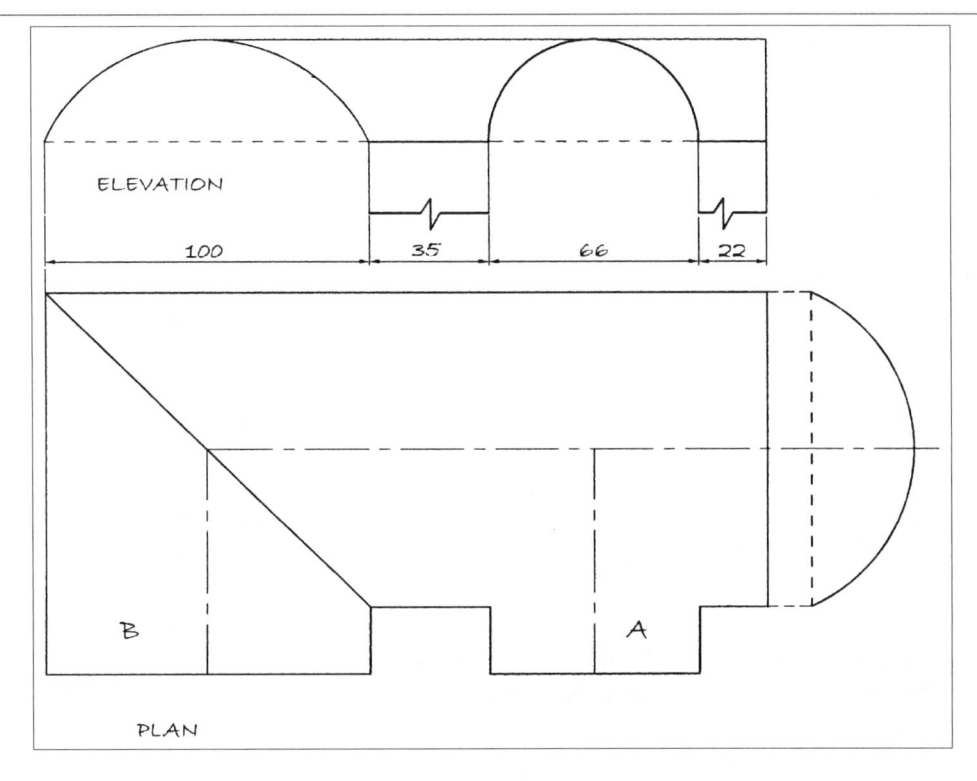

Fig. 15.7

16   OBLIQUE SOLIDS

In Chapter 4 on geometric solids, the solids dealt with are more accurately described as right solids, meaning that the axis of each type is at 90° to its base. A right cone, for example, has its axis at an angle of 90° to its base circle and consequently the apex is directly above the centre of the base. A right cylinder resting on its base will have its top circle directly above.

An oblique solid is one which has the axis inclined at a given angle to the base, that is, an angle other than 90°, so the solid is skewed to one side. In **Fig. 16.1**, an oblique square-based prism is shown. The section of the square base is constant throughout the height of the prism. This is illustrated with a section at **A-B** in elevation and plan. The sides of the prism are developed square-off the elevation, as shown, and it should be noted how the dimension **x**, which represents the length of the side of the square base, is stepped off in the development. The length of the top and bottom perimeters of the development must be four times line **x**.

The height (**H**) of the solid is indicated and it is important to mention here that the height of such solids is measured vertically from the base to the top. It would be inconceivable to measure height along the incline because height is altitude.

An oblique pyramid with a square base is given in plan and elevation in **Fig. 16.2**. The incline of the axis is evident and, again, the height is indicated. A section of the pyramid on line **E-F** will result in a smaller square in the plan.

Some of the sloping sides of the pyramid are developed using varying methods. The surface **o-b-d** is developed by hinging it on line **b-d** in the plan and knocking it flat in the elevation. The developed surface results in **O-b-d** (in the plan view).

Surface **o-a-b** is similarly hinged (from **a-b**). The development is completed by making **O-b** equal in both developments.

From the elevation two surfaces are developed as follows:

Method

- Pivot **o-c** in the plan to a horizontal position and project to **C** in the elevation.
- Join **C** to **o₁**. This is the true length of edge **o-c**.
- Swing **o₁-C** from **o₁** and make **C-A** equal to **c-a** (the length of the square side from the base); **o₁-A-C** is **o-a-c** developed.
- Pivot **o-b** in the plan and project to **B**.
- Join **o₁-B**. This is the true length of **o-b**.
- Swing **o₁-B** from **o₁** and make **A-B** equal to **A-C**; **o₁-A-B** is **o-a-b** developed (this is already done from the plan using an alternative method).

Fig. 16.3a shows the plan and elevation of an oblique cone with six generators indexed **o-1** to **o-7**. In the elevation, the generators which show true lengths are: **o₁-1₁** and **o₁-7₁**. The same cone is reproduced in **Fig. 16.3b**, with the generators developed to show their true lengths in the elevation:

16 OBLIQUE SOLIDS

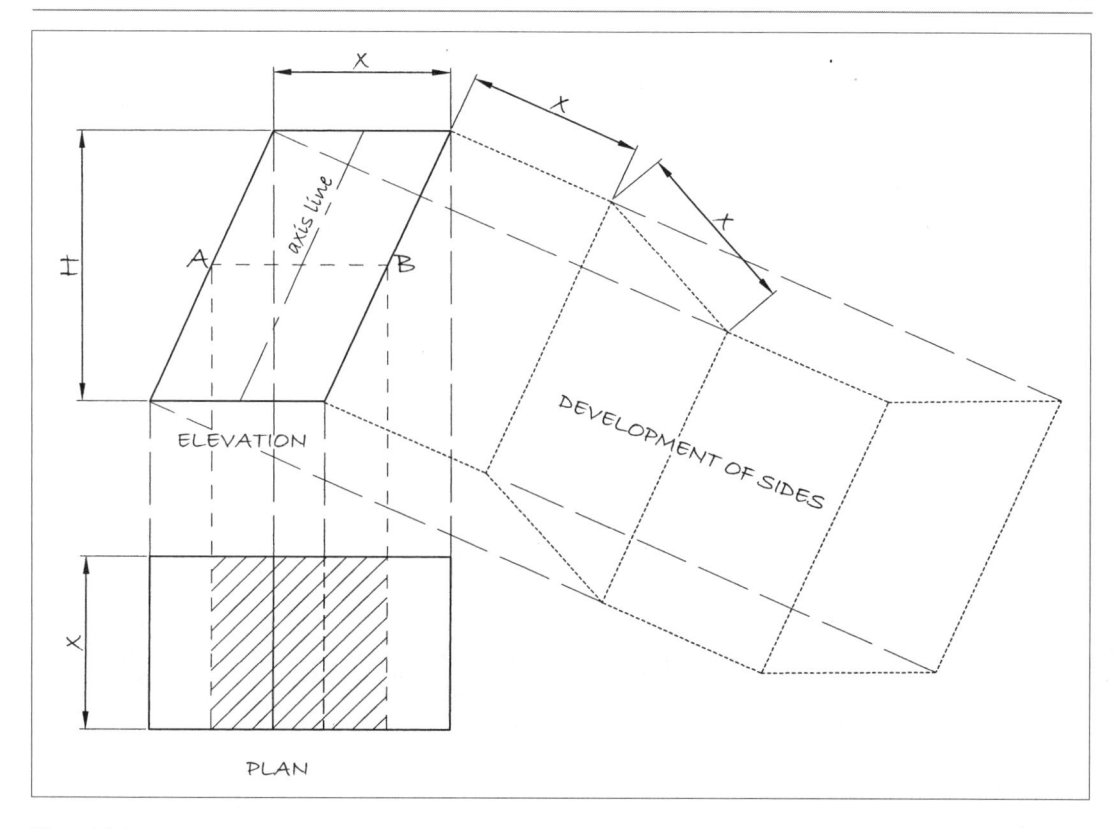

Fig. 16.1

Method

- In the plan, **o-2** to **o-6** are pivoted to a horizontal position along line **o-7**.
- Their new positions are found in the elevation (**2d** to **6d**) and joined to **o1**. They now reveal true lengths.
- From **o1**, pivot all of the true lengths as a series of arcs.
- From **7d**, swing distance **x** to intersect with the arc from **6d** and continue as shown, using **x** as a constant, so that the developed curve **7d-1d** equals half of the circle circumference.
- Join **1d** to **o1** to complete one half of the development of the curved surface of the cone.

The developed surface of an oblique cylinder is given in **Fig. 16.4a**. The cylinder is divided into a number of ordinates in the plan, **1** to **7** for example, and their positions in the elevation are found to run parallel with the axis line.

The development is hinged from ordinate **41-41**, and all other ordinates project square-off this line. The dimension **x** from the plan, is a constant measured from ordinate to ordinate in the development, as shown. The developed curved line **41-41** must equal the circumference of the circle.

Fig. 16.4b represents two elevations of a stack of coins. On the left the stack is upright like a right cylinder and on the right the stack is skewed like an oblique cylinder. This illustration

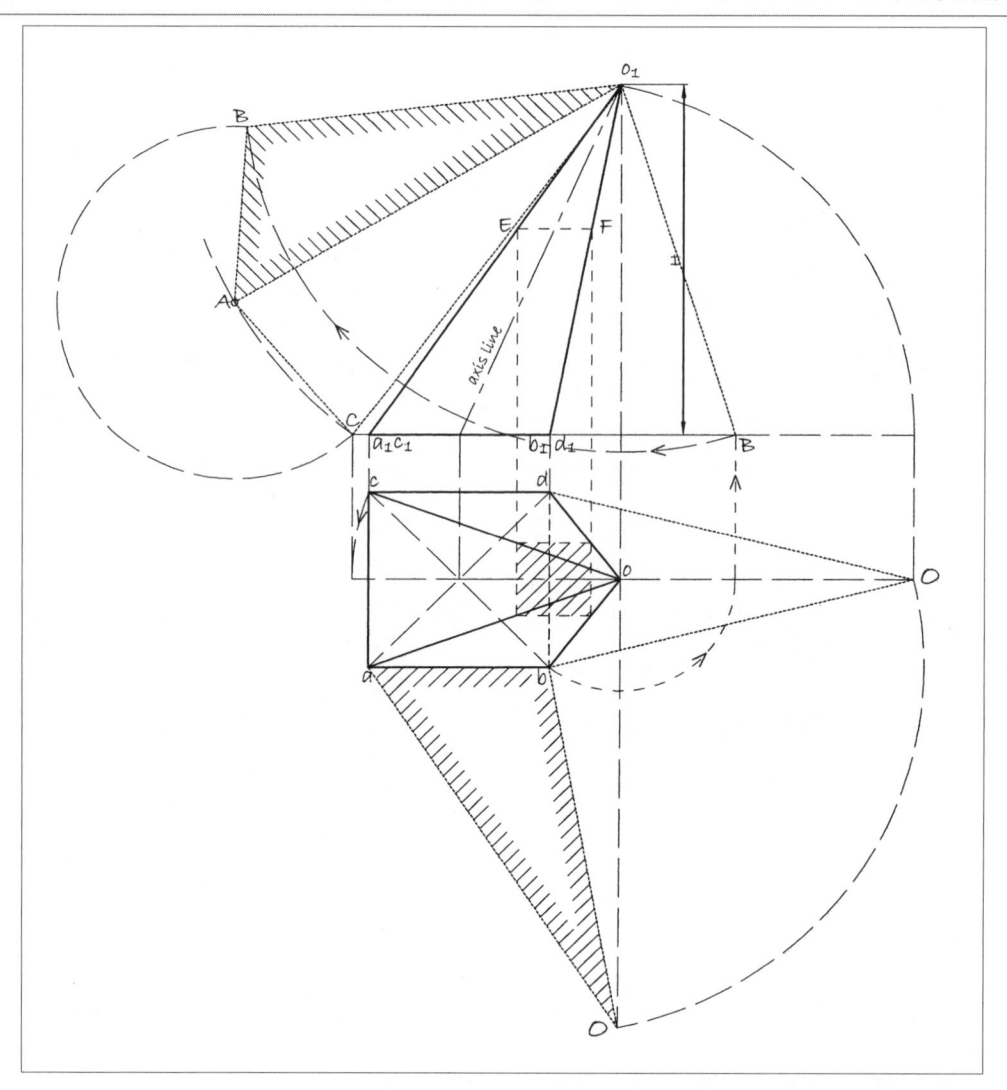

Fig. 16.2

is to show that both cylinders have the same height and that an oblique cylinder is a right cylinder pushed to one side. If the thickness of the coins is infinitely small, the fact that the cylinder has a serrated edge is barely visible. It can, therefore, be deduced from both examples that the surface area and volume does not alter in either case. Consequently, for both situations the formula ($\pi r^2 h$) for finding the volume of a cylinder is the same.

In **Fig. 16.5** an application of the oblique cylinder is given. A dormer roof, which is part of the cylinder, is used in the example. The elevation and end elevation of the dormer on a pitched roof is shown. The complete cylinder, on which the dormer is based, is indicated in dotted outline in the end elevation. The dormer roof surface is developed, and the shape of

16 OBLIQUE SOLIDS

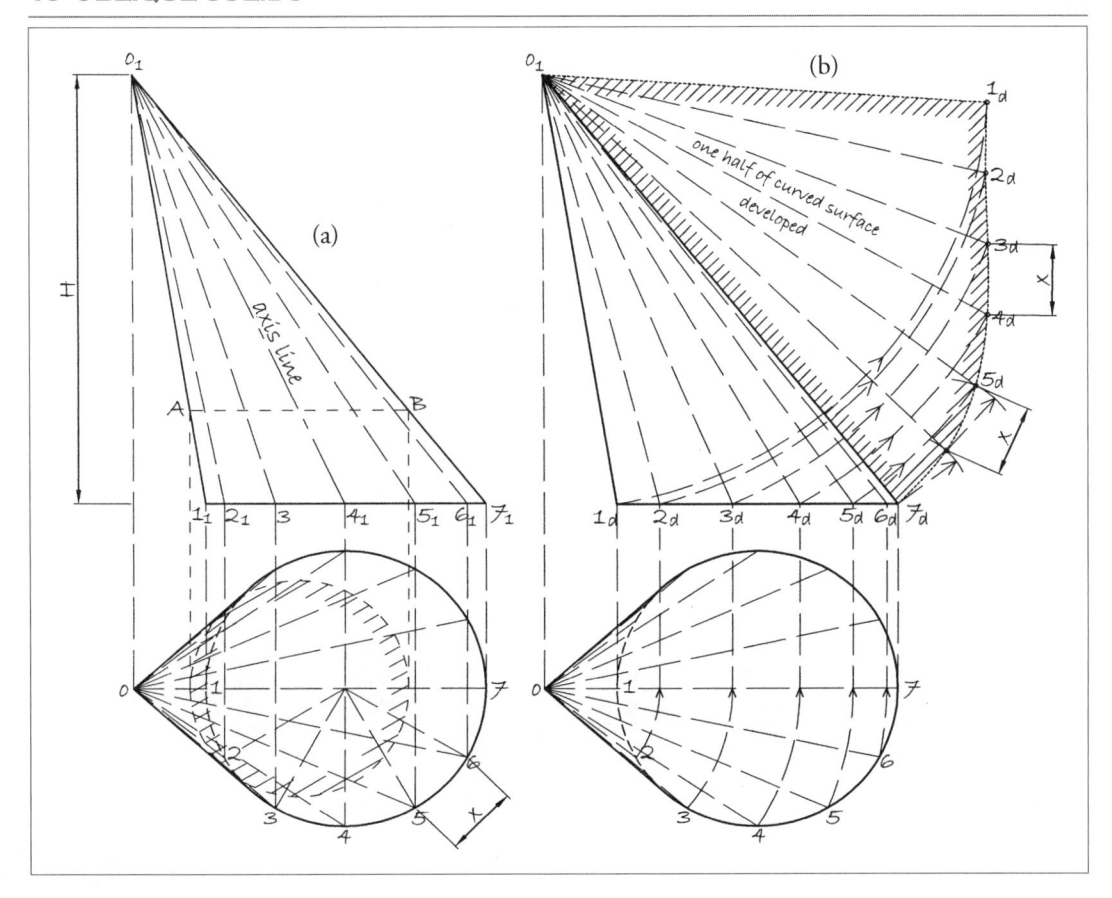

Fig. 16.3a
Fig. 16.3b

the lay board for the main roof is shown as an auxiliary plan set up on the line **c/l**. The widths **A**, **B** and **C** from the elevation are transferred to the development.

Exercise

Fig. 16.6 shows two oblique solids in plan and elevation: a hexagonal-based pyramid (left) and a half-cone (right). Reproduce **Fig. 16.6** and complete the exercise.

- Develop surfaces **A**, **B**, **C** and **D** of the pyramid.
- Develop surface **H** and the curved surface of the half-cone.

Fig. 16.4a
Fig. 16.4b

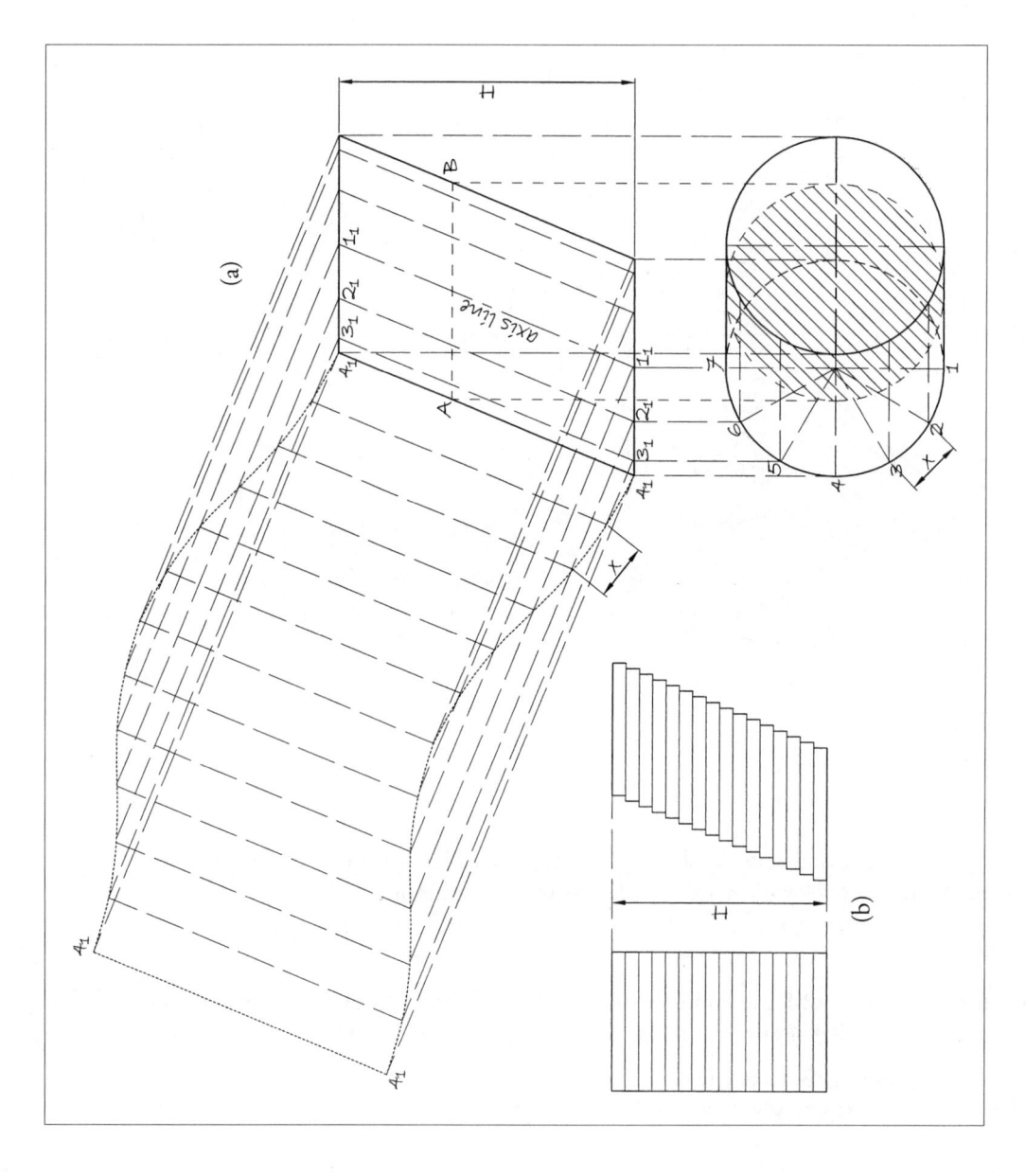

16 OBLIQUE SOLIDS

ELEVATION

END ELEVATION

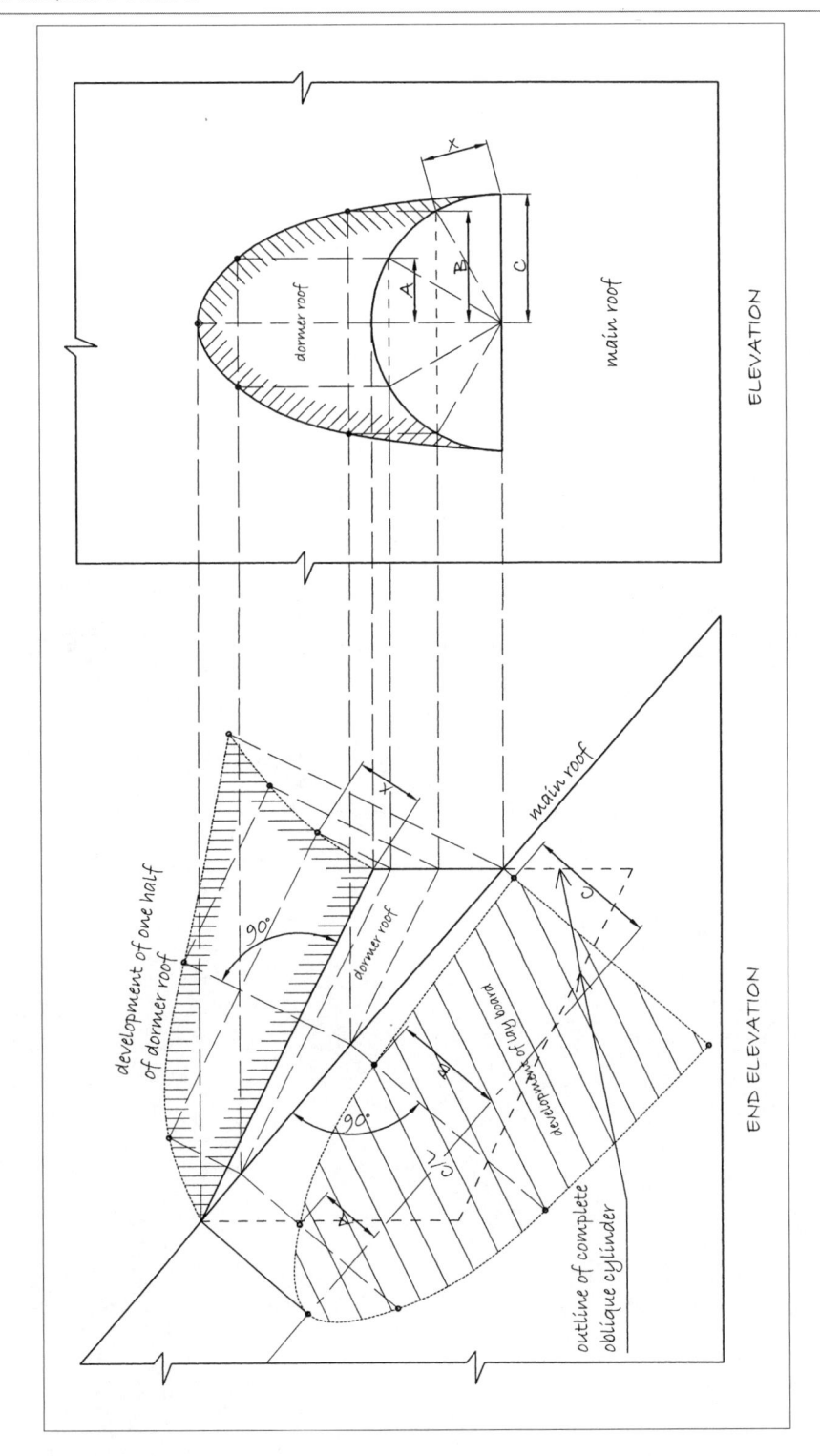

Fig. 16.5

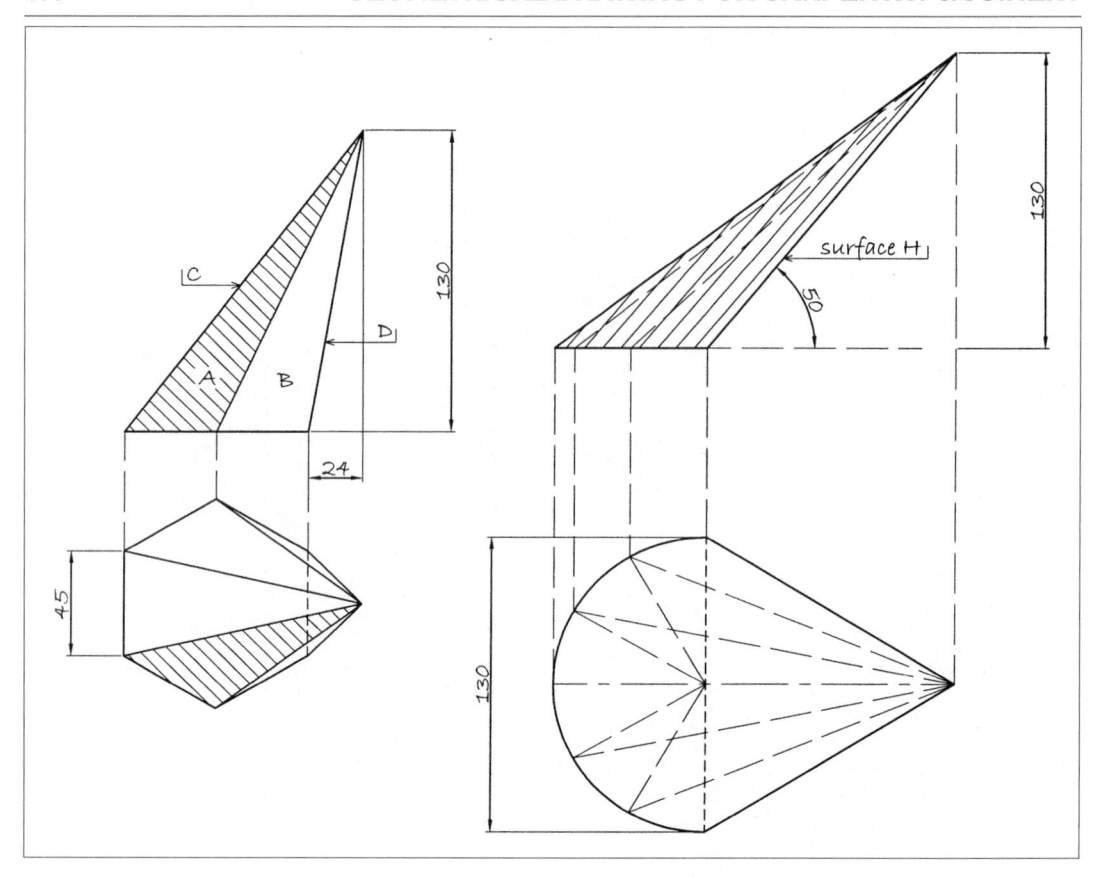

Fig. 16.6

17 CONSTRUCTIONS

SECTION I: ENLARGING AND REDUCING

This section covers the enlargement and reduction of areas using geometrical rather than mathematical principles. We shall deal with the method called *polar projection*, also know as *radial projection*.

Figs. 17.1 and **17.2** show *polar*, or *radial*, *projection*. This means that a figure is subjected to 'rays' from a point called the polar point (**p**). The figure is drawn, and the position of **p** determined anywhere convenient along the baseline. Radiating lines from **p** are projected through the relevant corners (intersections) of the figure and extended indefinitely. Between the radiating lines, the potential for an indefinite number of similar figures exists, from infinitely large to the size of point **p**. In **Fig. 17.1**, for instance, potentially there is no limit on the height of **a-b**.

It is important to remember here that with **Figs. 17.1** and **17.2** the enlargement and reduction is proportional, therefore, all the figures will be *similar* in shape. In similar figures, angles do not change. In **Fig. 17.1**, therefore, the slope in the roof-like shape will remain the same in all cases.

In **Figs. 17.3** and **17.4** the radial projection is done from point **p**, while it resides at the corner of the figure. Here, the enlarged or reduced shapes will also be similar, and an infinite number of figures between the smallest and largest are possible. Note: r = radius.

Fig. 17.5 shows a section of moulding where the relevant widths are swung down to the vertical so they lie on **a-b**. Radiating lines are then drawn through them and they enlarge to their new widths on line **a1-b1**. They are then swung back to the horizontal. This is an alternative method of transferring widths to the enlarged (or reduced) item.

In **Fig. 17.6** the polar point (**p**) is shown suspended in space to the rear of a try square, which is similarly suspended. A series of radiating lines through the corners of the try square show how an indefinite series of similar items may be produced.

Fig. 17.7 shows a moulding in section as well as a reduced version, with a different thickness. In this case, the thickness (or horizontal dimensions) is reduced by two-thirds, while the heights remain the same.

Method

- Divide **0-3** into three equal parts and swing two parts (**0-2**) to the vertical to become **0-21**.
- Draw horizontally from **21** to **4** and drop a vertical from **3** to **4**.
- Join **4** to **0** to find the *reducing* diagonal.

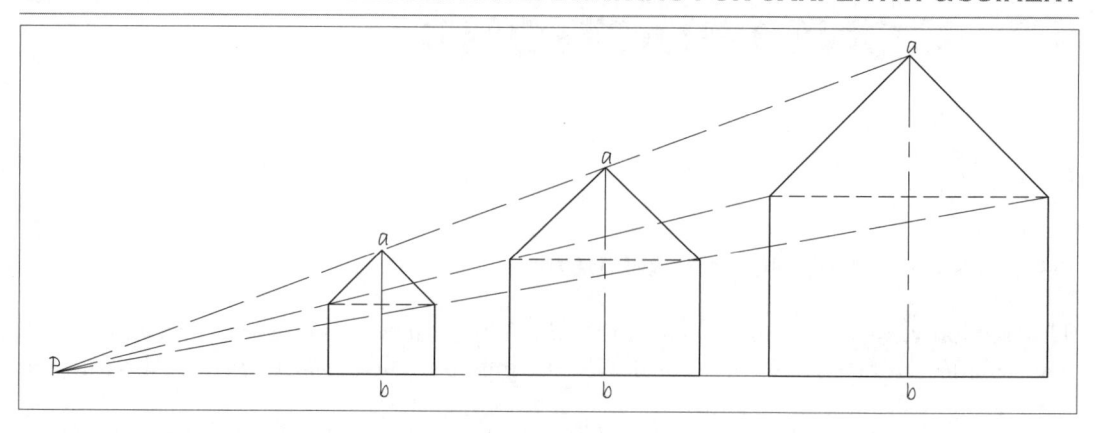

Fig. 17.1

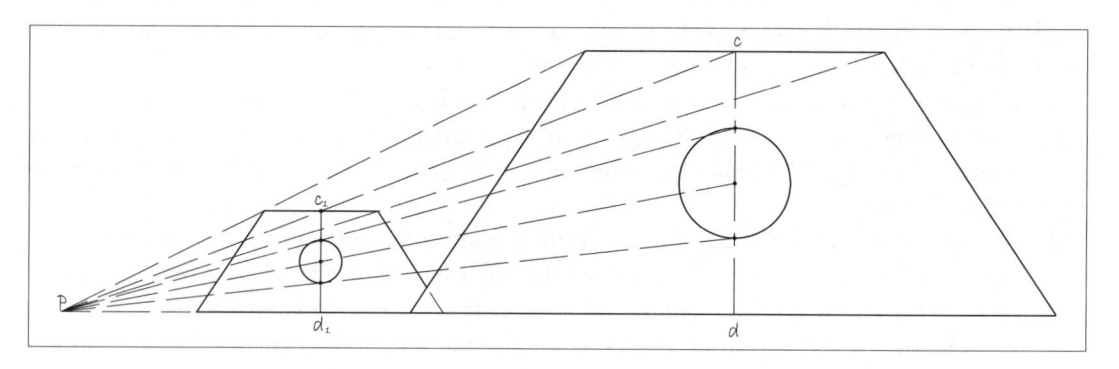

Fig. 17.2

- Project all of the thickness dimensions of the moulding vertically down to the diagonal **0-4** and project them horizontally to **0-21**. Here, they are now reduced by two-thirds.
- Swing them back to the horizontal and project them vertically to intersect with the horizontals from the given moulding.

When enlarging or reducing curves, as in this case, it is necessary to divide the curve into parts as shown at **x** and **y**. (Note: The more parts there are, the more accurate the final result. For clarity, only two parts have been used here.) In this example the reduction is not proportional, as with the previous figures.

In **Fig. 17.8** a section of architrave is shown. The task is to enlarge the height (**h1**) of the architrave to **h2**, while maintaining the same widths. Polar projection is employed for the height, while the widths are transferred geometrically as shown.

Fig. 17.3
Fig. 17.4

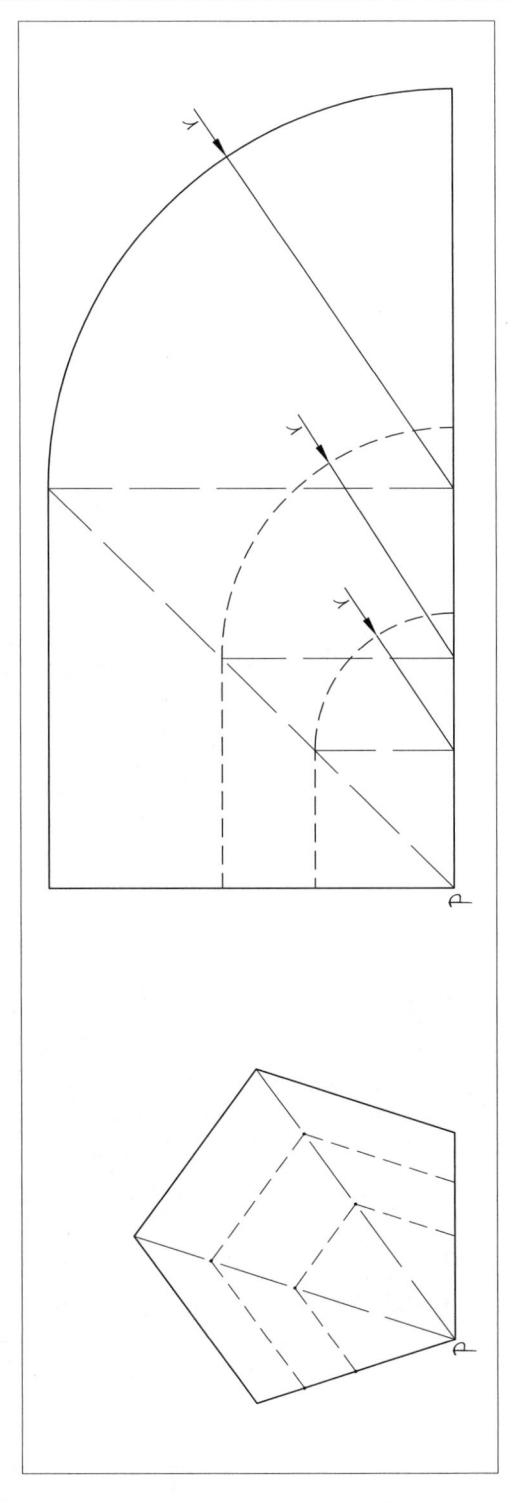

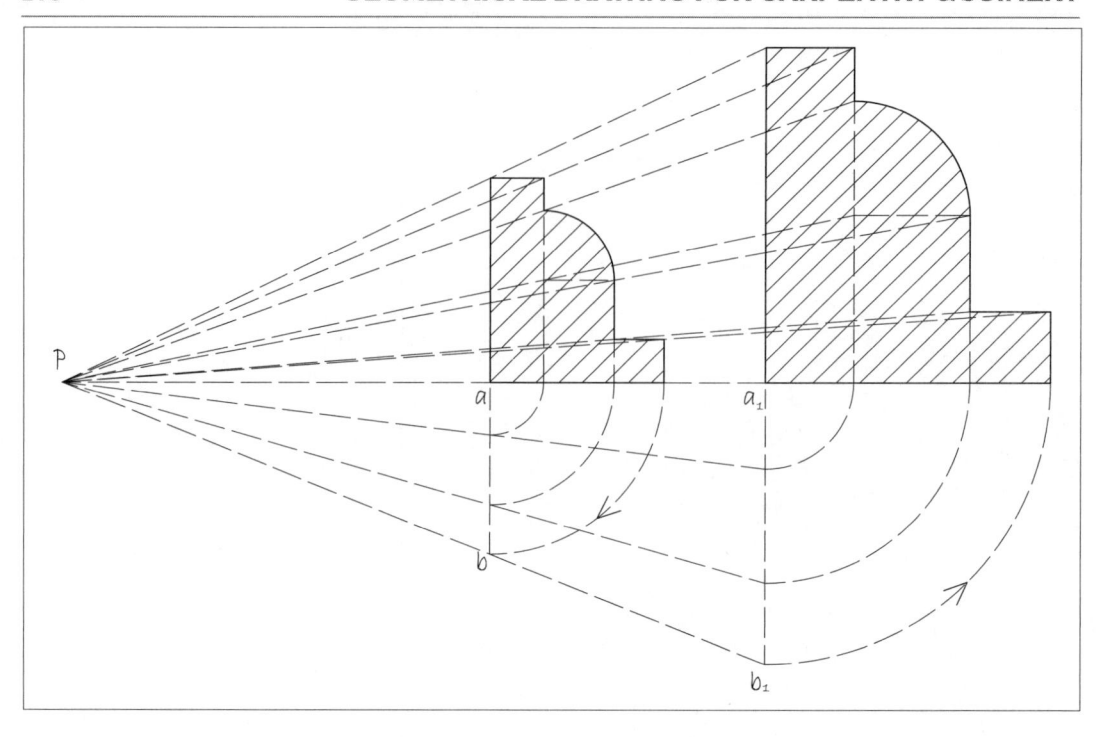

Fig. 17.5

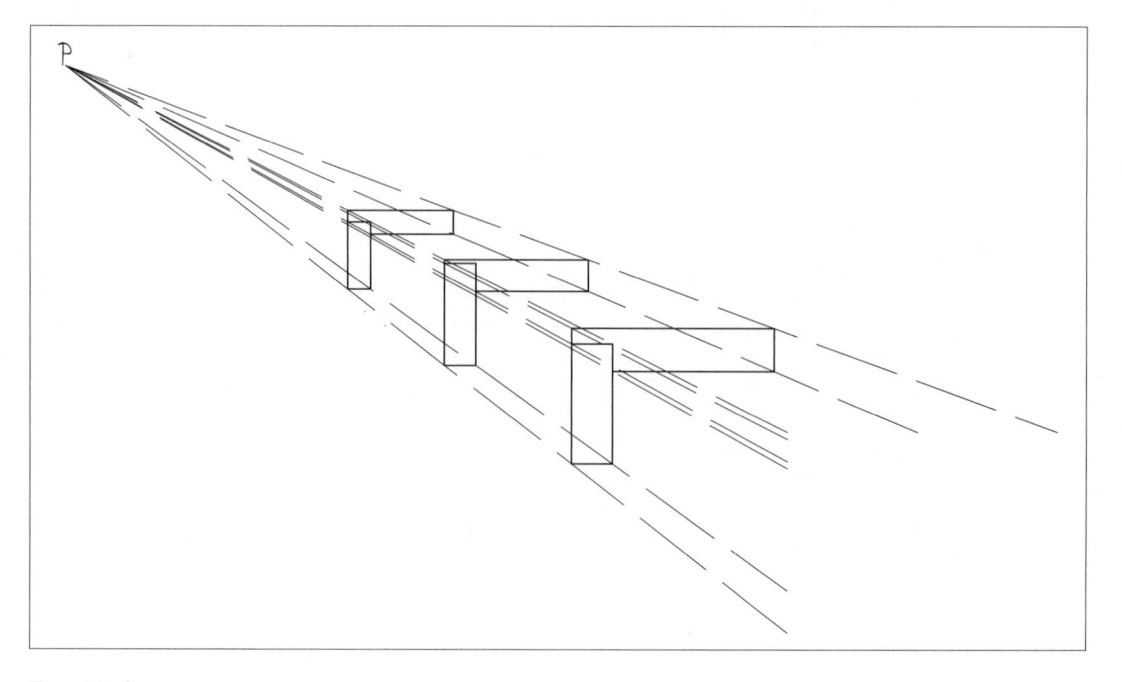

Fig. 17.6

17 CONSTRUCTIONS

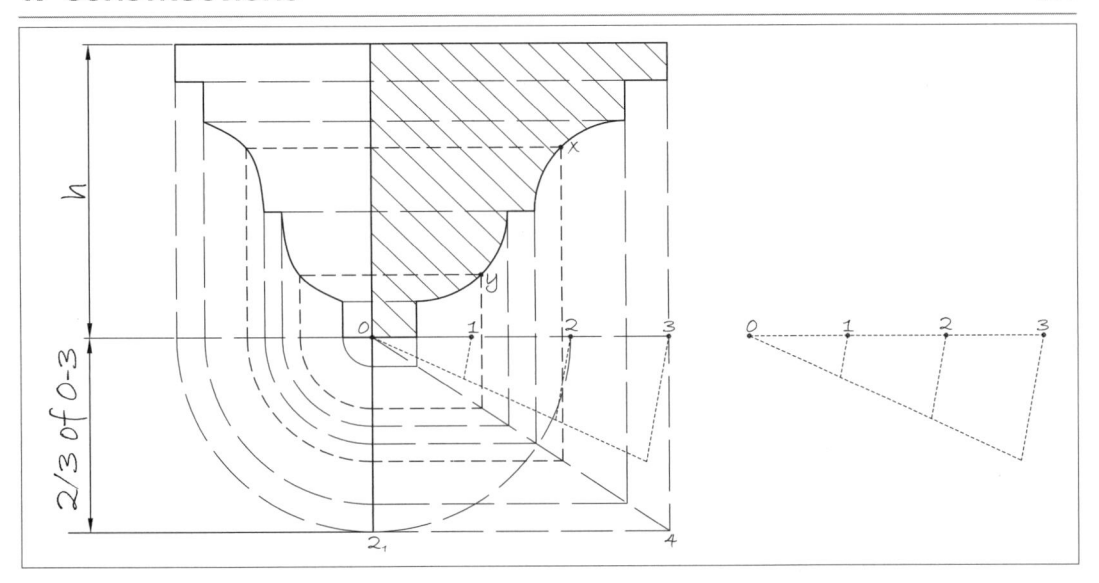

Fig. 17.7

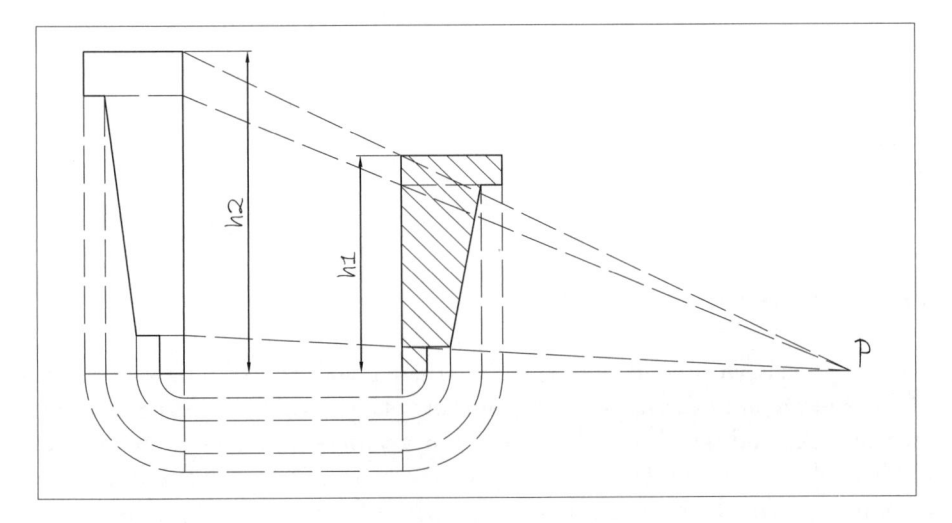

Fig. 17.8

Fig. 17.9 shows an ogee section moulding running horizontally and intersecting with a vertical moulding which has a different width. The mitre line for both mouldings is **a-b**.

- The thickness dimensions **a** to **g** are set off at 90° to the vertical moulding, where projection lines are drawn to intersect with their counterparts projected vertically from the mitre line.
- In this example, the mitre line **a-b** is the enlarging diagonal.

Mouldings are covered in greater detail in Chapter 13.

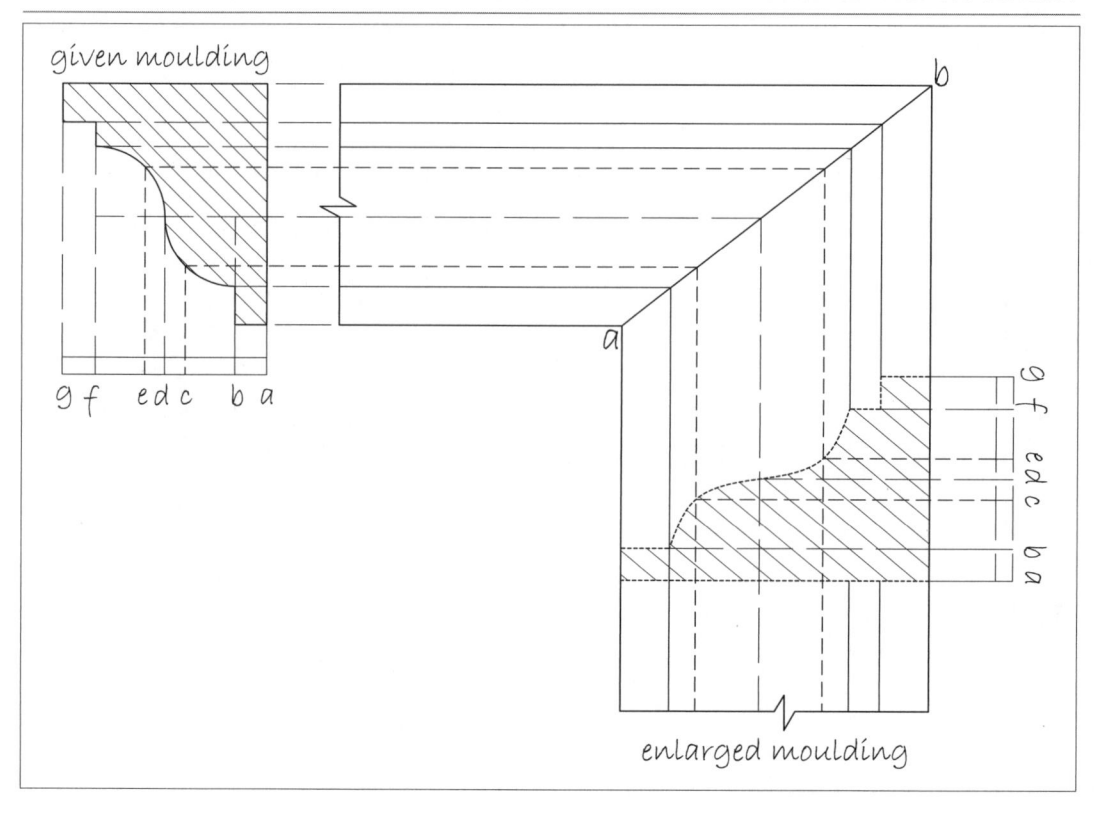

Fig. 17.9

SECTION II: GEOMETRIC PLANES

Solid geometry deals with solid objects rather than plane figures. Drawing these type of solids often requires an understanding of geometric planes, which are imaginary flat surfaces suspended in space and which may be inclined at a variety of angles to the horizontal and vertical. These planes have no thickness. To gain an understanding of planes, first try thinking of them as transparent, with the thickness of a sheet of paper, something like an extremely thin sheet of glass.

The basis for these planes is a horizontal and vertical which intersect at a line called **X-Y**, as shown in **Fig. 17.10a**. The planes make four quadrants, as indexed **1**, **2**, **3** and **4**, and are called, respectively, first, second, third and fourth angles of projection. What this means is that if an object is suspended in the first quadrant and its shape projected onto the horizontal and vertical planes, it is said to be drawn in first angle projection.

The shape projected onto the vertical plane is called the *elevation* and the shape projected onto the horizontal plane is called the *plan*. Subsequent *end elevations* may also be projected onto the end vertical plane.

17 CONSTRUCTIONS

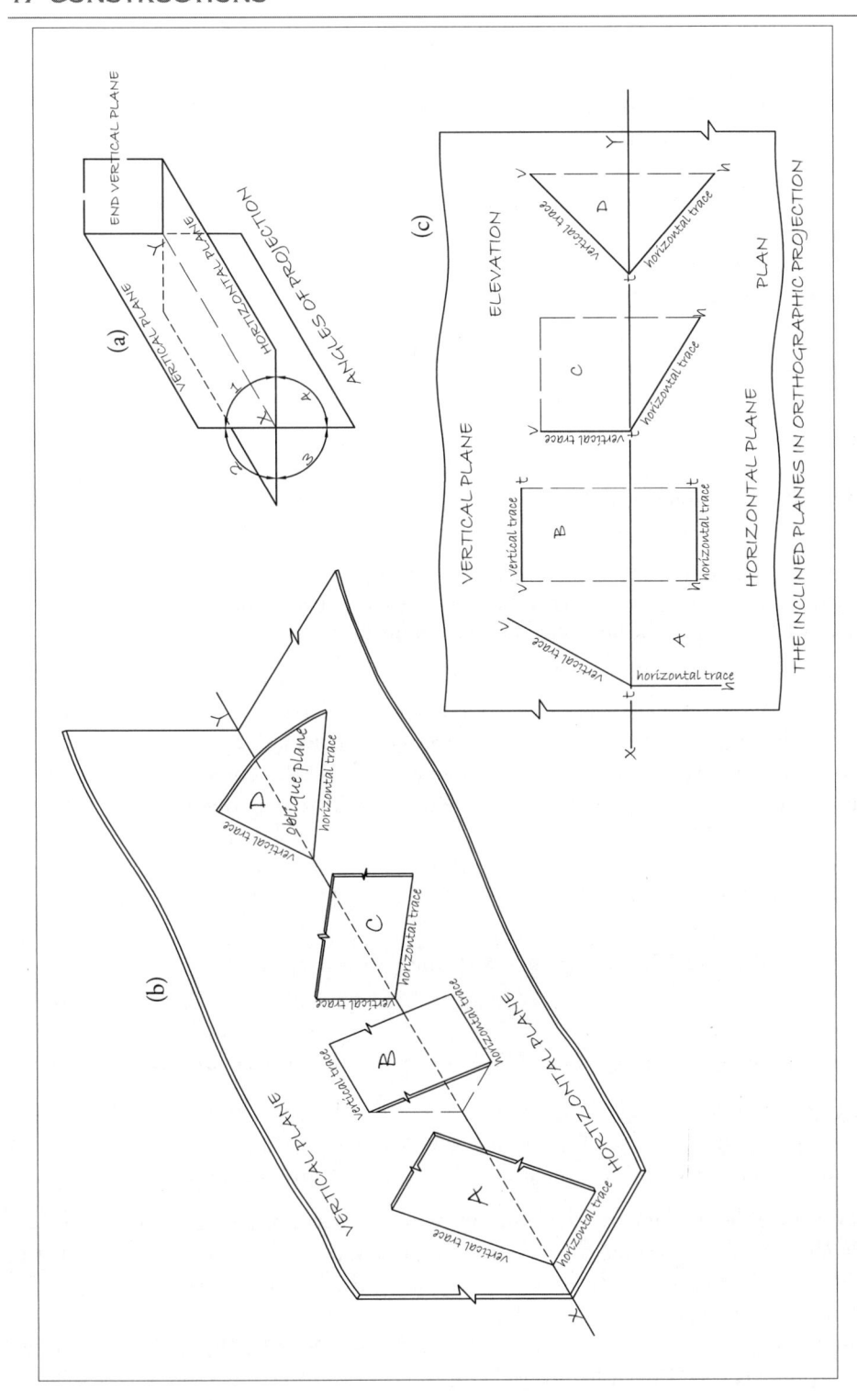

Fig. 17.10a
Fig. 17.10b
Fig. 17.10c

Simply inclined and oblique planes lean against the vertical and horizontal as shown in **Fig. 17.10b** and, where these touch the vertical and horizontal planes, they leave a *trace*, a line of intersection. **Fig. 17.10c** shows the vertical and horizontal planes 'flattened-out', with the traces as they would appear in plan and elevation.

When an object is drawn two-dimensionally on a flat surface with a variety of views called plans, elevations and end elevations, giving details and information about the object, this type of drawing is known as *orthographic projection*. Note: All of the orthographic projection in this book is done in first angle projection.

The planes shown in **Fig. 17.10b**, at **A**, **B**, **C** and **D** are as follows:

- **A** is simply-inclined (inclined to the horizontal plane).
- **B** is doubly-inclined (inclined to both the horizontal and vertical planes).
- **C** is an auxiliary vertical plane (may be used for projecting auxiliary elevations).
- **D** is an oblique plane (inclined to both the vertical and horizontal planes with the traces meeting on the **X-Y** line).

In **Fig. 17.11a** the first angle of projection is reproduced with a roof suspended from it. Projection lines from the roof are drawn to intersect with the vertical and horizontal planes.

- The front view projected onto the vertical plane gives the elevation.
- The end view projected onto the end vertical plane gives the end elevation.
- The top view projected onto the horizontal plane gives the plan.

Fig. 17.11b shows the planes 'flattened-out', with the projections as they would appear on a flat sheet. Simply inclined planes have been mentioned in Chapter 4 on geometric solids. They are used to illustrate the truncating of an object at a particular angle.

Knowledge of oblique planes is necessary to understand fully the geometry associated with handrail wreaths covered in Chapter 9. It is, therefore, useful to examine oblique planes to some extent here. In **Fig. 17.12a** a pictorial view of an oblique plane cutting a prism is shown. The cut surface is indexed **abcd**. The vertical and horizontal traces of the oblique plane are **VT** and **HT**. The prism base is parallel with the **X-Y** line and obviously the vertical plane **VP**. The horizontal plane is **HP**.

Look at **Figs. 17.12a** and **17.12b** together with the following procedure:

Method

- From the four corners of the prism (in plan), draw the base corners parallel with **HT** to touch the **X-Y** line.
- From the **X-Y** line project vertically to the **VT**.
- From the **VT**, project the points parallel with the **X-Y** line (**Fig. 17.12b**).
- Project verticals from **a**, **b**, **c**, and **d** in the plan to intersect at **a1**, **b1**, **c1** and **d1**.

Any point on the oblique plane can be found in plan and elevation using this method. It is known as the *level ordinates* method; for example, the line that contains **a** is a level ordinate: in other words, it is a horizontal line in reality. Drawing the line which contains **a** parallel with the **HT** in the plan ensures that it is a level line.

Note: In the orthographic presentation of the oblique plane, **Fig. 17.12b**, the **X-Y** line in the elevation also represents the **HT**, and conversely, in the plan the **X-Y** line represents the **VT**.

17 CONSTRUCTIONS

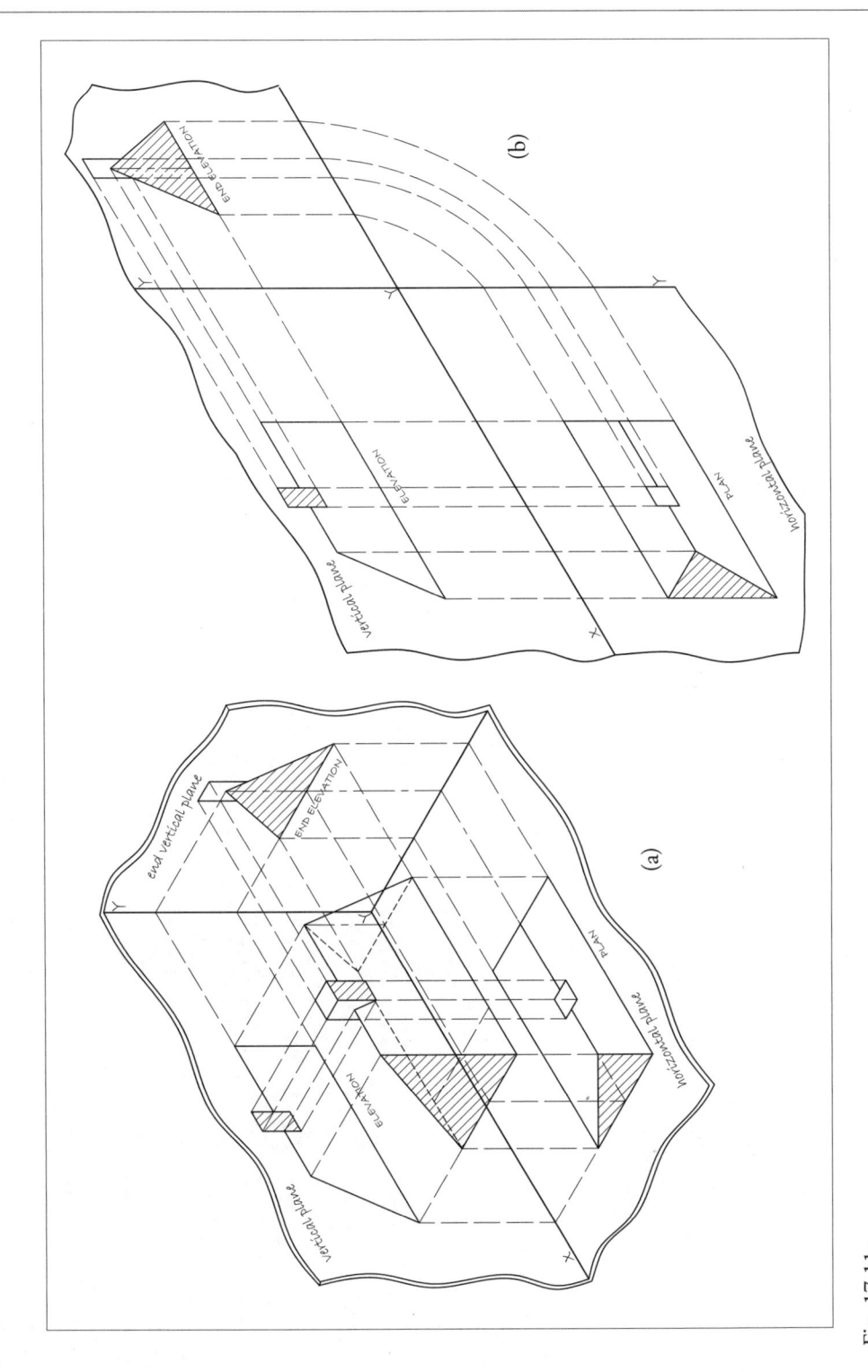

Fig. 17.11a
Fig. 17.11b

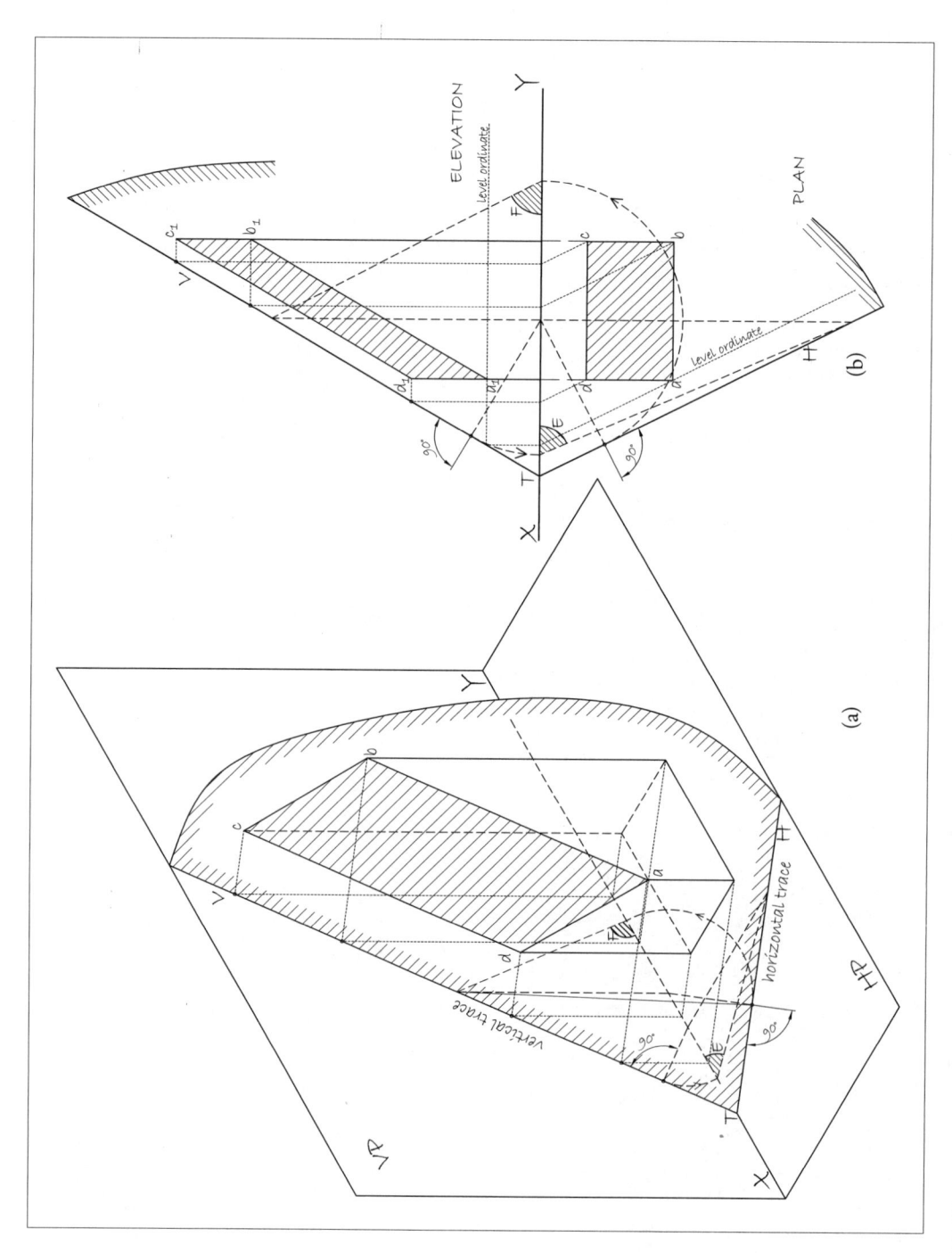

216　　　　　　　　**GEOMETRICAL DRAWING FOR CARPENTRY & JOINERY**

Fig. 17.12a
Fig. 17.12b

17 CONSTRUCTIONS

The (true) dihedral angles between the oblique plane and the vertical and horizontal planes are shown in **Fig. 17.12b** at **E** and **F**. These angles must be determined square-off the **VT** and **HT**. These are best understood in the pictorial view in **Fig. 17.12a**. If this 'wordy' explanation of the oblique plane seems complicated, do not despair, with a bit of study the drawing will reveal all.

The oblique plane is developed in **Fig. 17.13** where the figure from **Fig. 17.12b** is reproduced.

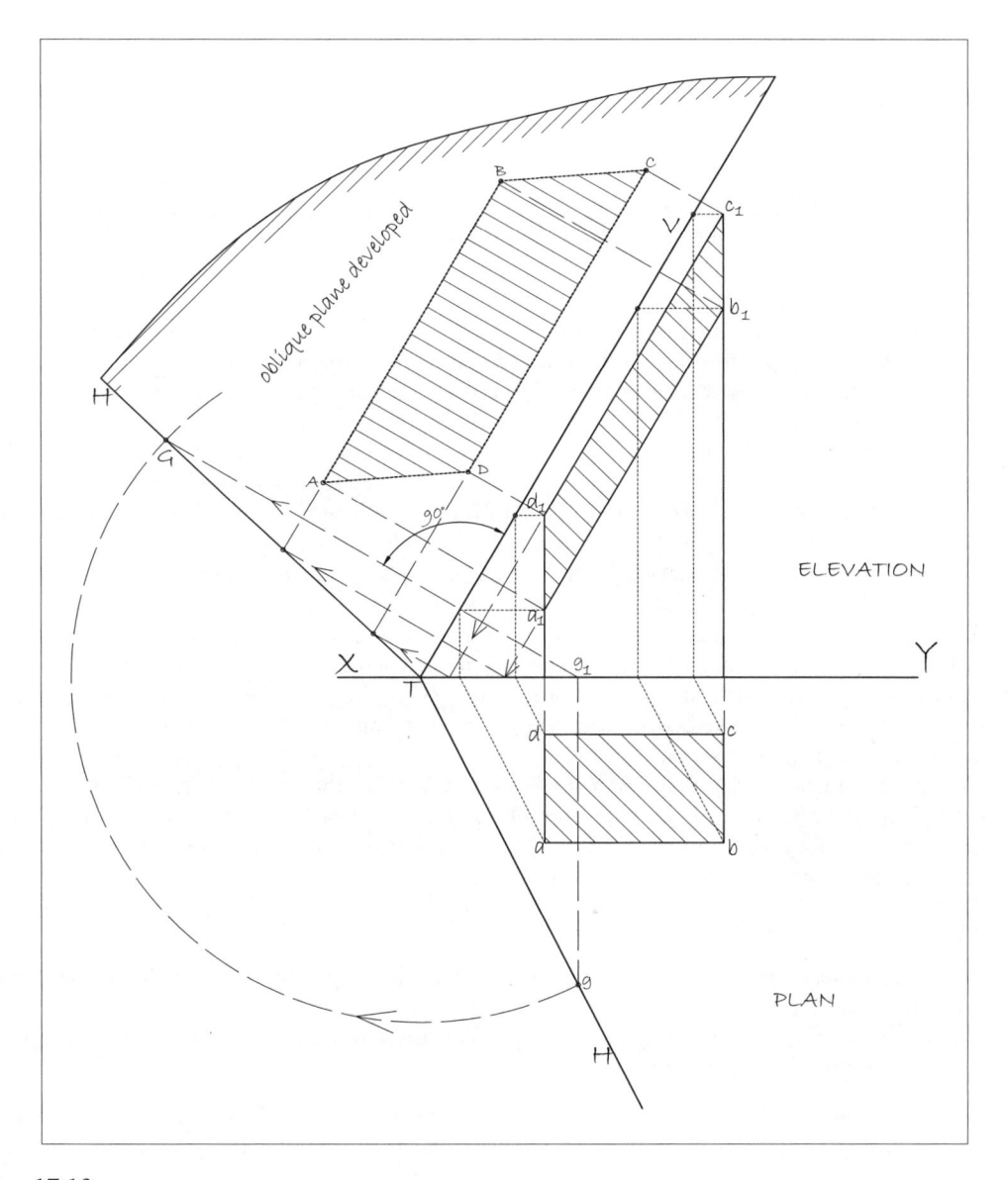

Fig. 17.13

Method

- On the **HT**, pick any point **g** and find its elevation at **g1**.
- Project **g1** square-off **VT** to touch the arc **T-g**, which has been struck from centre T, giving **G**.
- **TGH** is the developed horizontal trace of the oblique plane. **VT** stays where it is and is the 'hinge' for the development.
- Edge **c1-d1** is extended to the **X-Y** line and squared-off to touch the developed trace **TGH** from where it is drawn parallel with **VT**.
- Points **c1** and **d1** are drawn square-off to intersect at **C** and **D**.
- Points **a1** and **b1** are treated in the same way.
- **ABCD** is the true shape of the cut surface of the prism.

SECTION III: MISCELLANEOUS

In **Fig. 17.14** a selection of constructions are given.

A. A semi-circle is flattened to a straight line, which results in its true length and any sub-divisions of it. To facilitate this, an equilateral triangle is constructed on the diameter. Lines from **p** through the points on the semi-circle, and extended to the horizontal, give the developed sections and the overall length of the diameter.

B. A quick method of dividing a segment into equal parts is shown. Join **a** to **d** and square-off to the curve from the mid-point to give **b**. Join **b** to **d** and do the same to obtain **c**.

C. This shows that with any regular polygon the external angle is equal to the central angle.

D. A quick method of drawing a regular polygon within a circle is shown. Divide the diameter by the number of sides (in this case, seven) using the division of lines method. Draw arcs **0-7** and **7-0** from the extremities of the diameter to intersect at **k** and draw from **k** through point **2** to intersect the circumference of the circle at **m**. The length **0-m** is the length of the side of the polygon and can be stepped around the circle. Note: A seven-sided polygon is a heptagon.

E. A method of drawing a regular octagon within a square is shown. An arc of half of the diagonal of the square is drawn from the four corners to touch the sides and give the points needed.

F. A traditional boatwright's method of marking out an octagon on a squared baulk is shown. A distance of 12 inches is measured diagonally and marked across the face of the baulk. Three and a half inches are measured from each end of the line, giving the point on which the corner edge of the octagon lies. The square becomes an octagon, the octagon becomes a circle, the baulk becomes a mast and the boat goes to sea.

G. This is a rectangle known as the *golden rectangle,* which has the *golden ratio,* 1:1.618 … . This example of where geometrical drawing triumphs over mathematics is worthy of note. With a measuring tape it is not possible to measure 1.618 … , which is incomplete and goes on to infinity, but drawing the rectangle geometrically gives the exact ratio in linear form.

Method

- Construct a square of one unit (left of drawing). Join the diagonals and split the square in two vertically.
- Construct an arc equal to the diagonal of the half-square to terminate at **X** (right of drawing).
- The length of the rectangle is now acquired.

Within the rectangle, a spiral may be drawn as shown. This spiral is unique and can be found in nature, among other things, on seashells. The spiral is also used as a scroll form in handrailing.

17 CONSTRUCTIONS

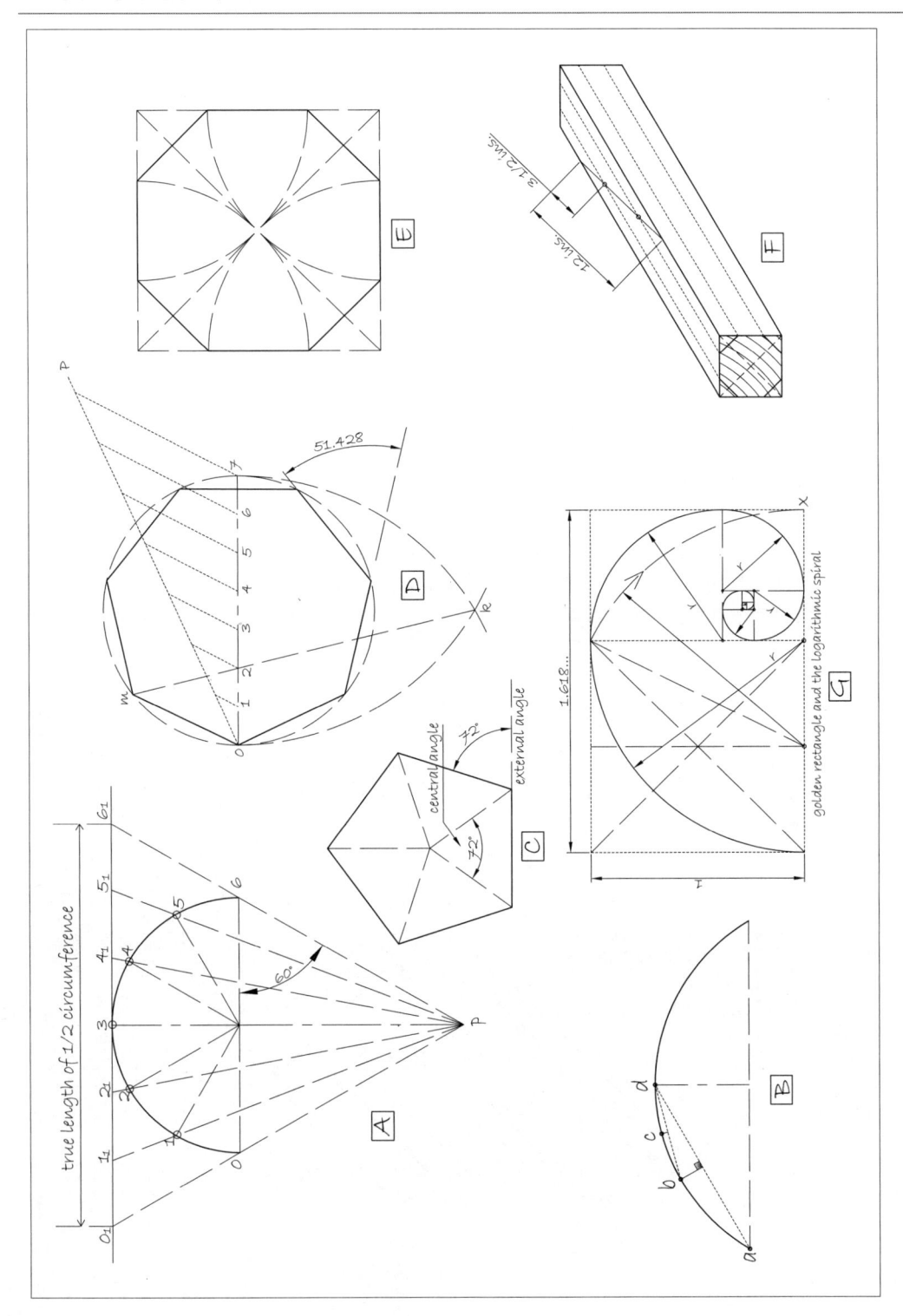

Fig. 17.14

AXONOMETRIC PROJECTION

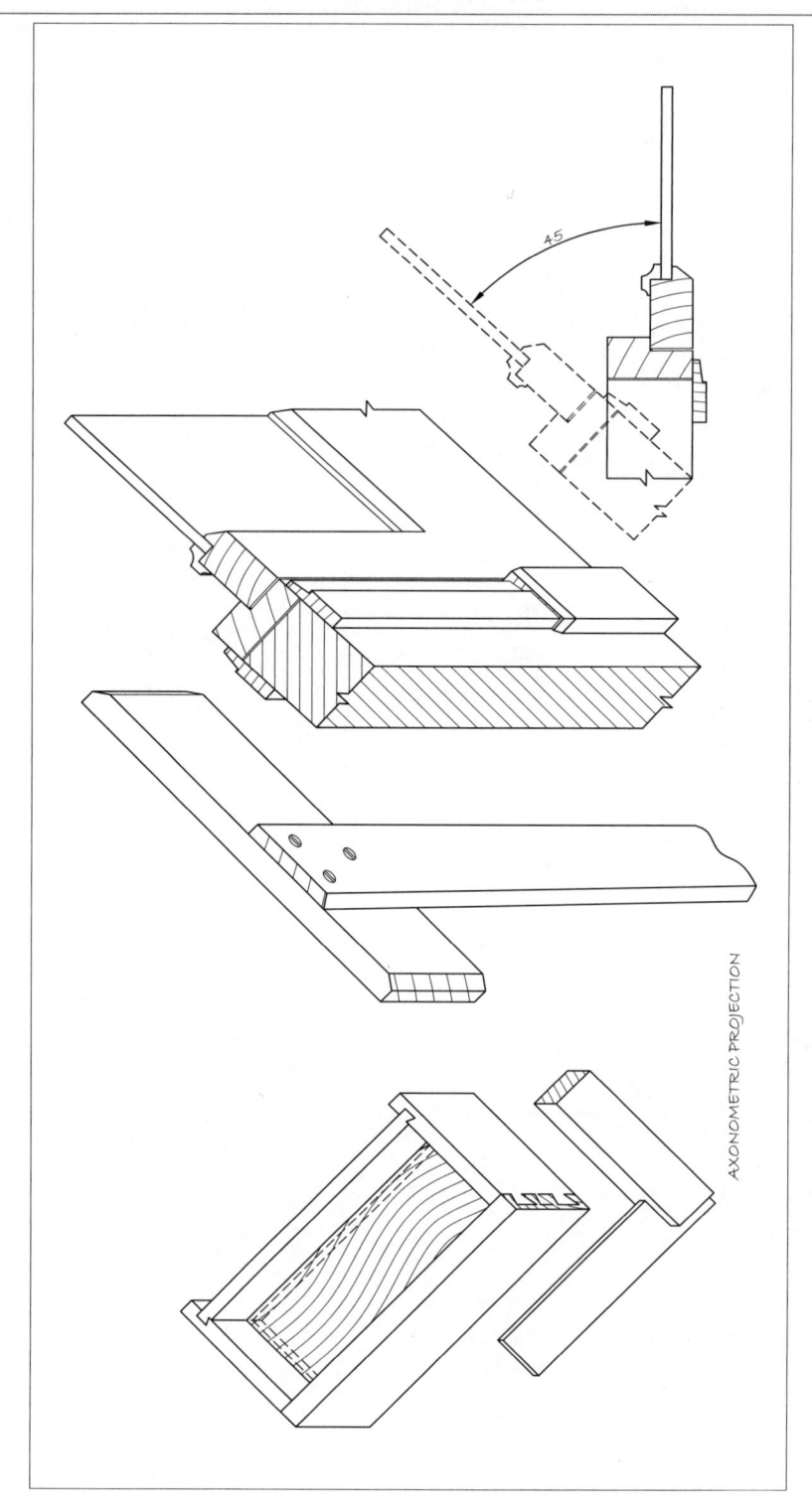

Fig. 17.15

17 CONSTRUCTIONS

The items in **Fig. 17.15** are common to the trade and drawn in *axonometric projection*, which is another form of pictorial projection useful to the carpenter or joiner. It differs from isometric projection in many respects and it comes into its own especially where sectional details have to be enhanced for better understanding.

Method

- The plan view of the item is drawn and rotated through an angle of 30 to 45°.
- Vertical lines are then drawn vertically from the rotated view as required. This is illustrated on the door frame section shown.

Exercise

1. **Figs. 17.16a** and **17.16b** represent solids being cut by an oblique plane. Reproduce **Figs. 17.16a** and **17.16b** and complete the exercise.

    - In each case, complete the elevation showing the cut surface.
    - Develop the shape of the cut surface.

2. An exercise on enlarging and reducing is given in **Figs. 17.17** (top) and **17.18** (bottom). Reproduce **Figs. 17.17** and **17.18** and complete the exercise.

    - Enlarge **Fig. 17.17** to an overall height of 90 mm.
    - Reduce **17.18** to an overall height of 65 mm.

Note: Use the given polar point (**P**) for each figure.

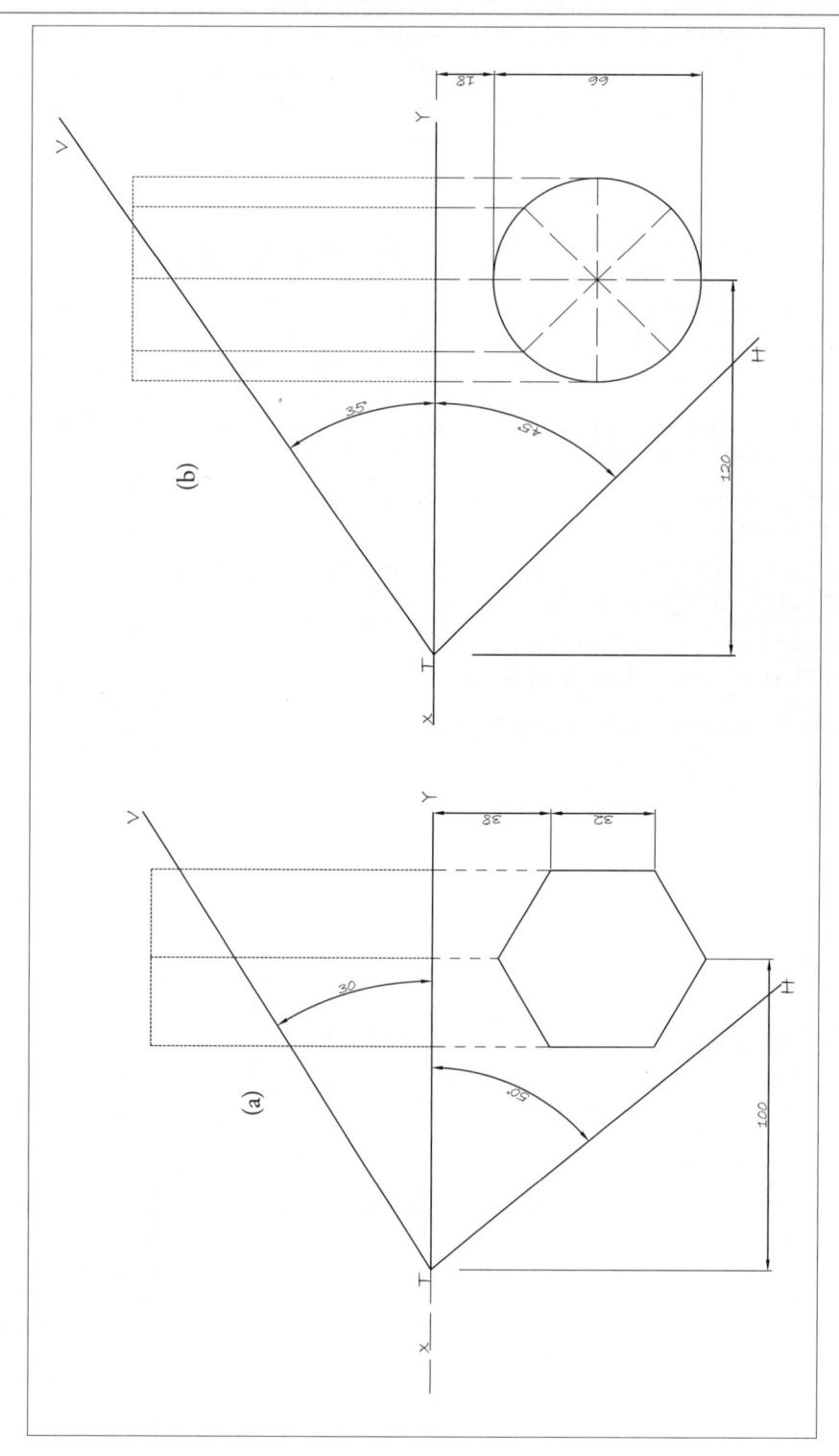

Fig. 17.16a
Fig. 17.16b

17 CONSTRUCTIONS

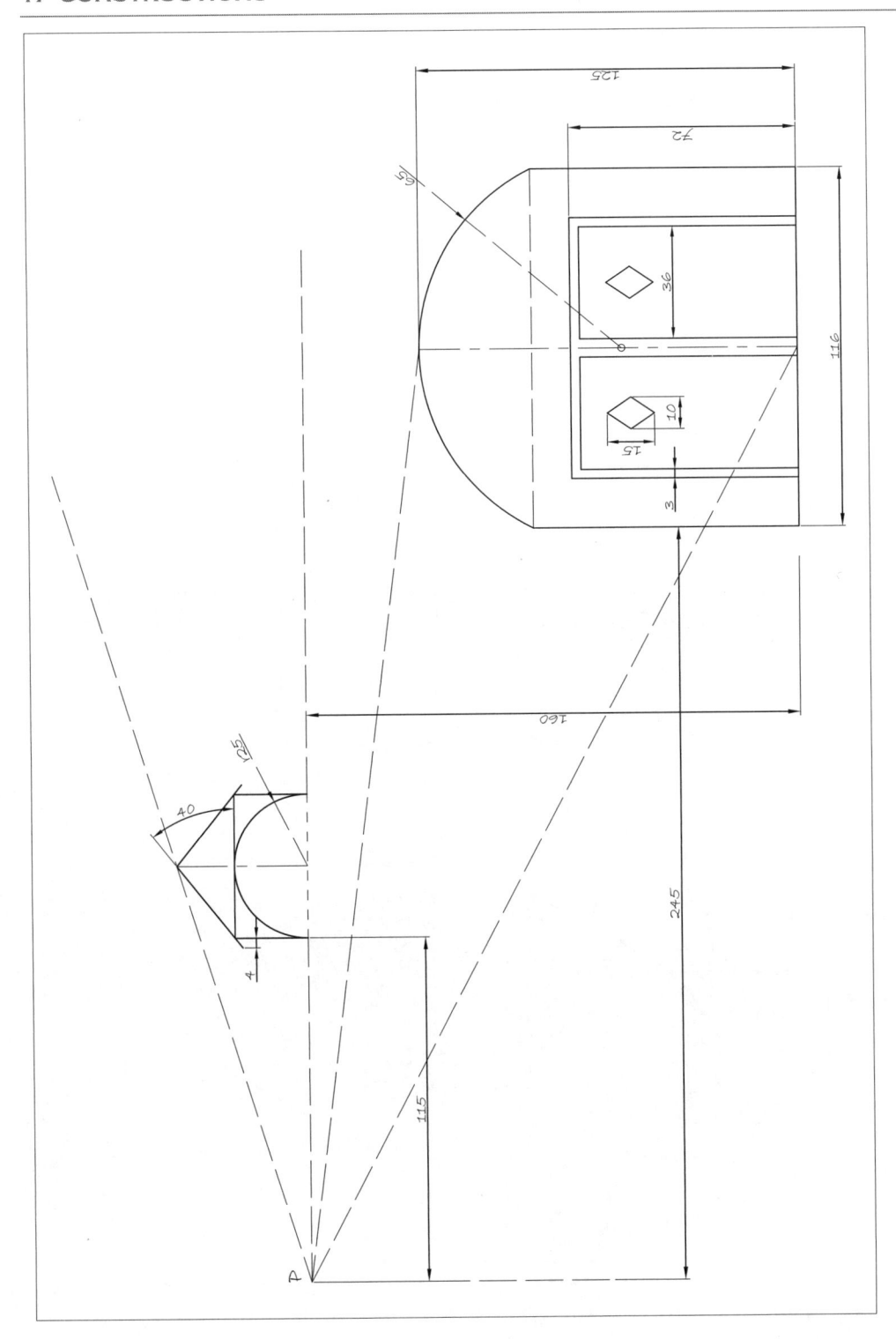

Fig. 17.17
Fig. 17.18

SOLUTIONS TO WORKED EXAMPLES

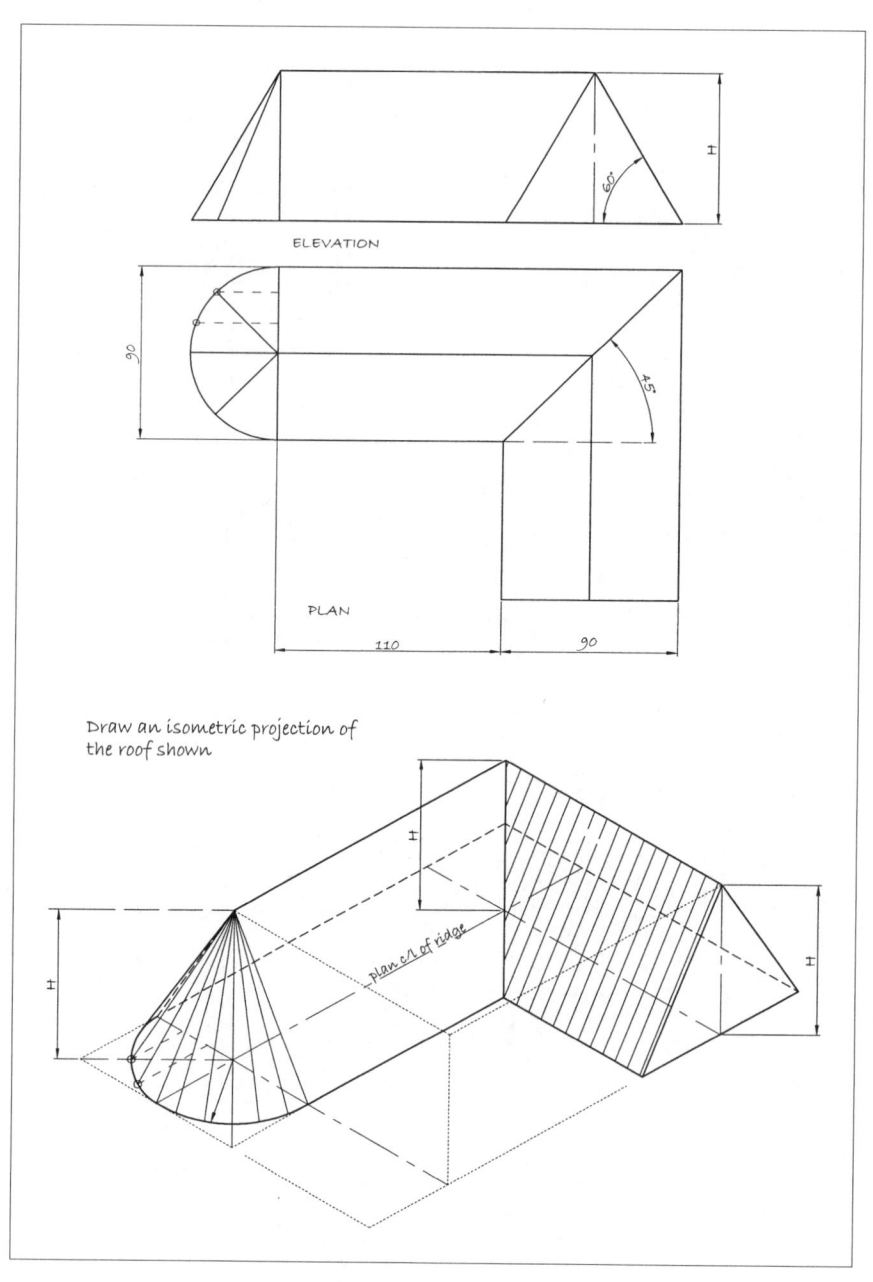

Chapter 1, Fig. 1.10

SOLUTIONS

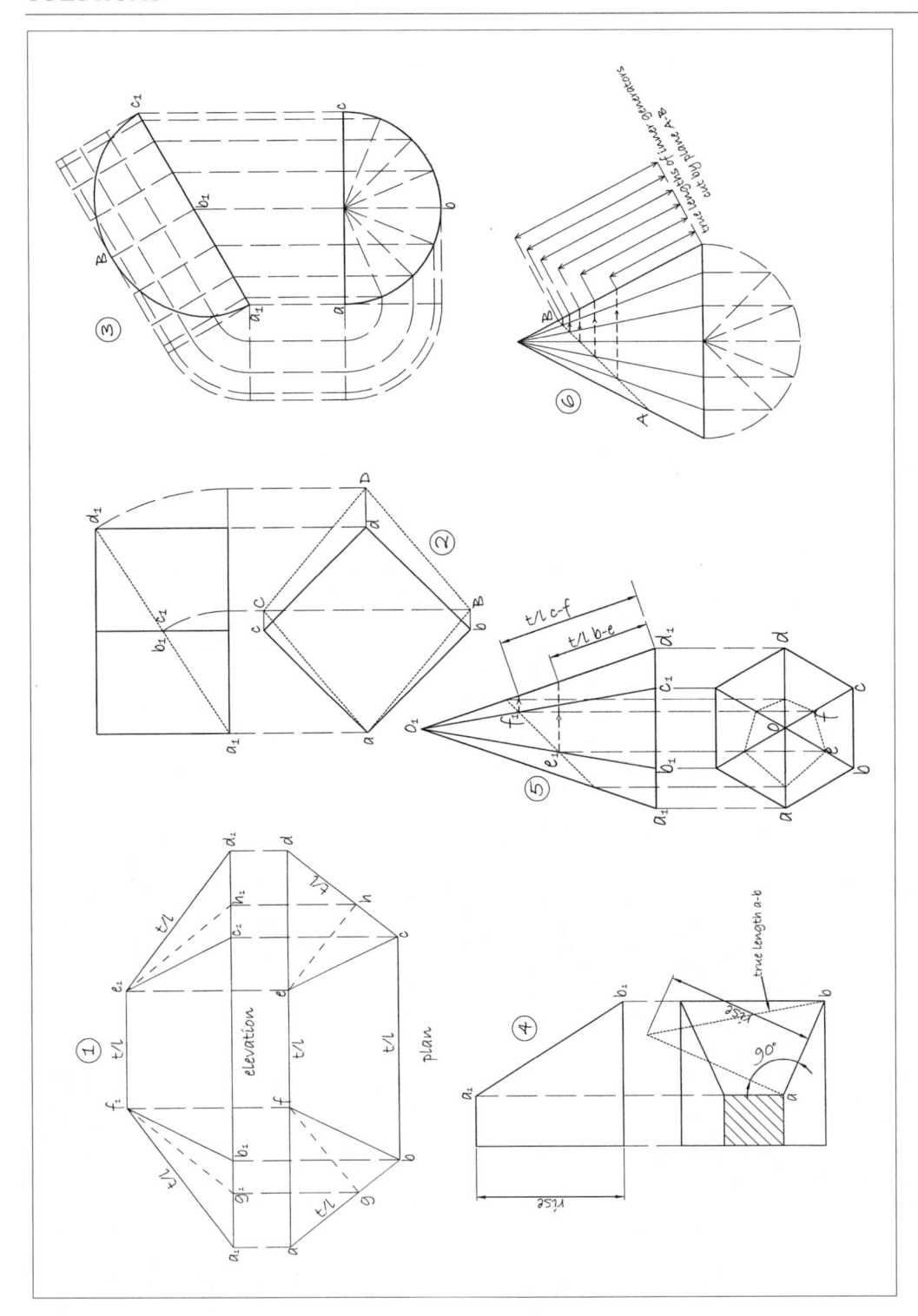

Chapter 2, Fig. 2.8

ELEVATION

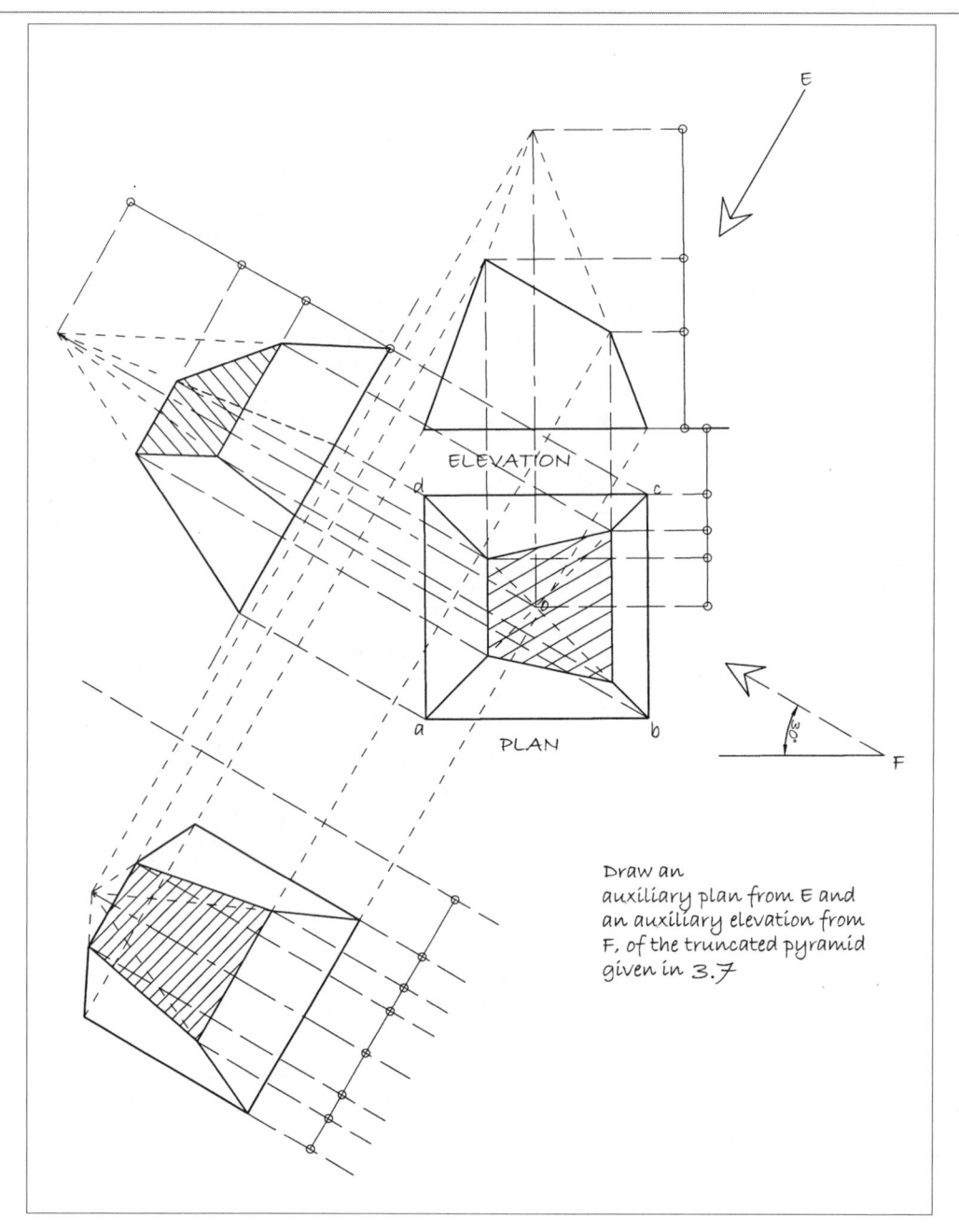

a PLAN b

Draw an auxiliary plan from E and an auxiliary elevation from F, of the truncated pyramid given in 3.7

30°

Chapter 3, Fig. 3.7

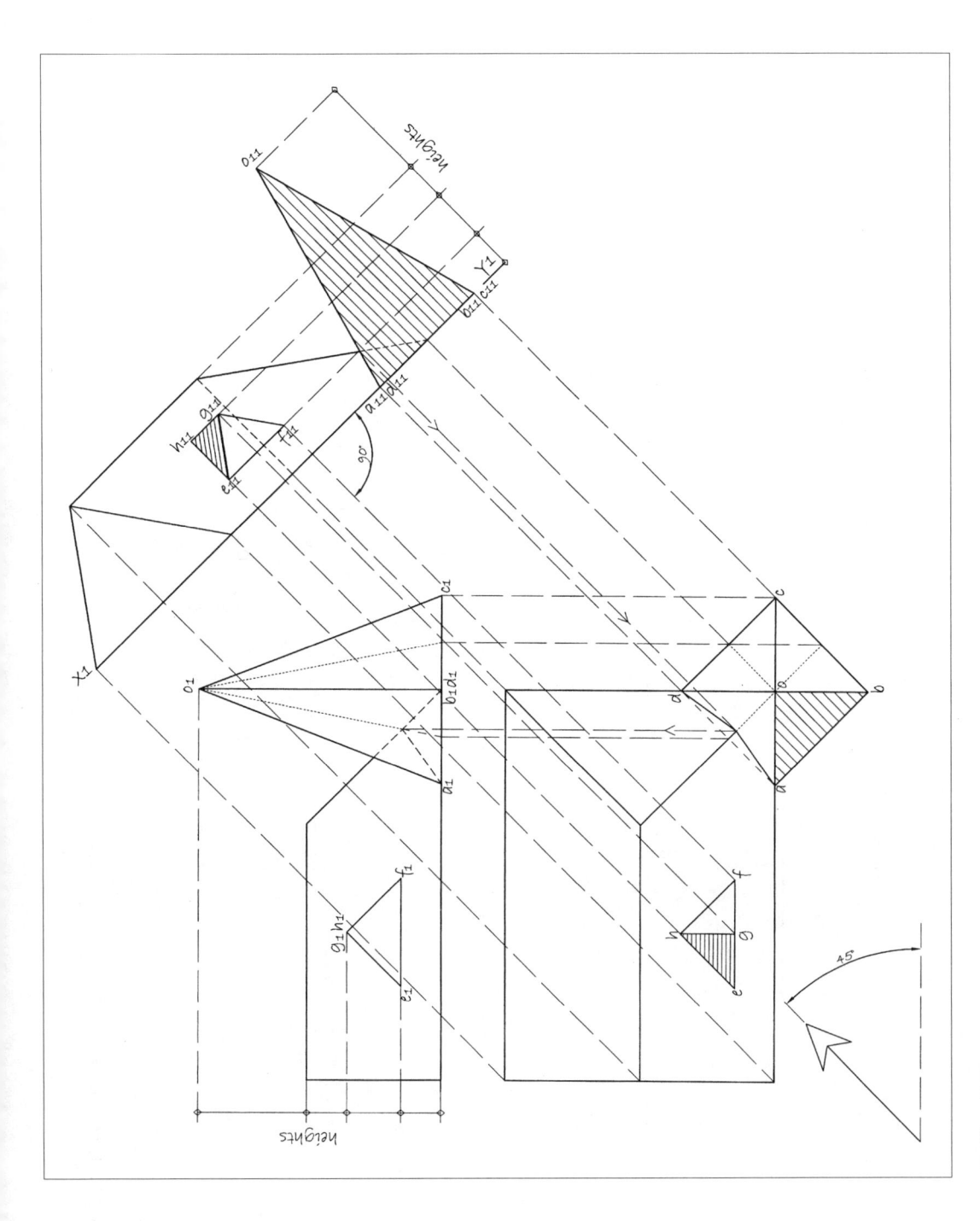

SOLUTIONS

Chapter 3, Fig. 3.8

GEOMETRICAL DRAWING FOR CARPENTRY & JOINERY

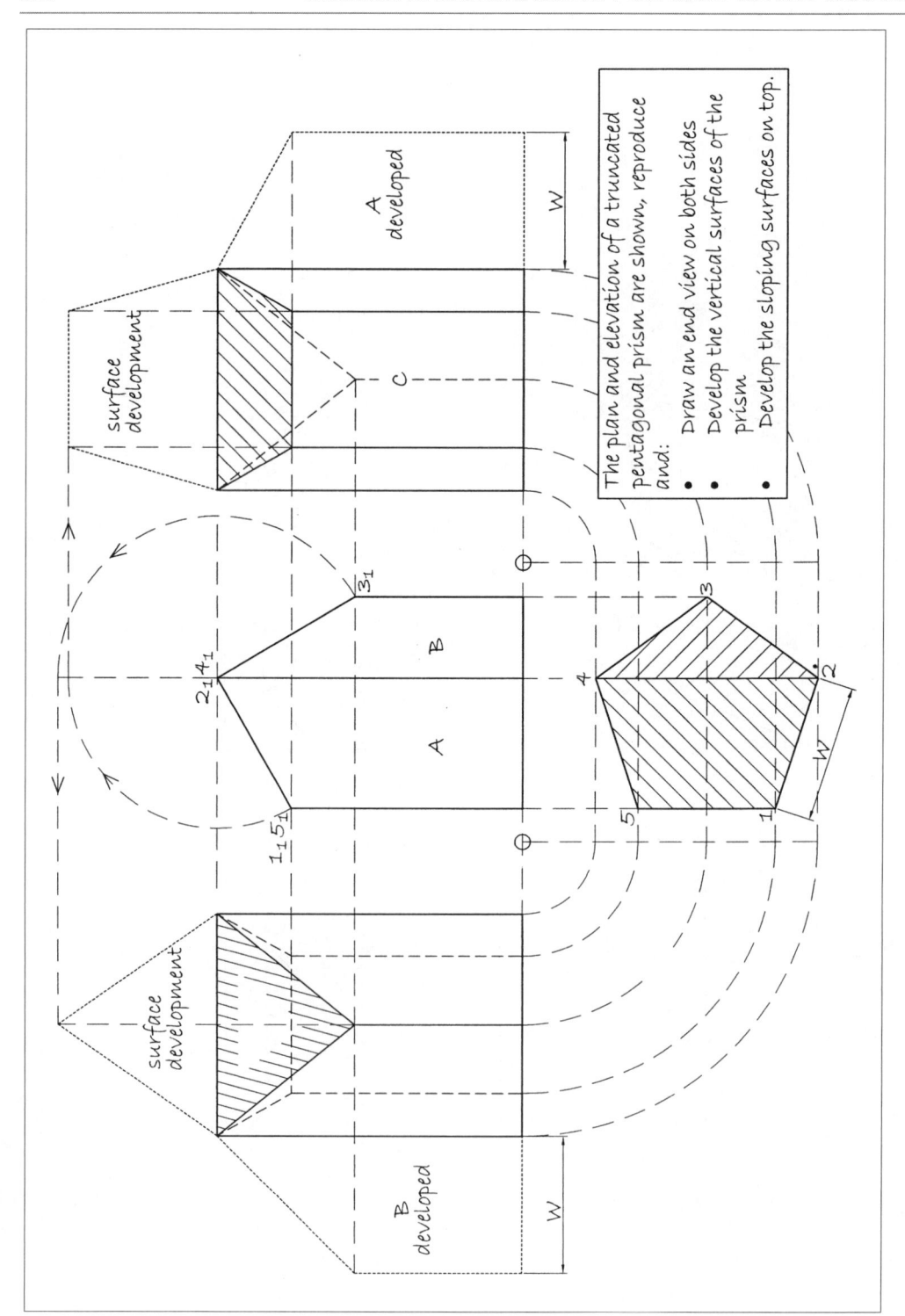

The plan and elevation of a truncated pentagonal prism are shown, reproduce and:
- Draw an end view on both sides
- Develop the vertical surfaces of the prism
- Develop the sloping surfaces on top.

Chapter 4, Fig. 4.11

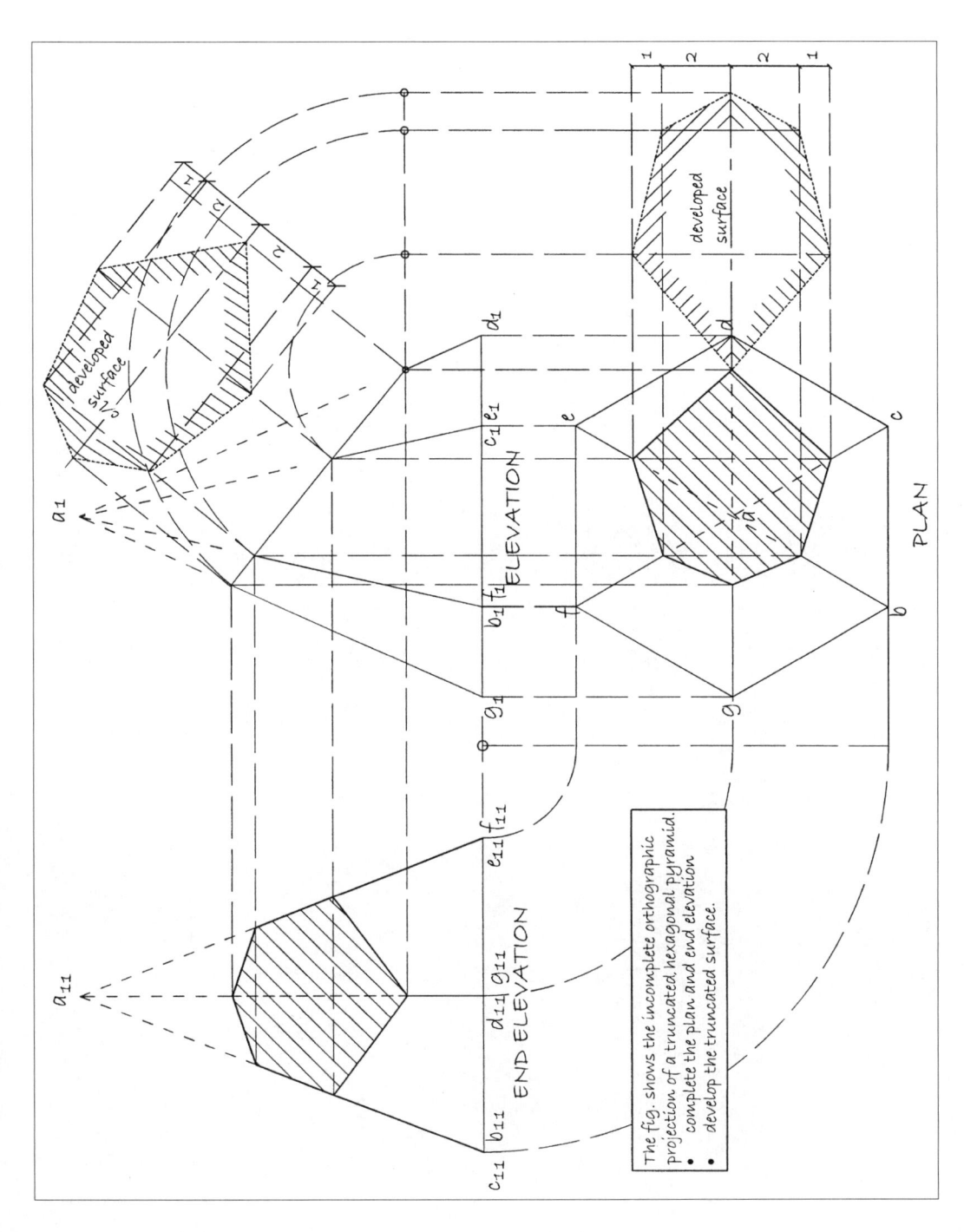

SOLUTIONS

The fig. shows the incomplete orthographic projection of a truncated, hexagonal pyramid.
- complete the plan and end elevation
- develop the truncated surface.

Chapter 4, Fig. 4.12

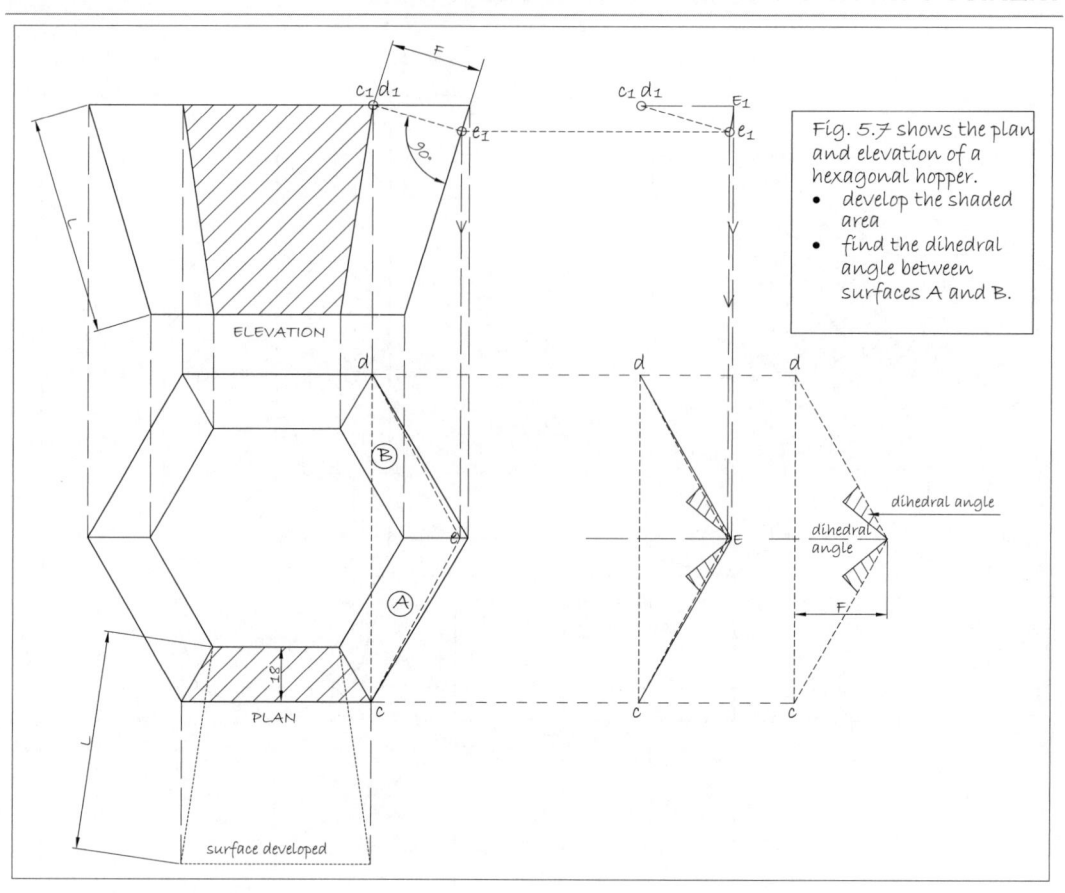

Fig. 5.7 shows the plan and elevation of a hexagonal hopper.
- develop the shaded area
- find the dihedral angle between surfaces A and B.

Chapter 5, Fig. 5.7

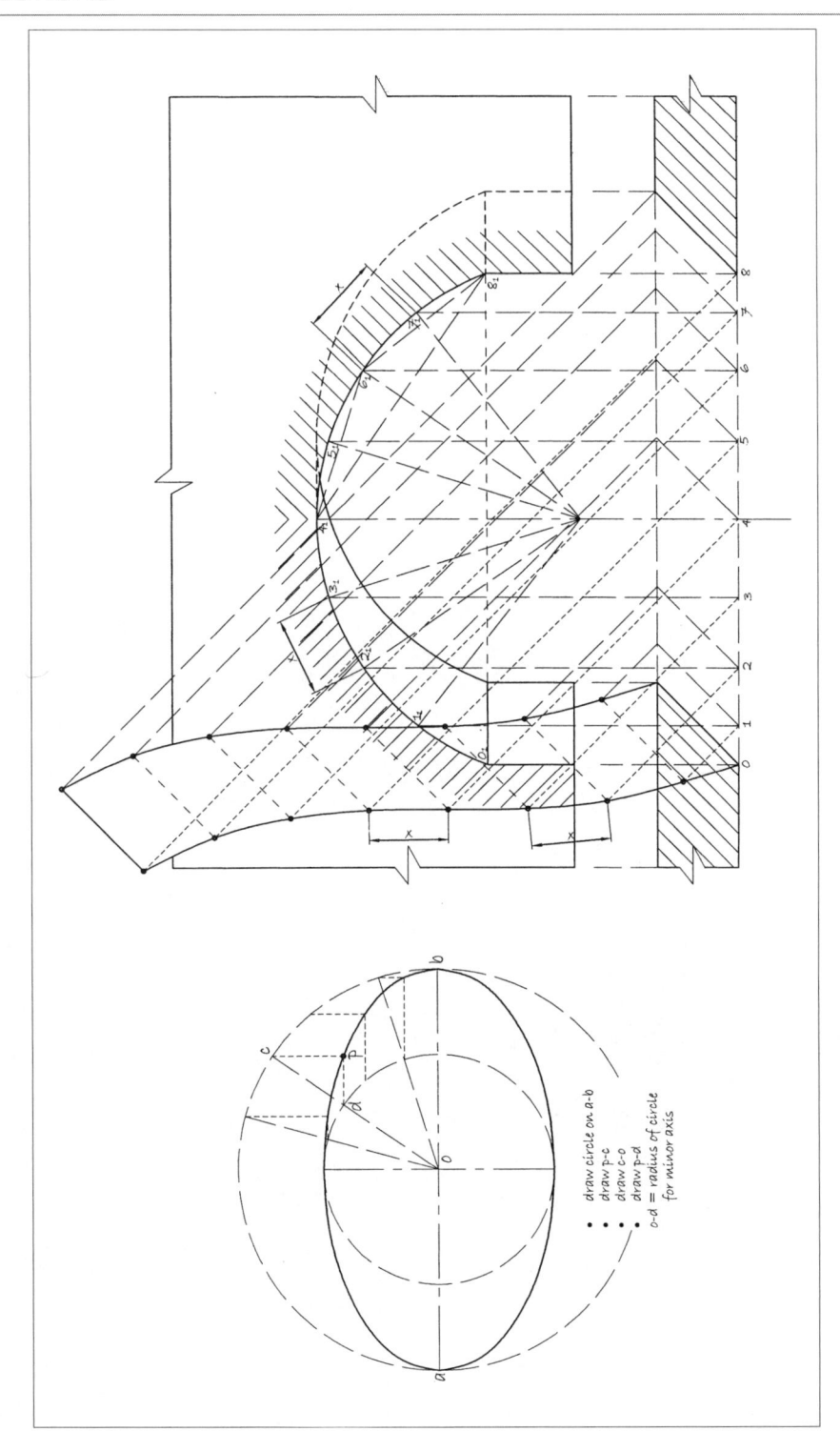

Chapter 6, Fig. 6.15 and 6.16

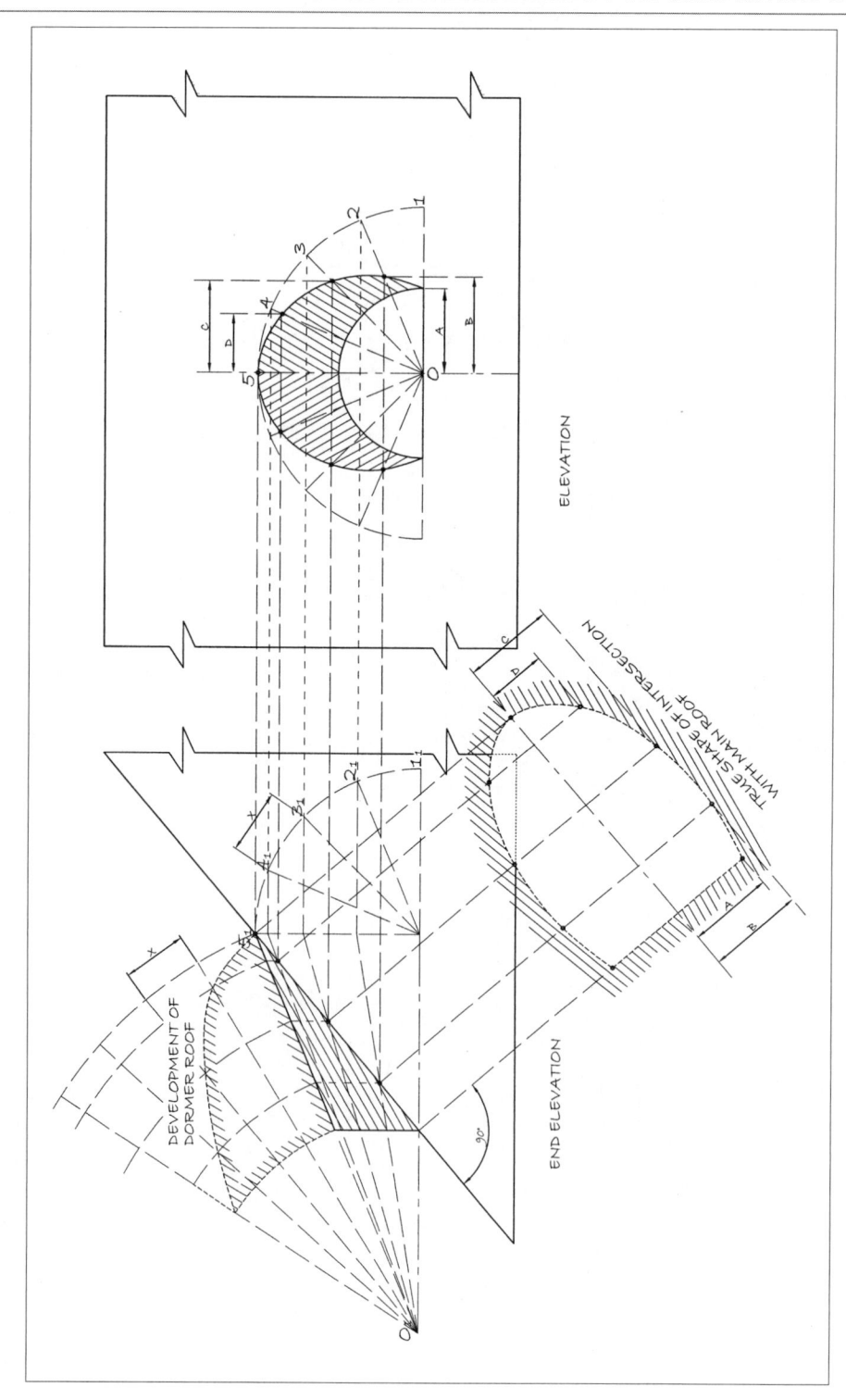

Chapter 7, Fig. 7.10

SOLUTIONS

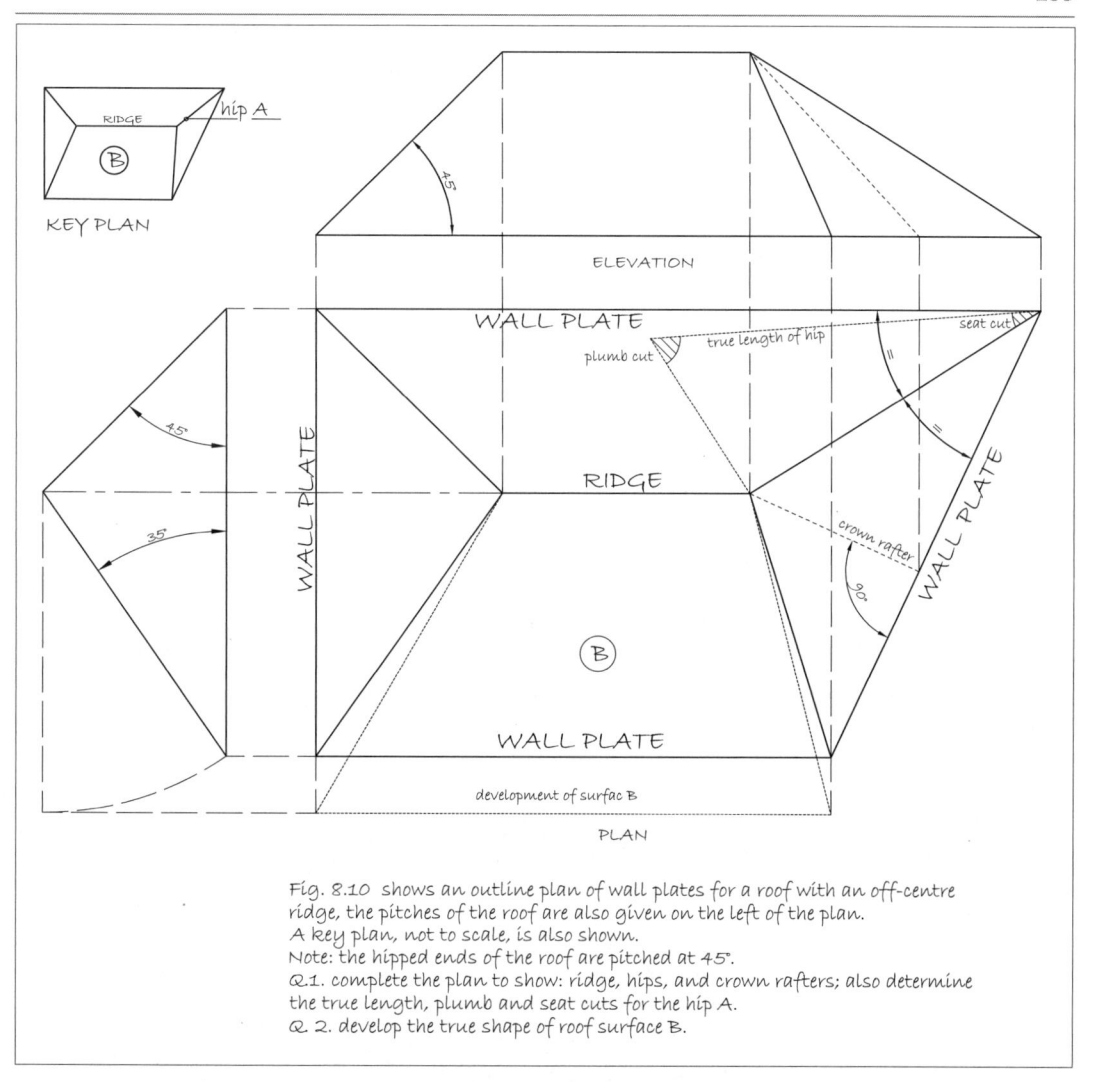

Fig. 8.10 shows an outline plan of wall plates for a roof with an off-centre ridge, the pitches of the roof are also given on the left of the plan.
A key plan, not to scale, is also shown.
Note: the hipped ends of the roof are pitched at 45°.
Q.1. complete the plan to show: ridge, hips, and crown rafters; also determine the true length, plumb and seat cuts for the hip A.
Q.2. develop the true shape of roof surface B.

Chapter 8, Fig. 8.10

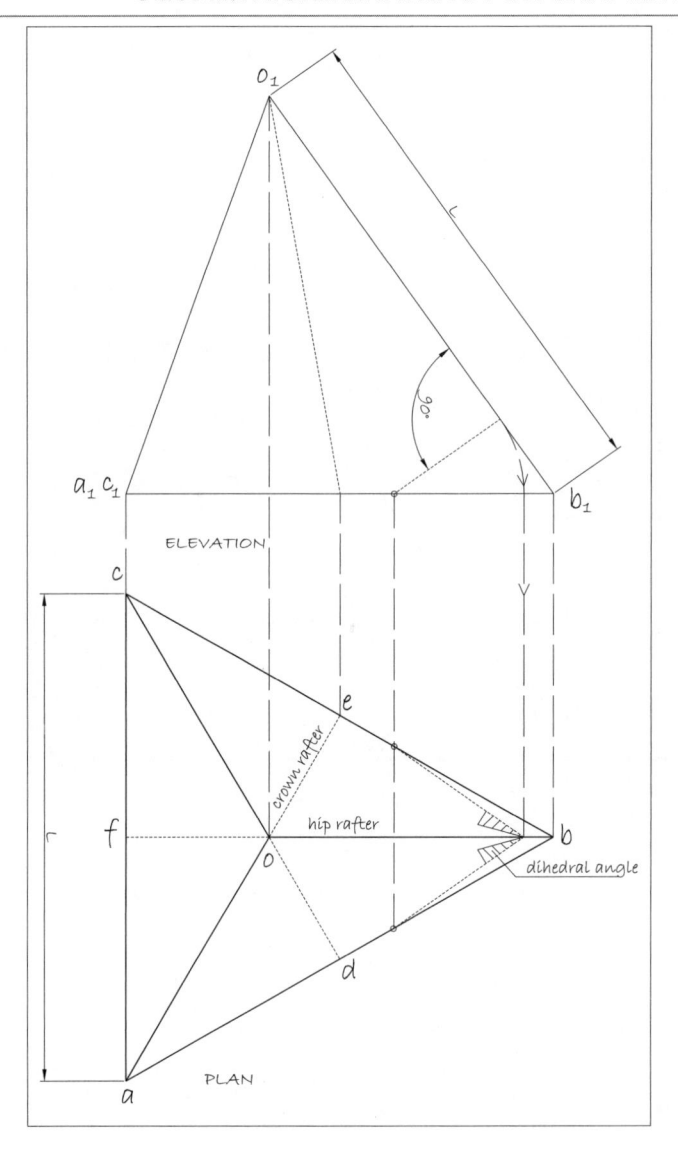

Chapter 8, Fig. 8.14b

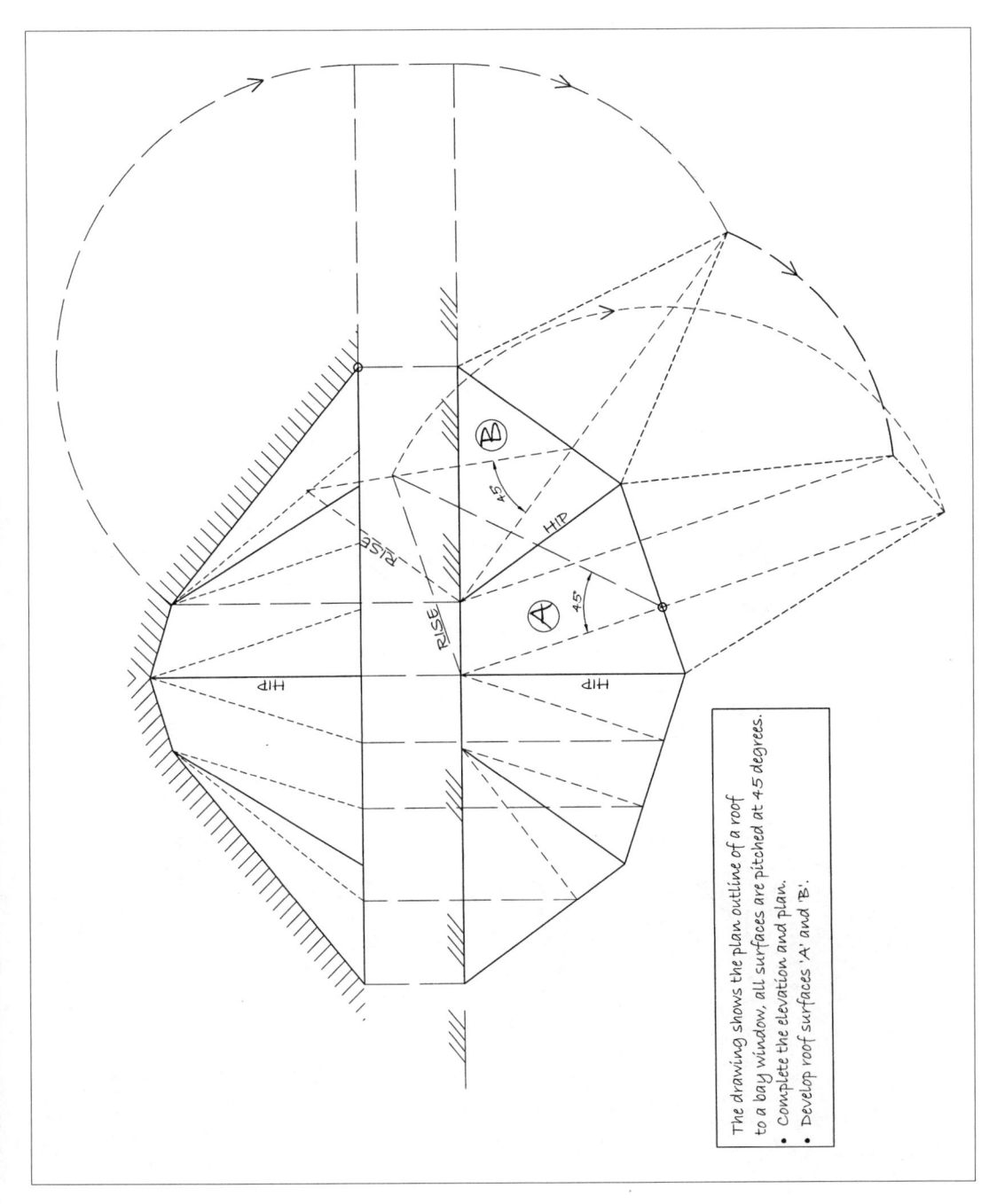

SOLUTIONS

Chapter 8, Fig. 8.20

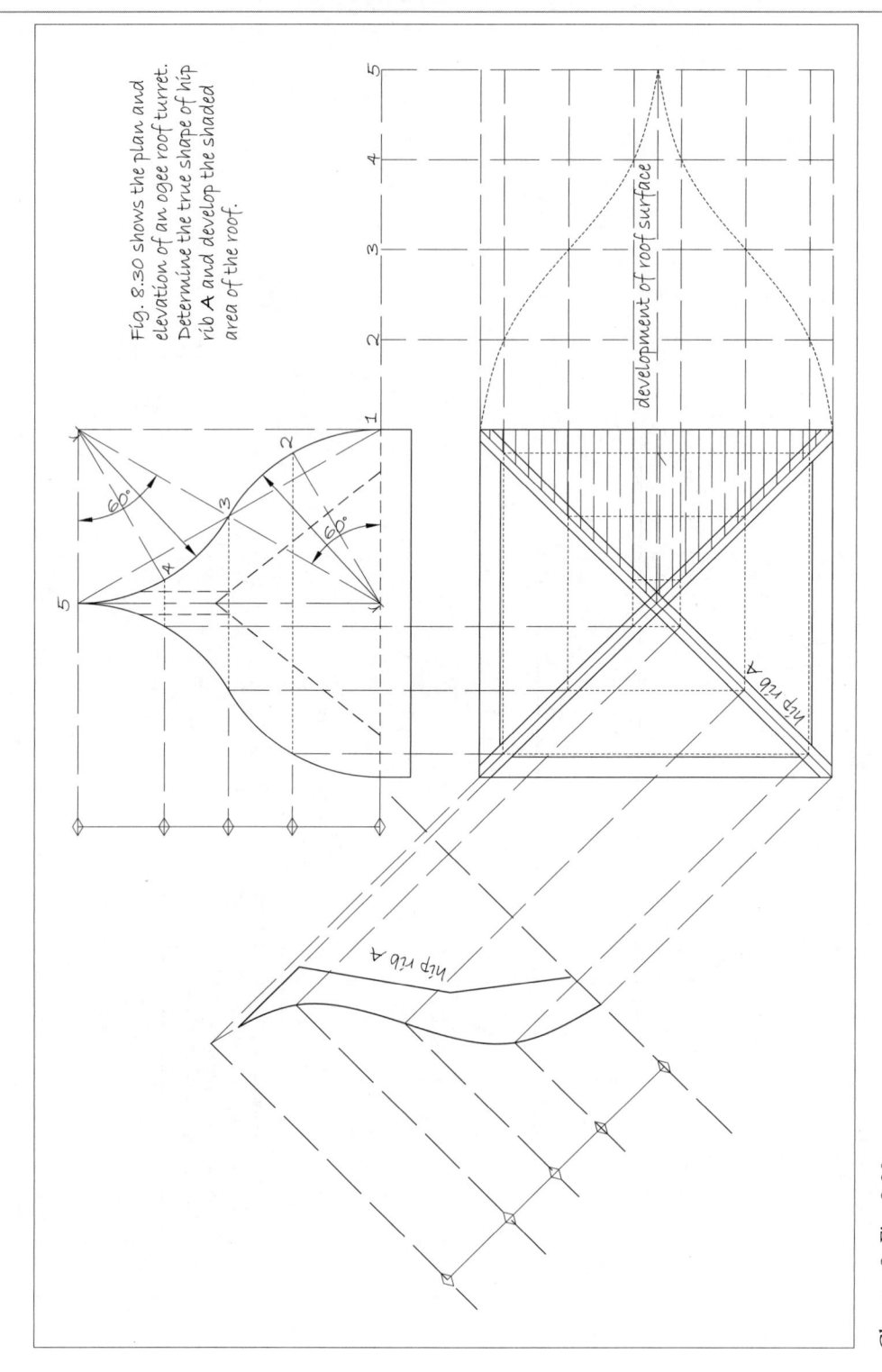

Fig. 8.30 shows the plan and elevation of an ogee roof turret. Determine the true shape of hip rib **A** and develop the shaded area of the roof.

Chapter 8, Fig. 8.30

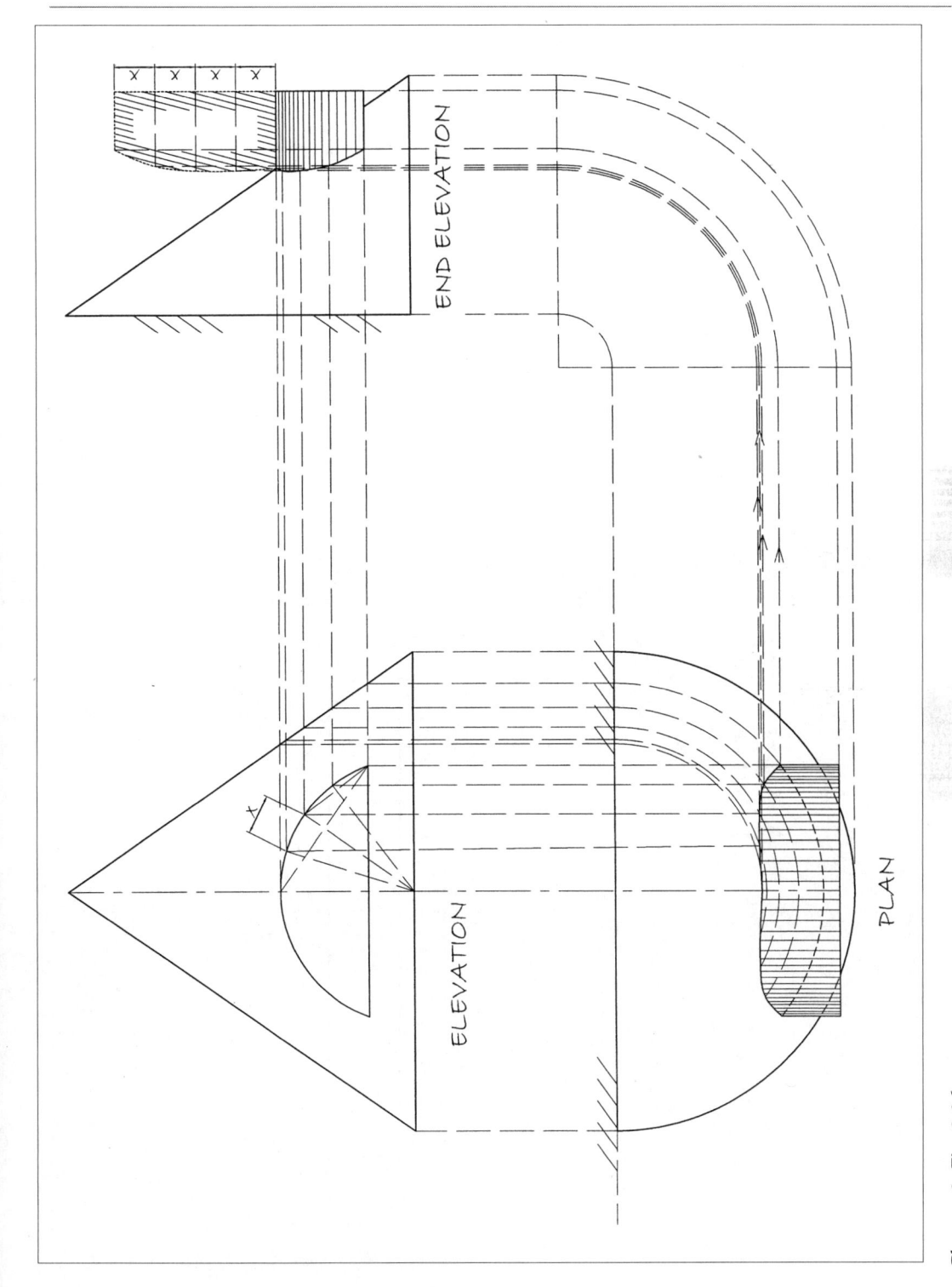

SOLUTIONS

Chapter 8, Fig. 8.36

GEOMETRICAL DRAWING FOR CARPENTRY & JOINERY

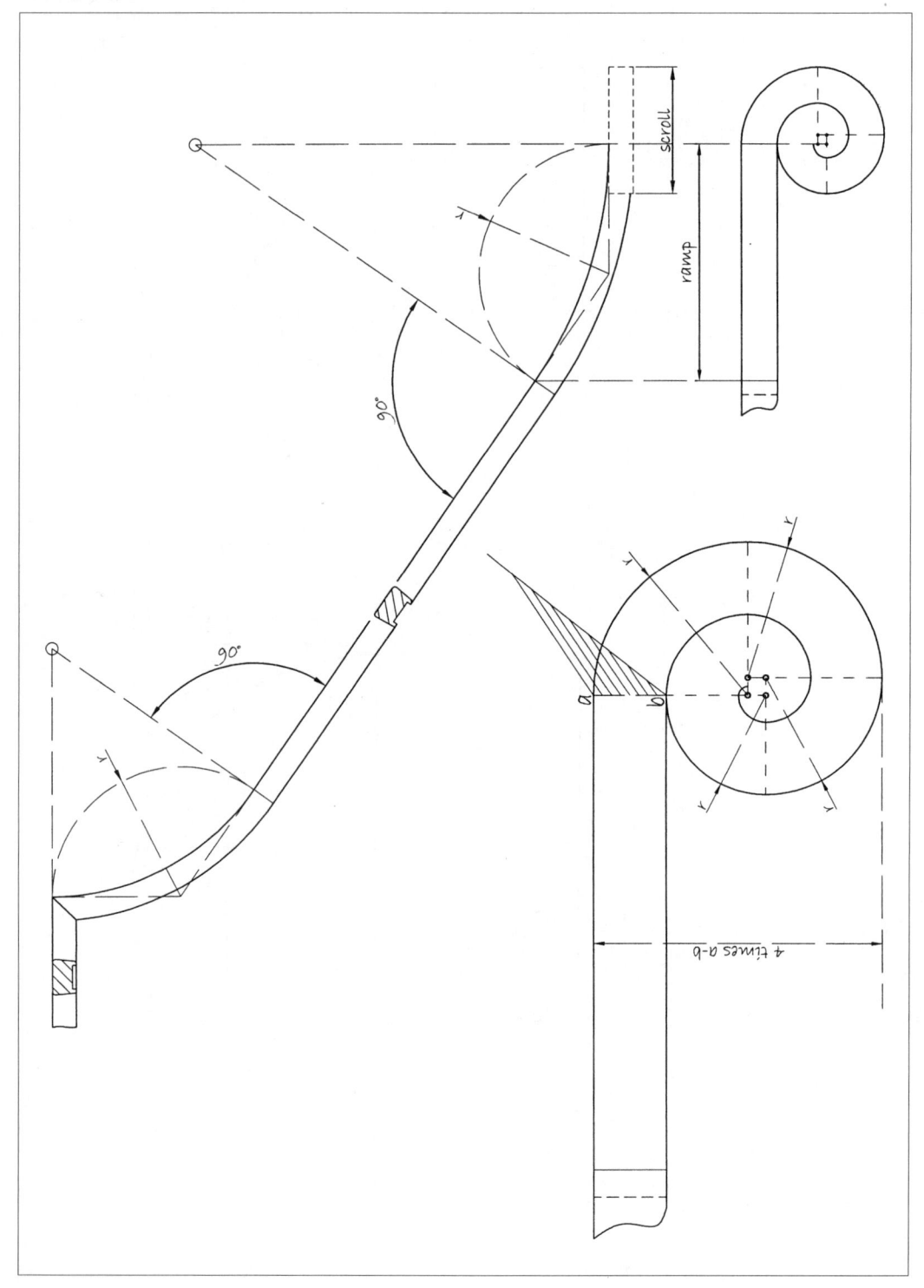

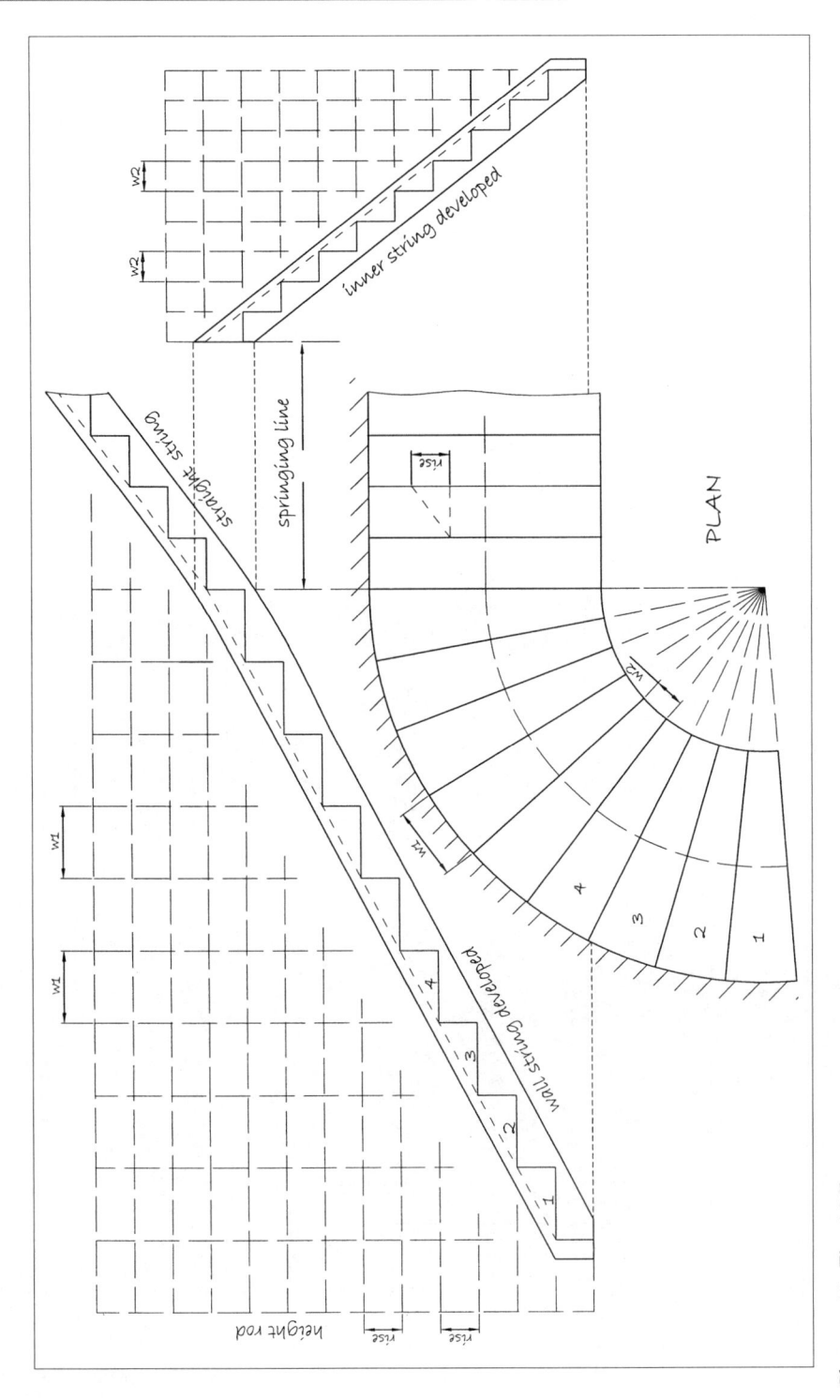

Chapter 10, Fig. 10.5

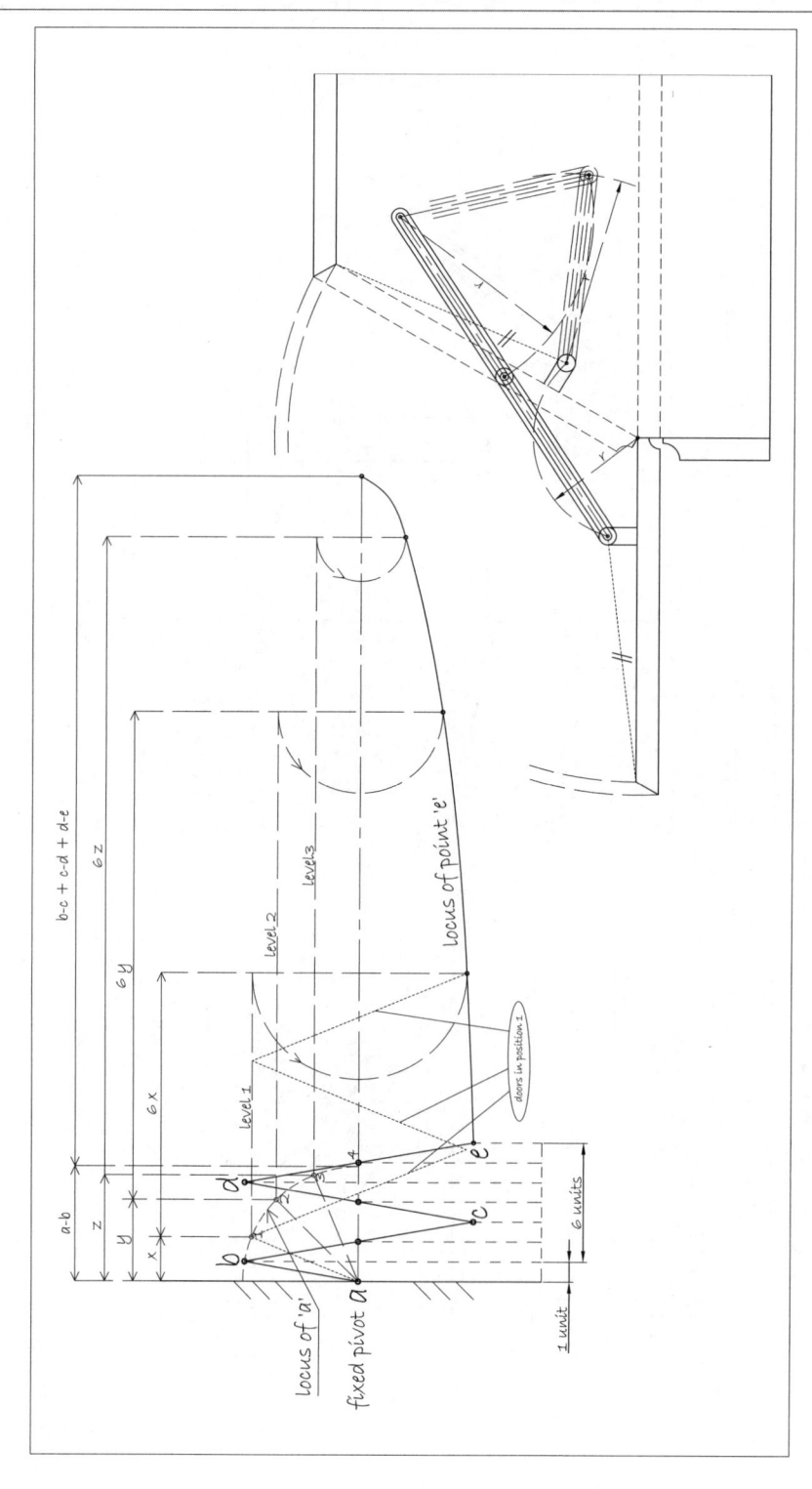

240   GEOMETRICAL DRAWING FOR CARPENTRY & JOINERY

Chapter 11, Fig. 11.13 and Fig. 11.14

SOLUTIONS

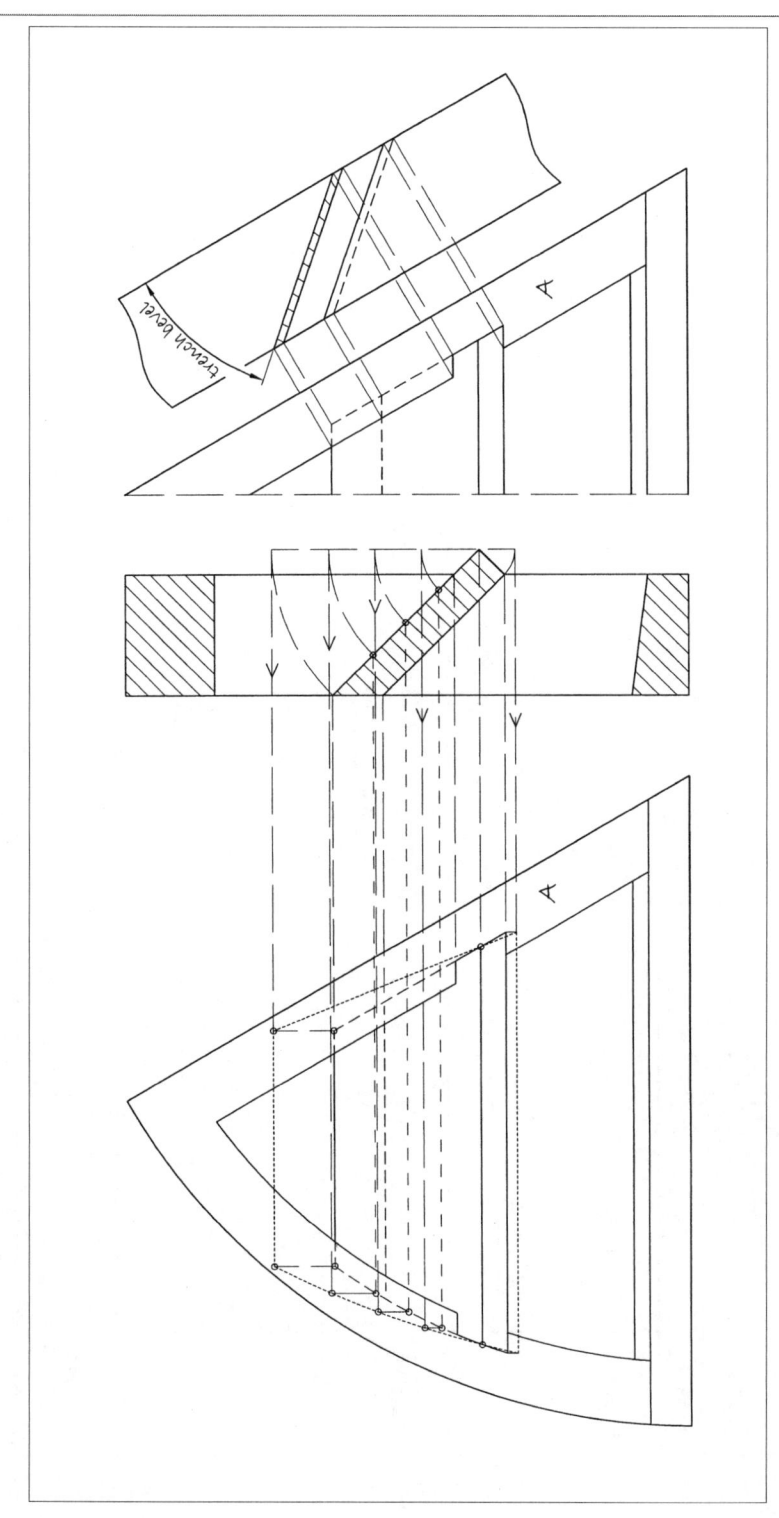

Chapter 12, Fig. 12.7

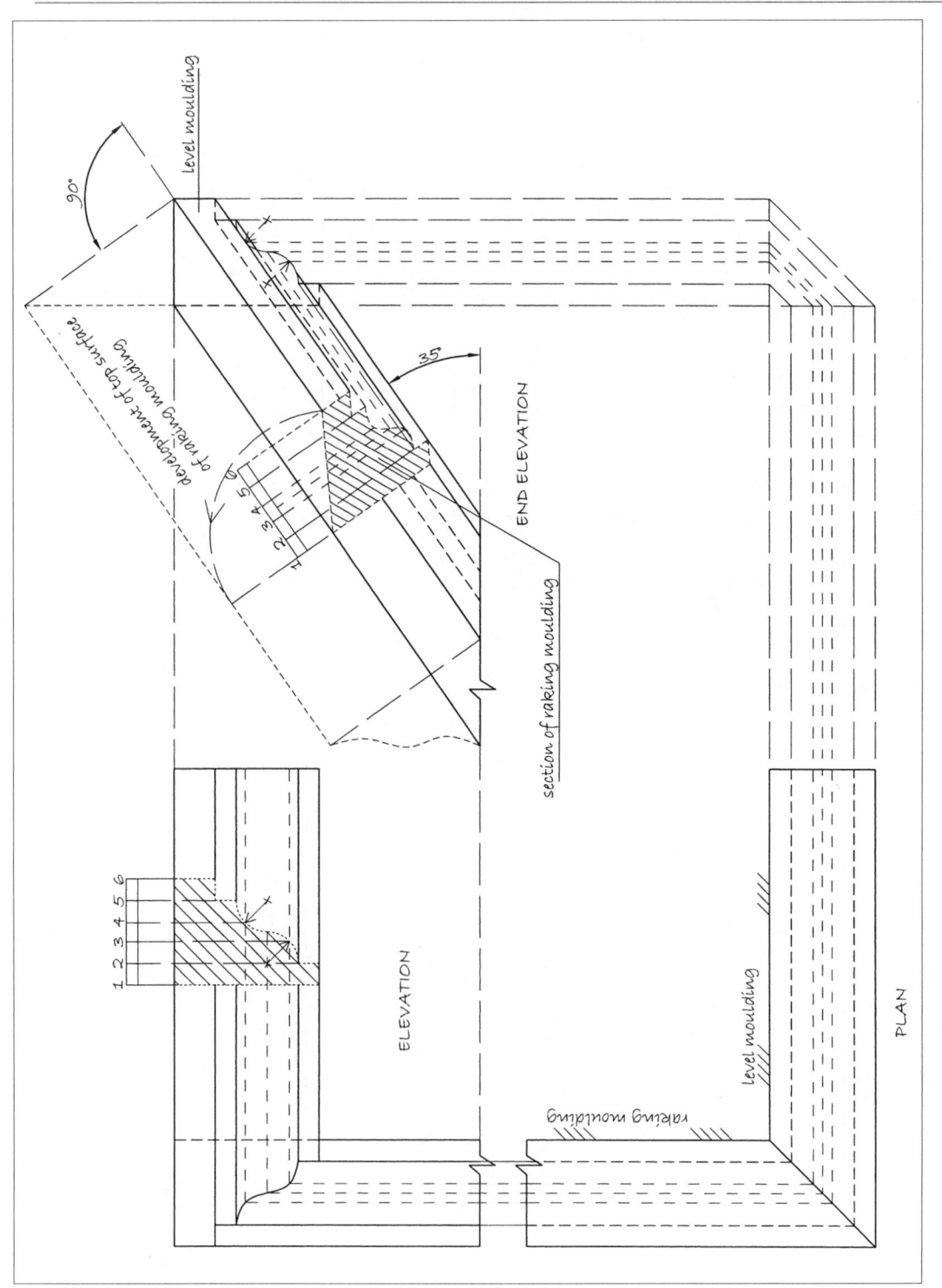

Chapter 13, Fig. 13.15

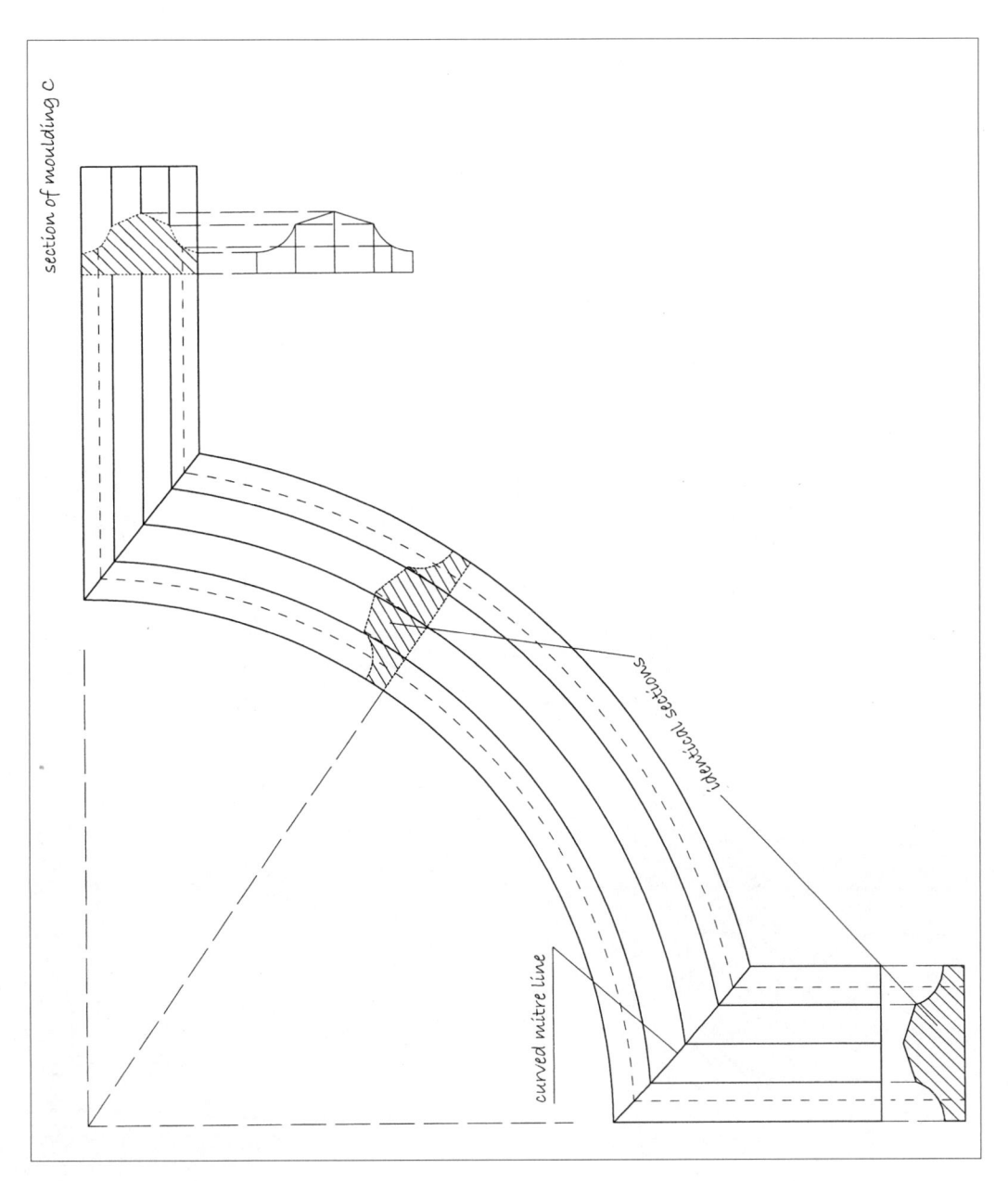

SOLUTIONS

Chapter 13, Fig. 13.16

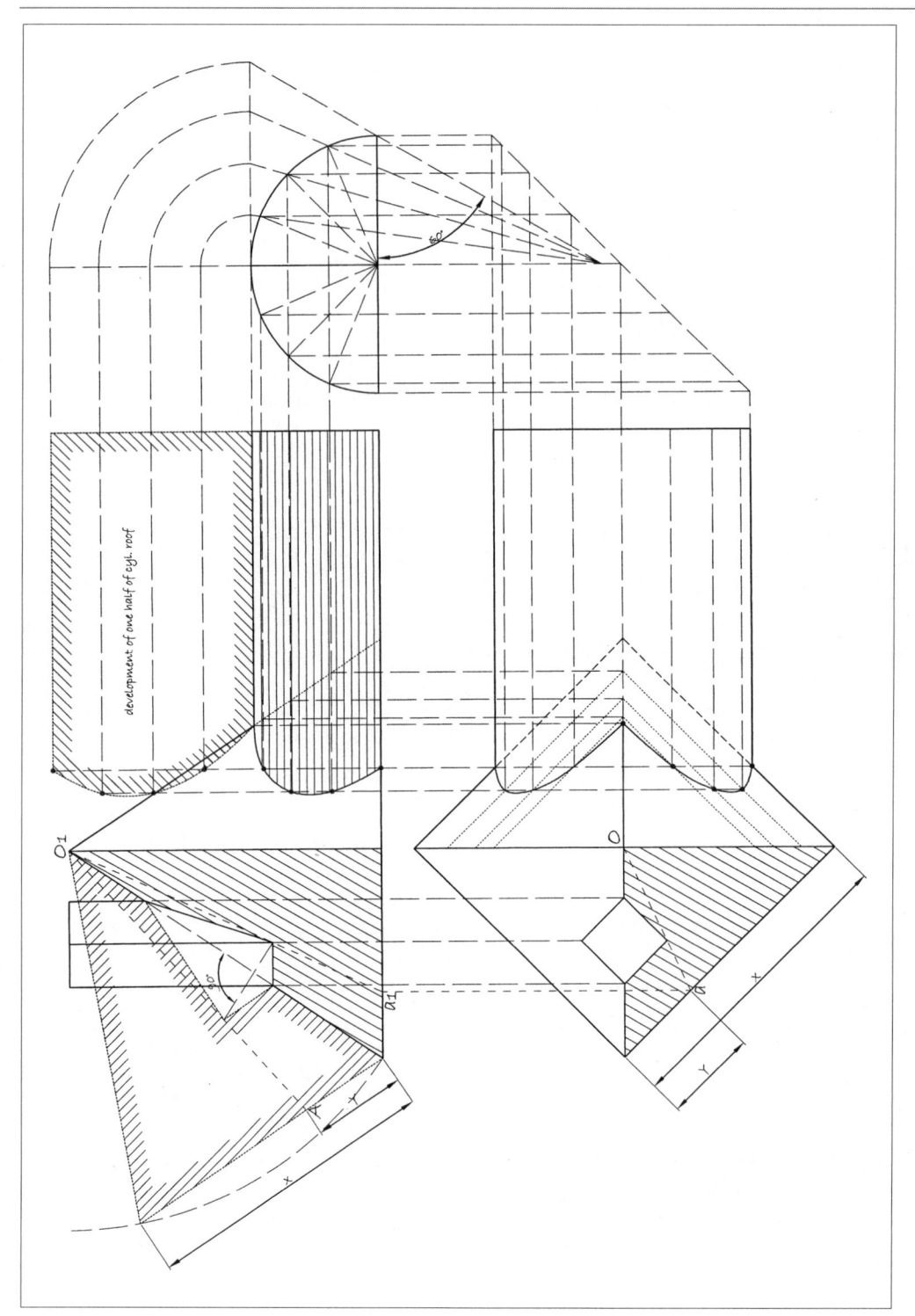

Chapter 14, Fig. 14.6

SOLUTIONS

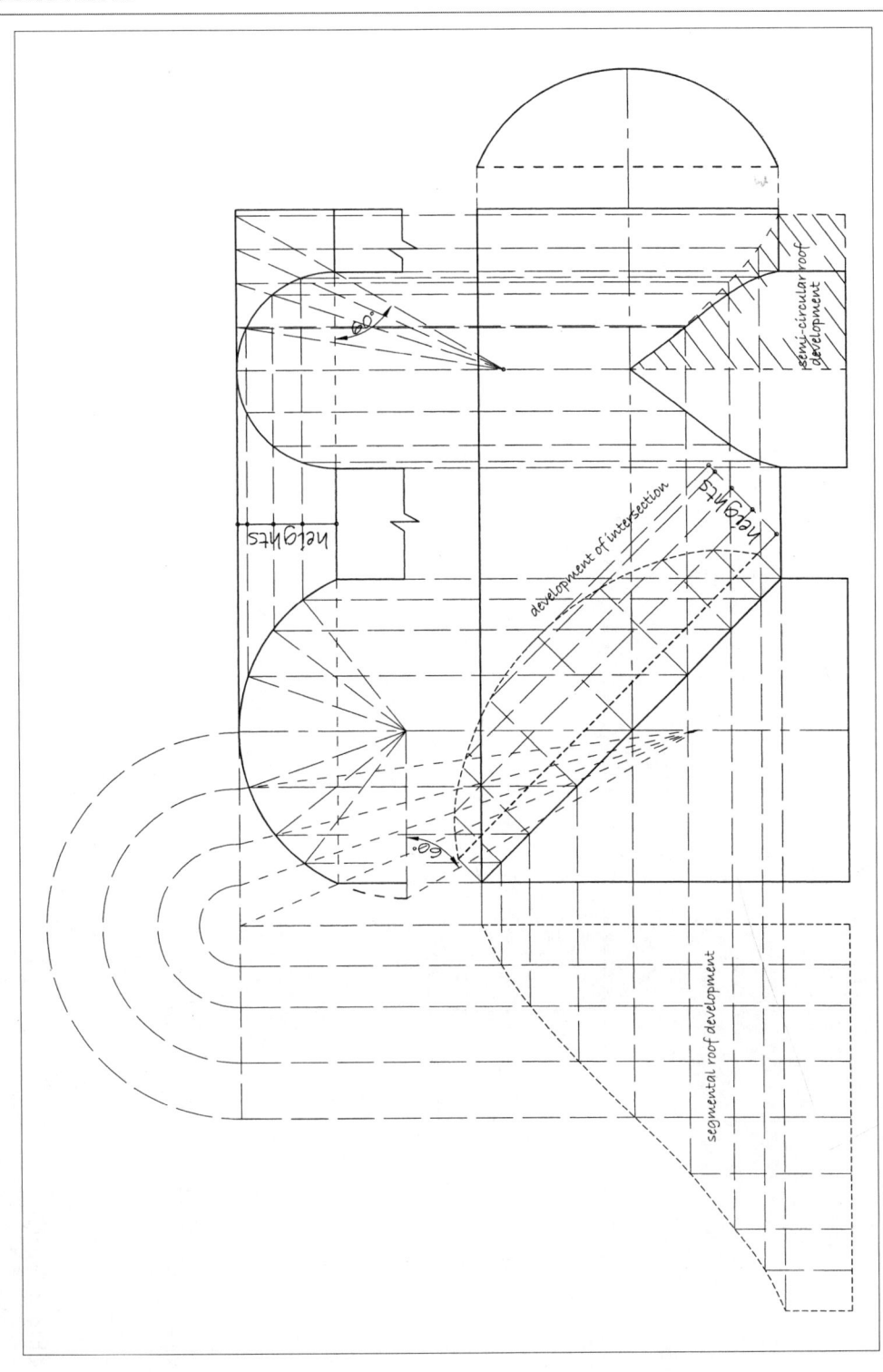

Chapter 15, Fig. 15.7

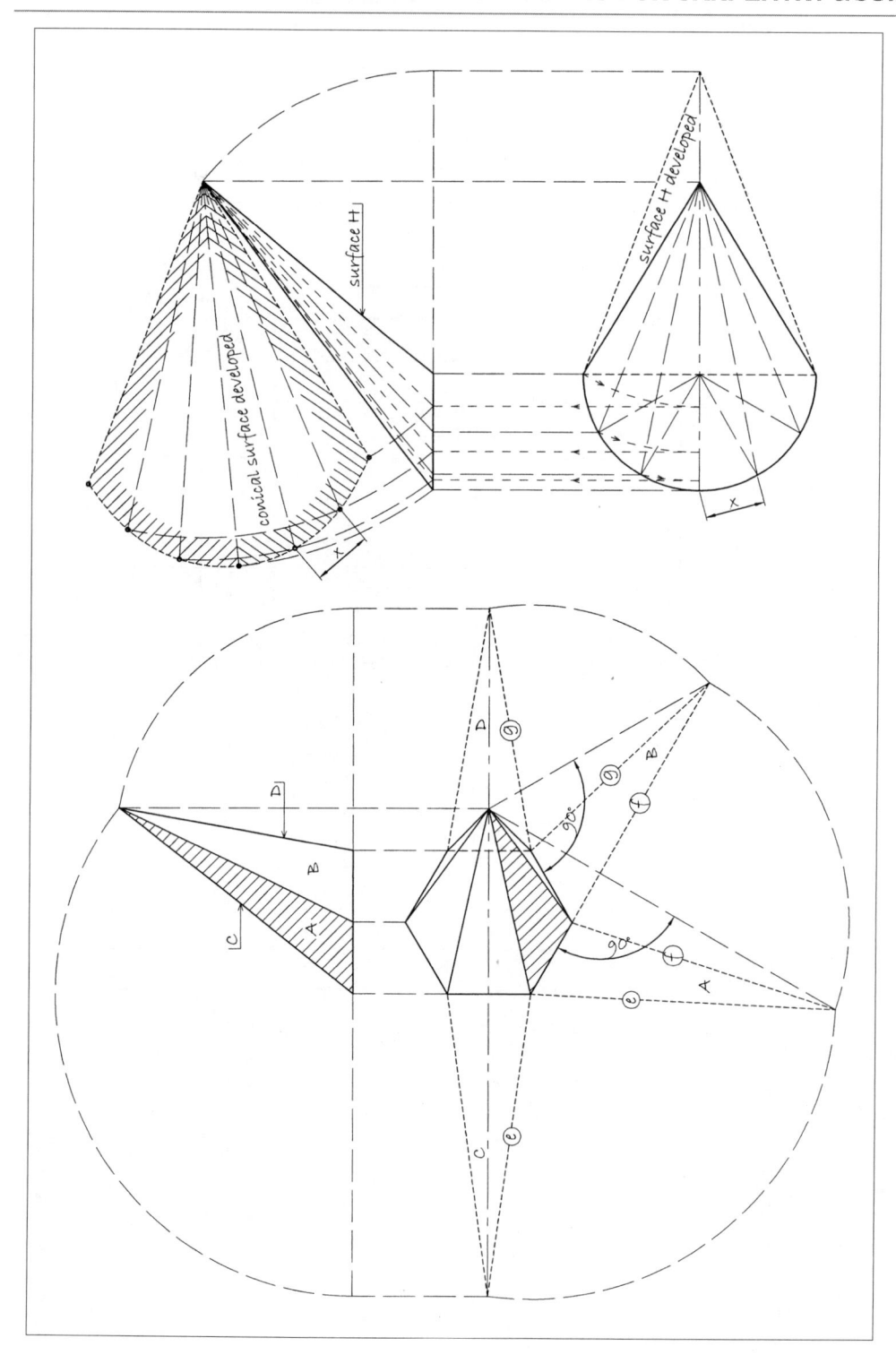

Chapter 16, Fig. 16.6

SOLUTIONS

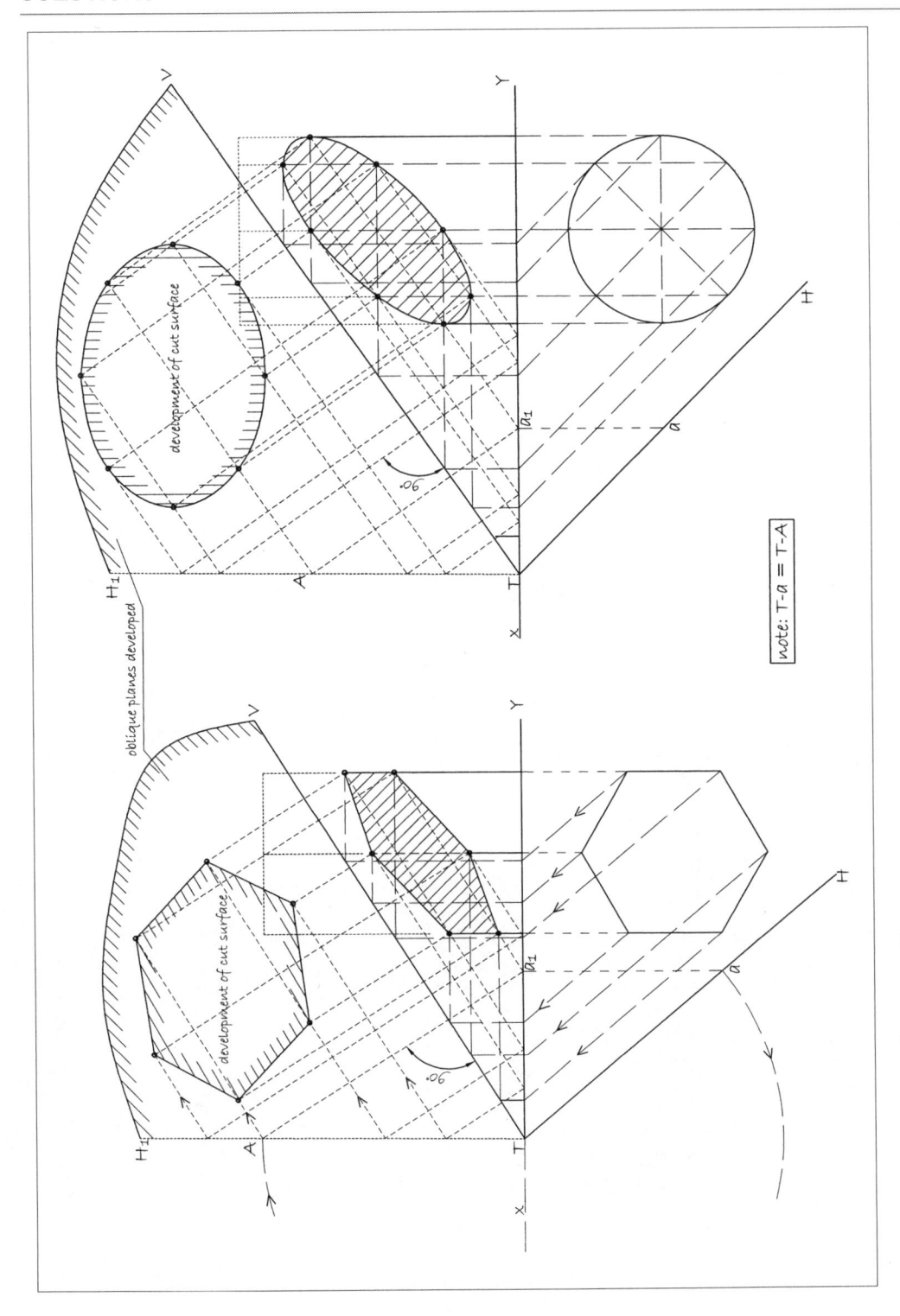

Chapter 17, Fig. 17.16a and Fig. 17.16b

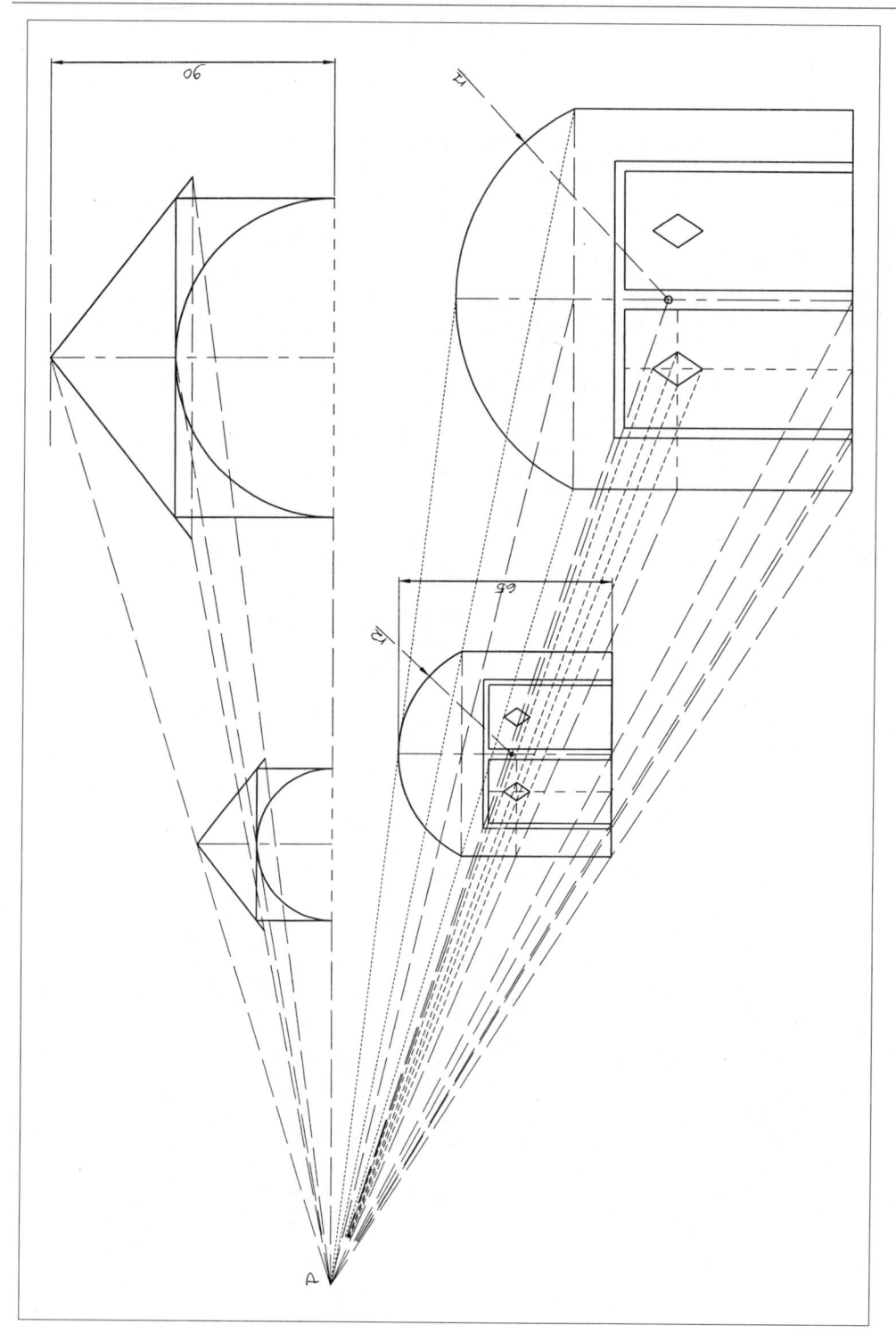

248 GEOMETRICAL DRAWING FOR CARPENTRY & JOINERY

Chapter 17, Fig. 17.17 and Fig. 17.18

INDEX

(Page numbers in italics refer to illustrations)

angles of projection, 212
apex, 27
aquatic centre, domed roof, 100, *102*
arc, diminishing, 132–5
arches, 50–65
    common normal of curves, 53, *54*, *55*
    four-centred, 53, *57*
    gothic, *56*
    gothic skew arch, 62, *64*
    ogee, *56*
    rise, self-determining, 53
    segmental, 50, *51*
    segmental skew arch, 62, *65*, *231*
    semi-circular, 50, *51*
    semi-circular and soffit, 57, *58*
    semi-circular skew arch, 61–2
    semi-elliptical, 50, *52*
    skew arches, 61–5, *231*
    soffit development, 57–60
    three-centred, 53, *54*, *55*
    three-centred and soffit, 59
    trefoil, *56*
    Tudor arch, 53, *57*
    see also vaults, intersecting
architrave
    intersection of mouldings, 165, *167*
    radial projection, 208, *211*
arrises, 122
auxiliary elevations and plans, 17–26
    conic section of roof, 72, *74*
    groins in vaults, 192, 194, *197*
    mitre line, 43–5, *46*
auxiliary vertical plane, 214
axis lines, isometric projection, 1, *2*
axonometric projection, 221
    intersecting vaults, 194, *197*

backing angle *see* dihedral angle
bathroom mirror, extendable, 145

bay window roofs, 95–8
    angled bay, 15, *16*, 95, 97, *98*, *225*, *235*
    segmental bay, 95
    square bay, 95, *96*
    triangular bay, *14*, 15, 95, *97*
    true lengths of lines, *14*, 15, *16*, *225*
bevels and purlins, in roofs, 84–7
bicycle, cycloidal curve, 145
boatwrights, 218
boxing the object, 1, *3*, *4*

cabinet with drop leaf, 150, *152*
cabinet with drop lid, 154, *155*, *240*
ceilings, sloping, and raking
    mouldings, 174, *176*
chimney, intersection with roof, 189–91
circle
    cycloid, 144–7, *148*
    involute and locus of a point,
        144, *146*
    semi-circle, dividing segment into equal
        parts, 218, *219*
    semi-circle, flattened to a straight
        line, 218, *219*
cone, 27, 185
    conic sections, 66–77
    curved surface, 34, 36, *38*
    dormer roofs, *75*, *76*, *77*, *232*
    elliptical conic sections, 66, *67*,
        68, *69*, *70*
    hyperbolic conic sections, 66, *67*, 68,
        69, 72, *73*, *75*, 77
    interpenetration with triangular
        prism, 185, *188*
    intersection of pitched/conical
        roofs, 185, *188*
    isometric projection, 5, *6*
    oblique, 200–1, *203*
    oblique half-cone, 203, *206*, *246*

cone (*continued*)
   parabolic conic sections, 66–8, 69, *70*, *75*, 77
   right cone, 200
   true lengths of lines, 15, *16*, *225*
counter, splayed work, 45, *47*
crane jib, 150, *151*
cross vaults, 192, *193*, 196, *198*
crown rafter, 31
   hexagonal pyramid roof, 89, *90*
curves
   isometric projection, 5–9
   true lengths of lines, 10, *12*
cutters, moulding, 180–4
cycloid, 144–7, *148*
cylinder, 27, 185
   auxiliary elevation and plan, 21, *23*
   handrail wreath, *120*, 121, 124, *125*
   interpenetration of half-cylinder with square prism and square-based pyramid, 191, *244*
   interpenetration with triangular prism and square prism, 185, *187*
   isometric projection, 5, *6*, *7*
   oblique, 201–3, *204*, *205*
   oblique, and dormer roof, 202–3, *205*
   oblique, and skew arch, 62, *63*
   right cylinder, 200, 201–2, *204*
   right cylinder and skew arch, 61–2
   truncated, 31, *32*, 33–4

dihedral angle, 43–5, *46*
   conventional pitched roofs, 80–4
   oblique planes, 217
   pyramid roofs, 88–91
disc, semi-elliptical
   true lengths of lines, 15, *16*, *225*
domes, 98–101
   hexagonal, 98–100
   interpenetration with triangular prism, 188, *189*
   pendentive, 196
domical elevation, 98–100, *101*
door, sliding/folding, 145

door frame
   axonometric projection, *220*, 221
   splayed work and jambs, 43, *45*
dormer roofs, 109–14
   conical roof, 111, *112*
   conic section, *75*, *76*, 77, *232*
   lean-to roof, 111, *113*
   oblique cylinder application, 202–3, *205*
   semi-conical roof, 111, *114*, *237*
   true shape, 33, *34*
drop leaf of cabinet, 150, *152*
drop lid of cabinet, 154, *155*, *240*

easings, 140
elevations, 212
elevations, auxiliary, 17–26
   conic section of roof, 72, *74*
   groins in vaults, 192, 194, *197*
   mitre line, 43–5, *46*
elevations, domical, 98–100, *101*
elevations, end, 212
ellipse, 31, 50–3
   drawing a normal to, 50, *52*
   elliptical conic sections, 66, *67*, 68, 69, *70*
   false/pseudo, 53, *54*
   locus of a point, 144, *146*
   skew arches, 62, *65*, *231*
end elevations, 212
enlarging areas, 207–12, 221, *223*, *248*
envelope, isometric projection, 1, *2*
exploded isometric projection, 5, *8*
extendable mirror, 145

face mould (handrail wreath), 118, *120*, 122–4, 127–9, *131*
falling lines, 122, *123*
folding doors, 145, 154, *240*

gable end, 78
golden rectangle/ratio, 218, *219*
gores, and domed roofs, 100, *101*
gothic arches, 56
   skew arch, 62, *64*
gothic tracery window, 168, *171*

INDEX

gothic vaults, 192–3, *195*
groin, intersecting vaults, 192–4
    waving groin, 193, *197*

handrails, 115–36
    oblique planes, 214
    ramps and knees, 115–17
    scrolls, 132–6, *238*
    swan neck, 115, *117*
    wreaths, 118–31, 214
height rod, 21
hemispherical elevation, 98–100, *101*
heptagon, 218
hexagon, involute and locus of a
    point, 144, *147*
hexagonal-based prism, 34, *37*, 185
hexagonal domes, 98–100
hexagonal hopper, 47, *49*, 230
hexagonal pyramid, 203, *206*, 246
    roofs, 89, *90*
    truncated, 37, *40*, 229
    turret roofs, 101, *104*, 105–6
hopper, splayed work, 43, *44*
    hexagonal hopper, 47, *49*, 230
horizontal planes, 212, *213*, 214
horizontal sections method, 185
hyperbolic conic sections, 66, *67*, 68, *69*, 72, *73*, *75*, 77

impost, vaults, 193
instructions for flat-packs, 5
interpenetration of solids, 185–91, *244*
isometric projection, 1–9, *224*
    curves, 5–9
    exploded isometric, 5, *8*

jib of crane, 150, *151*

knees (handrails), 115–17

lazy tongs, 145, 149–50
level ordinates method, 129, *131*, 214
lines and true lengths, 10–16, *225*
lip bevel, 84, *85*, 86–7

loci, 144–55, *240*
louvers, 156–64
    in curved frames, 161–4, *241*
    in straight frames, 156–60

mirror, extendable, 145
mitre line, 43–5, *46*, 165
mouldings, 165–84, *242–3*
    intersection, 165–71, 211, *212*
    moulding cutters, 180–4
    radial projection, 207–12
    raking mouldings, 171–80

name board, splayed work, 43, *46*

oblique cone, 200–1, *203*
    half-cone, 203, *206*, 246
oblique cylinder, 201–3, *204*
    dormer roof application, 202–3, *205*
    skew arch, 62, *63*
oblique planes, 87, 214–18, 221, *222*, 247
    handrail wreaths, 124, *127*, 129, *131*
oblique solids, 200–6
oblique square-based prism, 200, *201*
oblique square-based pyramid, 200, *202*
octagon
    drawing within a square, 218, *219*
    marking on squared baulk, 218, *219*
octagonal-based prism
    interpenetration with sloping
        plane, 189–90
    truncated, 34, *35*, 36
octagonal pyramid, 1, *4*, 5
octagonal wall plates, 88, *89*
ogee arch, *56*
ogee roof turret, 107, *108*, *236*
ogee section moulding, radial
    projection, 211, *212*
orthographic projection, 1, 214
overhang, 78

parabolic conic sections, 66–8, 69, *70*, *75*, 77
pendentives, 196

pentagon, involute and locus of a
    point, 144, *147*
pentagonal-based prism, 36, *39*, *228*
pitch of roof, 78
plan, 212
plan and auxiliary elevations, 17–26
planes, 212–18
    auxiliary vertical plane, 214
    doubly inclined, 214
    oblique, 87, 214–18, 221, *222*, 247
    oblique, handrail wreaths, 124, *127*, 129, *131*
    simply inclined, 86, 118, 214
plinth, true lengths of lines, 15, *16*, *225*
polar projection, 207–12, 221, *223*, *248*
polygon
    drawing within a circle, 218, *219*
    external/central angle, 218, *219*
post, splayed work, 45–7, *48*
prism, 27
    oblique plane cutting prism, 214, *216*, 217
    true length of lines, *14*, 15, *16*, *225*
prism, hexagonal-based, 34, *37*, 185
prism, octagonal-based
    interpenetration with sloping plane, 189–90
    truncated, 34, *35*, *36*
prism, pentagonal-based, 36, *39*, *228*
prism, square-based
    handrail wreaths, 124, *126*
    interpenetration with half-cylinder and square-based pyramid, 191, *244*
    interpenetration with triangular prism and cylinder, 185, *187*
    interpenetration with triangular prism and square-based pyramid, 185, *186*
    oblique, 200, *201*
    true lengths of lines, 15, *16*, *225*
prism, triangular-based, 185
    interpenetration with cone, 185, *188*
    interpenetration with dome, 188, *189*
    interpenetration with square prism and cylinder, 185, *187*
    interpenetration with square prism and square-based pyramid, 185, *186*
prism, truncated, 27
    isometric projection, 1, *3*
    octagonal base, 34, *35*, *36*
    true shape of sloping surface, 27, *28*
projection, angles of, 212
projection, axonometric, 221
    intersecting vaults, 194, *197*
projection, isometric, 1–9, *224*
    curves, 5–9
projection, orthographic, 1, 214
projection, polar/radial, 207–12, 221, *223*, *248*
pseudo ellipse, 53, *54*
purlins, roofs, 84–7
pyramid, 27
    rebatement, 40, *41*
    true shapes of sloping surfaces, 29–31
pyramid, hexagonal, 203, *206*, *246*
    truncated, 37, *40*, *229*
    turret roofs, 101, *104*, 105–6
pyramid, octagonal, 1, *4*, 5
pyramid roofs, 88–95
    hexagonal, 89, *90*
    intersection with semi-circular roof, 191, *244*
    octagonal wall plates, 88, *89*
    plan and elevation, 25, *26*, *227*
    square-based, 93–5
    tetrahedron, 91–3, *234*
    turret, 101, *104*, 105–6
pyramid, square-based
    auxiliary elevation, 17, *19*, 21, *22*
    interpenetration with half-cylinder and square prism, 191, *244*
    interpenetration with triangular prism and square prism, 185, *186*
    isometric and orthographic projection, 1, *4*
    oblique, 200, *202*
    truncated, auxiliary elevation and plan, 21, *22*

INDEX

pyramid, square-based (*continued*)
   truncated, true shape of cut surface, 27–8, *29*
pyramid, truncated
   elevation and plan, 25, *226*
   hexagonal, 37, *40*, *229*
   square-based, auxiliary elevation and plan, 21, *22*
   square-based, true shape of cut surface, 27–8, *29*
pyramid turret
   framed louvers, 158–9, *160*
   roofs, 101, *104*, 105–6
   true length of hips, 12, *13*

quadrants, diminishing, 132–5

radial projection, 207–12, 221, *223*, *248*
rafter, 78
   crown rafter, 31
   crown rafter in hexagonal pyramid roof, 89, *90*
   curved rafters (ribs), 98–108
   right-angled triangle and, 78
   valley rafter, 80, *81*
raking mouldings, 171–80
   not square-off, 175, *178*
ramps (handrails), 115–17
ratio, golden, 218, *219*
rebatement, 40–2
   raking mouldings, 174
reception counter, splayed work, 45, *47*
rectangle, golden, 218, *219*
reducing areas, 207–12, 221, *223*, *248*
ribbed vaults, 193–4, *196*, *197*
right cone, 200
right cylinder, 200, 201–2, *204*
right solids, 200
rise of arch, self-determining, 53
rise of roof, 78
road route, loci, 144, *145*
roofs, 78–114
   auxiliary elevation and plan, 21, *24*, 25, *26*, 72, *74*, *227*

bay windows *see* roofs to bay windows
bevels and purlins, 84–7
chimney intersection, 189–91
conic sections and development, 66–77
domes and turrets, 98–108
dormer, 109–14
dormer, conic section, *75*, *76*, 77, *232*
dormer, oblique cylinder application, 202–3, *205*
dormer, true shape, 33, *34*
hip and valley rafter, 80, *81*
hipped end, 78
hipped end, conic section, 69, *70*
hipped with splayed wall plate, 78–80
intersection of pitched/conical roofs, 185, *188*
intersection of pyramid/semi-circular roofs, 191, *244*
isometric projection, 9, *224*
off-centre ridge, 87, *88*, *233*
pitches, equal, 80, *82*
pitches, unequal, 83–7
planes and first angle of projection, 214, *215*
pyramid roof, 88–95, *234*
pyramid roof, intersection with semi-circular roof, 191, *244*
pyramid roof, plan and elevation, 25, *26*, *227*
pyramid roof, square-based, 93–5
rebatement, 40, *42*
segmental extension to building, 72, *73*
terminology, 78
roofs to bay windows, 95–8
   angled bay, 15, *16*, 95, 97, *98*, *225*, *235*
   segmental bay, 95
   square bay, 95, *96*
   triangular bay, *14*, 15, 95, *97*
   true lengths of lines, *14*, 15, *16*, *225*
route of road, loci, 144, *145*
run of roof, 78

sash of window, 150, *153*, 154
scratch stock, 181

scrolls, handrails, 132–6, *238*
segment, dividing into equal parts, 218, *219*
semi-circle
    dividing segment into equal parts, 218, *219*
    flattened to a straight line, 218, 219
set square, isometric projection, 1, *2*
shanks (handrail wreath), 118, *120*, 121
simply inclined planes, 86, 118, 214
skew arches, 61–5, *231*
slide mechanism, 150, *152*
sliding doors, 145
soffit of arches, 57–60
solids, 27–42, 185
    interpenetration, 185–91, *244*
    oblique, 200–6
    right solids, 200
span of roof, 78
spherical triangles, 196
spile batten, 161
spiral, 218, *219*
spiral stairs, 137–43, *239*
splayed work, 43–9, *230*
springing line, 50
    vaults, 193
springing plane (handrail wreath), 118, *128*
square-based prism *see* prism, square-based
square-based pyramid *see* pyramid, square-based
square-based turret, 101, *103*
stairs, spiral, 137–43, *239*
stilted vaults, 193
swan neck, 115, *117*

tangent, 50
tangent plane (handrail wreath), 118, 121, *128*
tetrahedron, roof based on, 91–3, *234*
trace (geometric planes), 214
traffic cone, 34, *38*
trammel, 53

trefoil arch, *56*
triangle, right-angled, and roof geometry, 78
triangle, spherical, 196
triangular-based prism *see* prism, triangular-based
triangular bay window roofs, *14*, 15, 95, *97*
try square, radial projection, 207, *210*
Tudor arch, 53, *57*
turret, pyramid
    framed louvers, 158–9, *160*
    roofs, 101, *104*, 105–6
    true length of hips, 12, *13*
turret roofs, 98–108
    ogee, 107, *236*
turret, square-based, 101, *103*
twist bevel (handrail wreath), 118, 119, *120*, 121, *122*, 127

valley rafter, 80, *81*
vaults, intersecting, 192–9, *245*
    cross vaults, 192, *193*, 196, *198*
    gothic vaults, 192–3, *195*
    ribbed vaults, 193–4, *196*, *197*
    stilted vaults, 193
verge of roof, 78
vertical planes, 212, *213*, 214
    auxiliary vertical plane, 214

waving groin, 193, *197*
window, bay *see* bay window roofs
window, gothic tracery, 168, *171*
window sash, 150, *153*, 154
wreathed strings, in spiral stairs, 137–43, *239*
wreaths, in handrails, 118–31
    level ordinates method, 129, *131*
    oblique planes, 214
    wreath centerline plane, 118, *120*, 121, 122

zones, and domed roofs, 100, *101*